Edited by
I. Gohberg

Birkhäuser
Boston · Basel · Stuttgart

I.S. Iohvidov

Hankel and Toeplitz Matrices and Forms

Algebraic Theory

Translated by
G. Philip A. Thijsse

1982

Birkhäuser
Boston • Basel • Stuttgart

Author:
Iosif Semenovich Iohvidov
VGU, Matematichesky fakultet
Kafedra matematicheskogo analiza
Universitetskaya ploshchad, 1
Voronezh, 394693, USSR

Library of Congress Cataloging in Publication Data

Iohvidov, I.S. (Iosif Semenovich)
Handel and Toeplitz matrices and forms : algebraic theory

Translation of: Gankelevy i teplitsevy
matritsy i formy.
Bibliography: p.
Includes index.
1. Matrices. 2. Forms (Mathematics)
I. Title.
QA188.I5713 512.9'434 82-4434
ISBN 3-7643-3090-2 AACR2

CIP-Kurztitelaufnahme der Deutschen Bibliothek

Iohvidov, Iosif S.:
Hankel and Toeplitz matrices and forms:
algebraic theory / I.S. Iohvidov. Transl.
by G. Philip A. Thijsse. (Ed. by I. Gohberg).
Boston ; Basel ; Stuttgart : Birkhäuser, 1982.
Einheitssacht.: Gankelevy i teplicevy matricy
i formy (engl.)
ISBN 3-7643-3090-2
Iohvidov, I.S.

ISBN: 3-7643-3090-2
Printed in USA

CONTENTS

EDITORIAL INTRODUCTION

This book is a valuable introduction to the theory of finite Hankel and Toeplitz matrices. These matrices are characterized by the property that in one of them the entries of the matrices depend only on the sum of indices, and in the other only on the differences of the indices.

The book is dedicated in general to the algebraic aspect of the theory and the main attention is given to problems of: extensions, computations of ranks, signature, and inversion. The author has succeeded in presenting these problems in a unified way, combining basic material with new results.

Hankel and Toeplitz matrices have a long history and have given rise to important recent applications (numerical analysis, system theory, and others).

The book is self-contained and only a knowledge of a standard course in linear algebra is required of the reader. The book is nicely written and contains a system of well chosen exercises. The book can be used as a text book for graduate and senior undergraduate students.

I would like to thank Dr. Philip Thijsse for his dedicated work in translating this book and Professor I.S. Iohvidov for his assistance and cooperation.

I. Gohberg

NOTE OF THE TRANSLATOR

The text of this edition is, but for some minor corrections, identical with that of the 1974 Russian edition. In order to inform the readers of new developments a list of additional literature was added, and at the end of the Chapters II, III and IV a Remark leads the reader to this list. For technical reasons all footnotes were replaced by notes at the end of the sections. For the convenience of the readers these notes have been listed separately in the table of contents.

The production of this translation would have been impossible without the invaluable help of Mrs. Bärbel Schulte, who typed the manuscript with much skill, and showed much patience during the process, and of Professor I.S. Iohvidov, who corrected the typescript with extreme diligence, and had to endure the critical remarks of the translator, from which he would have been spared if the job had been done by a non-mathematician.

PREFACE

The theory of Hankel and Toeplitz matrices, as well as the theory of the corresponding quadratic and Hermitian forms, is related to that part of mathematics which can in no way be termed non-prolific in the mathematical literature. On the contrary, many journal papers and entire monographs have been dedicated to these theories, and interest in them has not diminished since the beginning of the present century, and in the case of Hankel matrices and forms, even since the end of the previous century. Such a continuous interest can be explained in the first place from the wide range of applications of the mentioned theories - in algebra, function theory, harmonic analysis, the moment problem, functional analysis, probability theory and many applied problems.

Besides the mentioned regions of direct application, there is still one more section of mathematics in which Toeplitz and Hankel matrices play the role of distinctive models. The point is, that the continual analogues of systems of linear algebraic equations, in which the matrices of coefficients are Toeplitz matrices (i.e., the entries of these matrices depend only on the difference of the indices of the rows and the columns), are integral equations with kernels, which depend only on the difference of the arguments, including, in particular, the Wiener-Hopf equations, a class which is of such importance for theoretical physics. Not infrequently facts, discovered on the algebraic level for the mentioned linear systems, lead instantly to analoguous new results for integral equations (a quite recent example is the paper [25]; in § 18 of this book the reader will become partially acquainted with its contents). The analoguous situation holds for Hankel matrices (i.e. matrices in which the entries depend only on the sum of the indices) and kernels which depend on the sum of the arguments.

This makes it all the more paradoxical, that at least in the Russian language no monograph has been dedicated to Toeplitz and Hankel matrices and forms in a purely algebraic setting. Moreover, although some information on Hankel matrices and forms can be obtained from the monograph "Theory of Matrices" of F.R. Gantmaher ([3],Ch.X,§ 1o and Ch.XVI,§ 1o), practically all known Russian or translated courses on linear algebra and matrix theory make no mention of Toeplitz matrices and forms, except for the literally few lines devoted to them in the book of R. Bellmann [2]. As to the well-known monograph of U. Grenander and G. Szegö "Toeplitz forms and their applications" [7], that is on the whole devoted to analytic problems. The term "Toeplitz form" itself is, in spite of the general definition given to it by the authors in the preface, used in this book almost exclusively in the sense in which it entered in the literature following the works of C. Carathédory, O. Toeplitz, E. Fischer, G. Herglotz and F. Riesz (in the years 19o7-1915). Namely, they deal basically with forms with coefficients which are connected with certain power series, Laurent series or Fourier series, and not at all with forms of general shape and their purely algebraic properties.

To date, a large number of results relating to the algebra of Hankel and Toeplitz matrices and the corresponding forms has been accumulated in the journal literature, and these results combine already to form a sufficiently well-structured theory. It originated in the memoires of G. Frobenius [19,2o] (from the years 1984 and 1912), but further results, which enter into the present book, were only found in our days.

Highly remarkable in our view, are the deep analogies and also direct relations, which were discovered only in the later years between the two classes of matrices (and forms) to which this book is dedicated. These analogies and connections, namely, were the orientation which enabled us to clear up many questions which remained, until now, in the shadow, in spite of the venerable age of the considered theory.

The reasons delineated above constitute, in all probability, sufficient justification for the purpose adopted in the writing of this book: *to restrict it in particular to the algebraic aspect of the theory, but to reflect this, if possible, completely.* We note, that the first part of this formula, to set aside all kinds of applications, as long as these are presented in other monographs in a sufficiently complete way, is (just as the second part) not wholly sustained with due consequence - we could not resist the temptation to adduce if only the simplest application of Hankel and Toeplitz forms in the theory of the separation of the roots of algebraic equations, which, besides, does not really violate the algebraic character of the book, mentioned in its subtitle. The special Appendix I is dedicated to this matter, whereas Appendix II touches, albeit also only in the same elementary way, the deep connection between our subject and the classical moment problem.

As to the basic text of the book, it is, with the exception of Chapter I, entirely devoted to the algebraic properties of Hankel (Ch. II) and Toeplitz (Ch. III) matrices and forms, and also to the various transformations of these subjects, among them the mutual transformations of matrices and forms of each of these two classes to matrices and forms of the other class (Ch. IV).

Let us linger in some detail on the contents of Chapters II - IV. The core of the whole theory is the so-called method of singular extension of Hankel and Toeplitz matrices (§§ 9 and 12 respectively) and the notions of characteristics which are developed on this basis. These notions which allow, respectively in §§ 11 and 15, to establish comparatively rapidly fundamental theorems on the rank of Hankel and Toeplitz matrices separately, are then combined in § 17 to one single systems of characteristics, covering both considered classes of matrices. In §§ 12 and 16, respectively, signature rules are established - the well-known rule of Frobenius for Hankel forms and a new rule for Toeplitz forms; this and the other are obtained by the same method of singular extensions and characteristics. Section 18 is entirely devoted to the problem of inversion of Toeplitz and Hankel matrices, and § 19 to transformations which transfer into each other the forms of the two classes which interest us.

Chapter I plays an auxiliary role. In it information from the general theory of matrices and forms, which is necessary for the subsequent chapters, is gathered. Some of this material is presented in traditional form but another part, however, had to be presented in a new way in order to make the reading of the book, if possible, independent of the direct availability of other texts. This relates in particular to §§ 6 and 8, which deal with truncated forms and the signature rule of Jacobi (and its generalizations), respectively. Somewhat distinct is § 3, which contains purely technical but, for the construction of the entire theory, very important material - a lemma on the evaluation of one special determinant and its consequences.

Of the reader is required the knowledge of the elements of mathematical analysis and algebra, and also the knowledge of a basic course in linear algebra and matrix theory to the extent of, for example, the first ten chapters of the treatise of F.R. Gantmaher [3], to which this book is, actually, presented as a supplement. Such minimal general preparatory requirements of the reader has forced us to exclude from the book the theory of infinite extensions of Hankel and Toeplitz forms with a fixed number of squares with a certain sign. This theory, developed in the papers [3o, 42, 33, 43, 31] and others, is in this book represented only in two exercises in §§ 12 and 16 respectively, since it requires the application of tools from functional analysis (operators on Hilbert spaces with an indefinite metric). In addition the study of the asymptotics of the coefficients of the mentioned infinite extensions necessitates the engagement of the appropriate analytical apparatus.

The original text of the book was completed as a manuscript of special courses which the author held in the years 1968 - 197o at the Mathematics Department of the Voronež State University and at the Department of Physics and Mathematics of the Voronež Pedagogical Institute. Subsequently this text was significantly extended by the inclusion of new results, both published and unpublished, and also in favour of examples and exercises, which conclude each section of the basic text and both appendices. The range of these exercises is sufficiently wide - from elementary numerical examples, provided either with detailed calculations or with answers, to little propositions, and sometimes also important theorems, not occurring in the basic text. The most difficult among the exercises are accompanied by hints.

In the book continuous numeration of the sections is adopted; the propositions, and also the examples and exercises are numerated anew in each section; the items, lemmata and theorems, and also the individual formulae have double numbers (of which the first number denotes the section). The references to the literature in brackets [] lead the reader to the list of cited literature at the end of the book.

For his initial interest in Toeplitz forms, and also in other problems of algebra and functional analysis, the author is obliged to his dear teacher Mark Grigorevič Kreĭn. In this book (especially in the Appendices I and II) the reader will repeatedly encounter some of his ideas and results, relating to our subject.

The text of this book reflects the valuable suggestions of

V.P. Potapov, expressed by him at the earlier stages of its preparation, when the idea of the book was barely thought out. At the final stage of the work the interest shown in this project by the collaborators of the chair of algebra of the Moscow State University, O.N. Golovin, E.B. Vinberg, E.S. Golod and V.N. Latyšev was a great stimulus for the author.

T.Ya. Azizov and E.I. Iohvidov, students in the special courses in which the book "originated", did indeed extend invaluable help to the author in the realization of the manuscript. In particular T.Ya. Azizov undertook the unenviable task of reading the complete text and verifying all exercises and calculations, which resulted in the insertion of numerous corrections and improvements. Useful remarks during the presentation and the reworking of the lecture courses were made by F.I. Lander.

To all those mentioned here the author wishes to express his sincere gratitude.

Chapter I

SOME INFORMATION FROM THE GENERAL THEORY OF MATRICES AND FORMS.

§ 1 THE RECIPROCAL MATRIX AND ITS MINORS

double brackets

1.1 We shall consider arbitrary square matrices $A = \|a_{ij}\|_{i,j=1}^{m}$ of complex numbers. If

$$i_1 < i_2 < \cdots < i_p;\ j_1 < j_2 < \cdots < j_p$$

are two sets of p indices ($1 \leqq p \leqq n$) from the indices $1,2,\cdots,n$, then we denote, as usually, through

$$A\begin{pmatrix} i_1 & i_2 \cdots i_p \\ j_1 & j_2 \cdots j_p \end{pmatrix}$$

the minor, consisting of the elements on the intersection of the rows with numbers $i_1,i_2,\cdots,i_p$ and the columns with numbers $j_1,j_2,\cdots,j_p$ of A, i.e.,

$$A\begin{pmatrix} i_1 & i_2 \cdots i_p \\ j_1 & j_2 \cdots j_p \end{pmatrix} = \det \|a_{i_\mu j_\nu}\|_{\mu,\nu=1}^{p}.$$

Evidently,

$$A\begin{pmatrix} 1 & 2 \cdots n \\ 1 & 2 \cdots n \end{pmatrix} \equiv |A|,$$

the determinant of the matrix A.

We agree to denote the n-p indices remaining after taking from the set $\{1,2,\cdots,n\}$ the indices $i_1,i_2,\cdots,i_p$ $(j_1,j_2,\cdots,j_p)$ through $k_1,k_2,\cdots,k_{n-p}$ $(\ell_1,\ell_2,\cdots,\ell_{n-p})$ (here the indices are always written in increasing order). Then

$$A\begin{pmatrix} k_1 & k_2 \cdots k_{n-p} \\ \ell_1 & \ell_2 \cdots \ell_{n-p} \end{pmatrix}$$

is, by definition, the *complementary* minor to the minor

$$A\begin{pmatrix} i_1 & i_2 \cdots i_p \\ j_1 & j_2 \cdots j_p \end{pmatrix}.$$

Evidently, complementary to the minors $a_{ij} = A\binom{i}{j}$ of the first order are the determinants

$$A\begin{pmatrix}1 & 2\cdots i-1 & i+1\cdots n\\ 1 & 2\cdots j-1 & j+1\cdots n\end{pmatrix} \equiv \tilde{a}_{ij} \quad (i,j=1,2,\cdots,n),$$

and the numbers $A_{ij} = (-1)^{i+j}\tilde{a}_{ij}$ represent the cofactors to the elements a_{ij} in the matrix A, respectively $(i,j=1,2,\cdots,n)$.

1.2 Diverging somewhat from more extended terminology, we shall, following [3], call the matrix

$$\tilde{A} = \|\tilde{a}_{ij}\|_{i,j=1}^{n}$$

consisting of the minors of order n-1 of A the *reciprocal* matrix with respect to the matrix A. We establish a rule for the computation of minors of the reciprocal matrix.

THEOREM 1.1. *For an arbitrary natural number* p $(1 \leqq p \leqq n)$ *one has*

$$\tilde{A}\begin{pmatrix}i_1 & i_2\cdots i_p\\ j_1 & j_2\cdots j_p\end{pmatrix} = |A|^{p-1} \quad A\begin{pmatrix}k_1 & k_2\cdots k_{n-p}\\ \ell_1 & \ell_2\cdots \ell_{n-p}\end{pmatrix}. \tag{1.1}$$

Here, in the case where $p = n$, *formula* (1.1) *should be understood as*

$$|\tilde{A}| = \tilde{A}\begin{pmatrix}1 & 2\cdots n\\ 1 & 2\cdots n\end{pmatrix} = |A|^{n-1}, \tag{1.2}$$

and for $p = 1$ *and* $|A| = 0$ *one should assume* $|A|^{p-1} = 1$.

PROOF. Without loss of generality one can restrict oneself to considering minors of the shape

$$\tilde{A}\begin{pmatrix}1 & 2\cdots p\\ 1 & 2\cdots p\end{pmatrix} \qquad (1 \leqq p \leqq n),$$

as the general case is obtained from this easily by appropriate permutation of rows and columns. Now formula (1.1) takes the form

$$\tilde{A}\begin{pmatrix}1 & 2\cdots p\\ 1 & 2\cdots p\end{pmatrix} = |A|^{p-1} \quad A\begin{pmatrix}p+1\cdots n\\ p+1\cdots n\end{pmatrix}. \tag{1.3}$$

In order to establish this, we multiply in the determinant

$$\tilde{A}\begin{pmatrix}1 & 2\cdots p\\ 1 & 2\cdots p\end{pmatrix} = \begin{vmatrix}\tilde{a}_{11} & \tilde{a}_{12}\cdots\tilde{a}_{1p}\\ \tilde{a}_{21} & \tilde{a}_{22}\cdots\tilde{a}_{2p}\\ \cdots\cdots & \cdots\cdots & \cdots\\ \tilde{a}_{p1} & \tilde{a}_{p2}\cdots\tilde{a}_{pp}\end{vmatrix}$$

the i-th row with $(-1)^i (i=1,2,\ldots p)$ and the j-th column with $(-1)^j$ $(j=1,2,\ldots,p)$ It is easy to understand that by such a transformation the value of the determinant doesn't change, and the determinant itself takes the form

$$\tilde{A}\begin{pmatrix}1 & 2\ldots p\\ 1 & 2\ldots p\end{pmatrix} = \det \| (-1)^{i+j}\tilde{a}_{ij} \|_{i,j=1}^{p} = \begin{vmatrix} A_{11} & A_{12}\ldots A_{1p}\\ A_{21} & A_{22}\ldots A_{2p}\\ \cdot & \cdot \;\ldots\\ A_{p1} & A_{p2}\ldots A_{pp}\end{vmatrix} .$$

It is convenient to assume that this minor has the shape of a determinant of order n (in the case p = n the next step is, clearly, not necessary):

$$\tilde{A}\begin{pmatrix}1 & 2\ldots p\\ 1 & 2\ldots p\end{pmatrix} = \left|\begin{array}{ccc|ccc} A_{11} & \ldots & A_{1p} & A_{1,p+1} & \ldots & A_{1n}\\ A_{21} & \ldots & A_{2p} & A_{2,p+1} & \ldots & A_{2n}\\ \cdot & \ldots & \cdot & \cdot & \ldots & \cdot\\ A_{p1} & \ldots & A_{pp} & A_{p,p+1} & \ldots & A_{pn}\\ \hline 0 & \ldots & 0 & 1 & \ldots & 0\\ \cdot & \ldots & \cdot & \cdot & \ldots & \cdot\\ 0 & \ldots & 0 & 0 & \ldots & 1\end{array}\right| .$$

We multiply both sides of this equality with the determinant

$$|A| = \left|\begin{array}{ccc|ccc} a_{11} & \ldots & a_{1p} & a_{1,p+1} & \ldots & a_{1n}\\ a_{21} & \ldots & a_{2p} & a_{2,p+1} & \ldots & a_{2n}\\ \cdot & \ldots & \cdot & \cdot & \ldots & \cdot\\ a_{p1} & \ldots & a_{pp} & a_{p,p+1} & \ldots & a_{pn}\\ \hline a_{p+1,1} & \ldots & a_{p+1,p} & a_{p+1,p+1} & \ldots & a_{p+1,n}\\ \cdot & \ldots & \cdot & \cdot & \ldots & \cdot\\ a_{n1} & \ldots & a_{np} & a_{n,p+1} & \ldots & a_{nn}\end{array}\right| ,$$

where we shall develop the product of the determinants on the right-hand side through the "row on row" method. Then, taking into account the well-known properties of the cofactors A_{ij}, we obtain

$$\tilde{A}\begin{pmatrix}1 & 2\ldots p\\ 1 & 2\ldots p\end{pmatrix} \cdot |A| =$$

$$= \left| \begin{array}{cccc|ccc} |A| & 0 & \ldots 0 & 0 & \ldots 0 \\ 0 & |A| & \ldots 0 & 0 & \ldots 0 \\ \cdot & \cdot & \ldots & \cdot & \ldots \\ 0 & & \ldots |A| & 0 & \ldots 0 \\ \hline a_{1,p+1} & a_{2,p+1} & \ldots a_{p,p+1} & a_{p+1,p+1} & \ldots a_{n,p+1} \\ \cdot & \cdot & \ldots \cdot & \cdot & \ldots \cdot \\ a_{1n} & a_{2n} & \ldots a_{pn} & a_{p+1,n} & \ldots a_{nn} \end{array} \right| =$$

$$= \quad |A|^p \cdot A\begin{pmatrix} p+1 & \cdots & n \\ p+1 & \cdots & n \end{pmatrix}. \tag{1.4}$$

We observe at once that, for p = n, it is simple to obtain that

$$\tilde{A}\begin{pmatrix} 1 & 2 \ldots n \\ 1 & 2 \ldots n \end{pmatrix} \cdot |A| = |A|^n. \tag{1.5}$$

If the matrix A is nonsingular ($|A| \neq 0$) then formula (1.3) (resp.(1.2)) follows immediately from (1.4) (resp.(1.5)). In the case where $|A| = 0$ the identity (1.3) (resp.(1.2)) is obtained by a standard limit transition. Namely, the matrix $A_\varepsilon = A + \varepsilon E$ - where E is the identity matrix on order n - is considered. The determinant $|A_\varepsilon|$ is a polynomial in ε. Therefore, in an arbitrarily small neighbourhood of zero there can be found values ε for which $|A_\varepsilon| \neq 0$. Having noted that for such ε the identity in formula (1.3)(resp.(1.2)) is valid for A_ε (strictly speaking for the minors of the reciprocal $\tilde{A}_\varepsilon$) we take the limit for $\varepsilon \to 0$ over those values of ε for which $|A_\varepsilon| \neq 0$. Hereby the minors of A_ε and $\tilde{A}_\varepsilon$ go to the respective minors of the matrices A and $\tilde{A}$ we obtain the identity (1.3)(resp.(1.2)).

EXAMPLES AND EXERCISES

1. Let

$$A = \left\| \begin{array}{ccc} 3 & 0 & -1 \\ 2 & 7 & -2 \\ -3 & 4 & 0 \end{array} \right\|.$$

Whithout constructing $\tilde{A}$ we evaluate $\tilde{A}\binom{1\ \ 2}{2\ \ 3}$. Here p = 2,
$|A| = 3.8+(-1)(29) = -5$,

$$|A|^{p-1} = -5, \quad A\binom{3}{1} = -3.$$

With formula (1.1)

$$\tilde{A}\begin{pmatrix}1 & 2\\ 2 & 3\end{pmatrix} = |A|^{p-1}\, A\begin{pmatrix}3\\ 1\end{pmatrix} = (-5)(-3) = 15. \qquad (1.6)$$

2. For the matrix A of example 1 we evaluate $|\tilde{A}|$. With formula (1.2)

$$|\tilde{A}| = |A|^{n-1} = (-5)^2 = 25.$$

We note, that the matrix $\tilde{A}$ itself has, in the given case, the form

$$\tilde{A} = \begin{Vmatrix} 8 & -6 & 29\\ 4 & -3 & 12\\ 7 & -4 & 21 \end{Vmatrix}.$$

Returning to example 1 we verify the result (1.6):

$$\tilde{A}\begin{pmatrix}1 & 2\\ 2 & 3\end{pmatrix} = \begin{vmatrix} -6 & 29\\ -3 & 12\end{vmatrix} = -72 + 87 = 15.$$

3. If the matrix A is "lower triangular",

$$A = \begin{Vmatrix} a_{11} & 0 & 0 & \cdots & 0\\ a_{21} & a_{22} & 0 & \cdots & 0\\ \cdot & \cdot & \cdot & \cdots & \cdot\\ a_{n1} & a_{n2} & a_{n3} & \cdots & a_{nn} \end{Vmatrix},$$

i.e., $a_{ij} = 0$ for $j > i$, then the reciprocal matrix $\tilde{A}$ is "upper triangular",

$$\tilde{A} = \begin{Vmatrix} \tilde{a}_{11} & \tilde{a}_{12} & \cdots & \tilde{a}_{1n}\\ 0 & \tilde{a}_{22} & \cdots & \tilde{a}_{2n}\\ \cdot & \cdot & \cdots & \cdot\\ 0 & 0 & \cdots & \tilde{a}_{nn} \end{Vmatrix}$$

i.e., $\tilde{a}_{ij} = 0$ for $i > j$.

4. Is the converse of the statement, for lated in example 3, correct?

§ 2 THE SYLVESTER IDENTITIES FOR BORDERED MINORS

2.1 We consider in the matrix $A = \|a_{ij}\|_{i,j=1}^{n}$ $(n \geqq 2)$ the minor

$$A\begin{pmatrix}1 & 2 & \cdots & p\\ 1 & 2 & \cdots & p\end{pmatrix} \qquad (1 \leqq p < n)$$

and we shall border it, adding any row and any column from the remaining n-p rows and n-p columns, i.e., forming the minors

$$b_{rs} \equiv A\begin{pmatrix}1 & 2 & \cdots & p & r\\ 1 & 2 & \cdots & p & s\end{pmatrix}.$$

With these minors we construct the matrix

$$B = \| b_{rs} \|_{r,s=p+1}^{n}$$

of order n-p and we set ourselves the aim to evaluate its determinant

$$B\begin{pmatrix} p+1 & \cdots & n \\ p+1 & \cdots & n \end{pmatrix} .$$

THEOREM 2.1 (SYLVESTER). *The following identity holds:*

$$B\begin{pmatrix} p+1 & \cdots & n \\ p+1 & \cdots & n \end{pmatrix} = |A| \left[A\begin{pmatrix} 1 & 2 & \cdots & p \\ 1 & 2 & \cdots & p \end{pmatrix} \right]^{n-p-1} . \quad (S)$$

PROOF. Since for n = p-1 formula (S) is trivial, we shall assume $1 \leqq p < n-1$. We consider the matrix $\tilde{A}$, reciprocal with respect to A, and some of its minors, namely

$$c_{rs} \equiv \tilde{A}\begin{pmatrix} p+1 & \cdots & r-1 & r+1 & \cdots & n \\ p+1 & \cdots & s-1 & s+1 & \cdots & n \end{pmatrix}$$

$$(r,s = p+1,\cdots,n). \quad (2.1)$$

According to Theorem 1.1 (formula (1.1)) we have

$$c_{rs} = |A|^{n-p-2} A\begin{pmatrix} 1 & 2 & \cdots & p & r \\ 1 & 2 & \cdots & p & s \end{pmatrix} = |A|^{n-p-2} b_{rs}$$

$$(r,s = p+1,\cdots,n). \quad (2.2)$$

Now setting up the matrix $C = \| c_{rs} \|_{r,s=p+1}^{n}$ of order n-p, we calculate from (2.2) its determinant $|C|$:

$$|C| = |A|^{(n-p-2)(n-p)} B\begin{pmatrix} p+1 & \cdots & n \\ p+1 & \cdots & n \end{pmatrix} . \quad (2.3)$$

The determinant (2.3) can be evaluated also in an alternative way taking advantage of the fact, that by the definition (2.1) of the number c_{rs}, the matrix C is the reciprocal for the matrix $\| \tilde{a}_{rs} \|_{r,s=p+1}^{n}$. Taking this into account, we have on the basis of (1.2)

$$|C| = \left[\tilde{A}\begin{pmatrix} p+1 & \cdots & n \\ p+1 & \cdots & n \end{pmatrix} \right]^{n-p-1} .$$

Now, having evaluated the minors $\tilde{A}\begin{pmatrix} p+1 & \cdots & 1 \\ p+1 & \cdots & 1 \end{pmatrix}$ standing between the brackets we obtain, again by formula (1.1),

$$|C| = \left[|A|^{n-p-1} A\begin{pmatrix} 1 & 2 & \cdots & p \\ 1 & 2 & \cdots & p \end{pmatrix} \right]^{n-p-1} =$$

$$= |A|^{(n-p-1)^2} \left[A \begin{pmatrix} 1 & 2 & \cdots & p \\ 1 & 2 & \cdots & p \end{pmatrix} \right]^{n-p-1} \quad (2.4)$$

It remains to equate the expressions (2.3) and (2.4) for C and in the case $|A| \neq 0$ we obtain, after cancellation, the Sylvester identity (S). If $|A| = 0$, then the proof is completed by means of the same method, which was exploited above for the analogous case in the ascertainment of Theorem 1.1.

2.2 We shall often have to use the Sylvester identity, mainly in two special cases, which we consider in detail.

Let $p = n-2$. Then formula (S) is reduced to the identity

$$|A| \; A \begin{pmatrix} 1 & 2 & \cdots & n-2 \\ 1 & 2 & \cdots & n-2 \end{pmatrix} = B \begin{pmatrix} n-1 & n \\ n-1 & n \end{pmatrix}$$

or, more explicitly,

$$|A| \; A \begin{pmatrix} 1 & 2 & \cdots & n-2 \\ 1 & 2 & \cdots & n-2 \end{pmatrix} =$$

$$= A \begin{pmatrix} 1 & \cdots & n-2 & n-1 \\ 1 & \cdots & n-2 & n-1 \end{pmatrix} A \begin{pmatrix} 1 & \cdots & n-2 & n \\ 1 & \cdots & n-2 & n \end{pmatrix} -$$

$$- A \begin{pmatrix} 1 & \cdots & n-2 & n-1 \\ 1 & \cdots & n-2 & n \end{pmatrix} A \begin{pmatrix} 1 & \cdots & n-2 & n \\ 1 & \cdots & n-2 & n-1 \end{pmatrix}. \quad (2.5)$$

It is easy to make clear to oneself the structure of the minors appearing in (2.5), having considered the diagram [1)]

$$|A| = \begin{vmatrix} A \begin{pmatrix} 1 & 2 & \cdots & n-2 \\ 1 & 2 & \cdots & n-2 \end{pmatrix} & \begin{matrix} n-1 & n \\ \vdots & \vdots \\ \vdots & \vdots \end{matrix} \\ \cdots\cdots\cdots\cdots\cdots\cdots\cdots & \begin{matrix} n-1 \\ n \end{matrix} \end{vmatrix} .$$

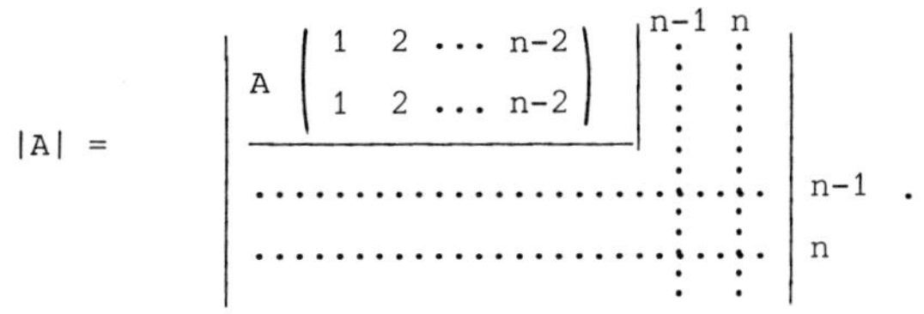

By a simple permutation of rows and columns one obtains from formula (2.5) another variant, which is very convenient in a series of computations

$$|A| \; A \begin{pmatrix} 2 & 3 & \cdots & n-1 \\ 2 & 3 & \cdots & n-1 \end{pmatrix} =$$

$$= A \begin{pmatrix} 1 & 2 & \cdots & n-1 \\ 1 & 2 & \cdots & n-1 \end{pmatrix} A \begin{pmatrix} 2 & 3 & \cdots & n \\ 2 & 3 & \cdots & n \end{pmatrix} -$$

$$- A \begin{pmatrix} 1 & \cdots & n-2 & n-1 \\ 2 & \cdots & n-1 & n \end{pmatrix} A \begin{pmatrix} 2 & \cdots & n-1 & n \\ 1 & \cdots & n-2 & n-1 \end{pmatrix}. \quad (2.6)$$

Here the related diagram is derived as

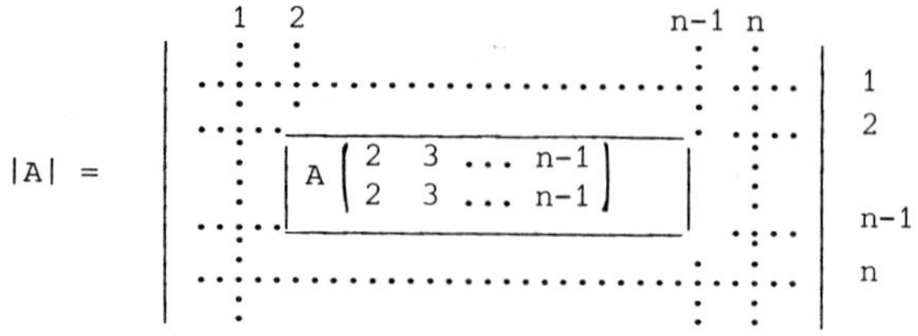

EXAMPLES AND EXCERCISES

1. Let

$$A = \begin{Vmatrix} 3 & -2 & 1 & 0 \\ 4 & -1 & 2 & 3 \\ 0 & 3 & 1 & 5 \\ 7 & -4 & 2 & 6 \end{Vmatrix} .$$

By rule (2.5)

$$|A| \begin{vmatrix} 3 & -2 \\ 4 & -1 \end{vmatrix} = \begin{vmatrix} 3 & -2 & 1 \\ 4 & -1 & 2 \\ 0 & 3 & 1 \end{vmatrix} \begin{vmatrix} 3 & -2 & 0 \\ 4 & -1 & 3 \\ 7 & -4 & 6 \end{vmatrix} - \begin{vmatrix} 3 & -2 & 0 \\ 4 & -1 & 3 \\ 0 & 3 & 5 \end{vmatrix} \begin{vmatrix} 3 & -2 & 1 \\ 4 & -1 & 2 \\ 7 & -4 & 2 \end{vmatrix} = (-1)\cdot 24-(-2)(-3) = -30,$$

i.e., 5 $|A| = -30$, $|A| = -6$.

By rule (2.6)

$$|A| \begin{vmatrix} -1 & 2 \\ 3 & 1 \end{vmatrix} = \begin{vmatrix} 3 & -2 & 1 \\ 4 & -1 & 2 \\ 0 & 3 & 1 \end{vmatrix} \begin{vmatrix} -1 & 2 & 3 \\ 3 & 1 & 5 \\ -4 & 2 & 6 \end{vmatrix} - \begin{vmatrix} -2 & 1 & 0 \\ -1 & 2 & 3 \\ 3 & 1 & 5 \end{vmatrix} \begin{vmatrix} 4 & -1 & 2 \\ 0 & 3 & 1 \\ 7 & -4 & 2 \end{vmatrix} = (-1)(-42)-0 = 42,$$

i.e, $|A| = -6$.

2. Let

$$A = \begin{Vmatrix} 3 & 1-i & 2+5i \\ 1+i & 3 & 1-i \\ 2-5i & 1+i & 3 \end{Vmatrix} .$$

By rule (2.6)

$$3\ |A| = \begin{vmatrix} 3 & i-1 \\ 1+i & 3 \end{vmatrix} \begin{vmatrix} 3 & 1-i \\ 1+i & 3 \end{vmatrix} - \begin{vmatrix} 1-i & 2+5i \\ 3 & 1-i \end{vmatrix} \begin{vmatrix} 1+i & 3 \\ 2-5i & 1+i \end{vmatrix} = (9-|1-i|^2)^2 -$$

$$- |(1-i)^2-3(2+5i)|^2 = 7^2-|1-2i-1-6-15i|^2 =$$

$$= 49-|-6-17i|^2 = 49-325 = -276,$$

i.e., $|A| = -92$.

3. Generalize identity (S) in the following way:

$$B\begin{pmatrix} i_1 & i_2 & \cdots & i_q \\ k_1 & k_2 & \cdots & k_q \end{pmatrix} = A\begin{pmatrix} 1 & 2 & \cdots & p & i_1 & i_2 & \cdots & i_q \\ 1 & 2 & \cdots & p & k_1 & k_2 & \cdots & k_q \end{pmatrix} \left[A\begin{pmatrix} 1 & 2 & \cdots & p \\ 1 & 2 & \cdots & p \end{pmatrix}\right]^{q-1} \tag{2.7}$$

where

$$p < \begin{matrix} i_1 < i_2 < \cdots < i_q \\ k_1 < k_2 < \cdots < k_q \end{matrix} \leq n$$

HINT. Apply identity (S) to the determinant

$$A\begin{pmatrix} 1 & 2 & \cdots & p & i_1 & i_2 & \cdots & i_q \\ 1 & 2 & \cdots & p & k_1 & k_2 & \cdots & k_q \end{pmatrix} .$$

4. (Kronecker [47]). Let (in the notation of § 1.1) the minor $A\begin{vmatrix} 1 & 2 & \cdots & p \\ 1 & 2 & \cdots & p \end{vmatrix} \neq 0$ for the matrix A, but let all bordered minors $b_{rs} = 0$ $(r,s = p+1,\cdots,n)$. Then the rank of the matrix A is equal to p.

HINT. In order to prove that all minors of order $m > p$ are equal to zero, use the Sylvester identity (S) (or (2.7)) in connection with the matrix

$$\mathring{A} = \left\| \begin{array}{ccc|ccc} a_{11} & \cdots & a_{1p} & a_{1k_1} & \cdots & a_{1k_m} \\ \cdot & \cdots & \cdot & \cdot & \cdots & \cdot \\ a_{p1} & \cdots & a_{pp} & a_{pk_1} & \cdots & a_{pk_m} \\ \hline a_{i_1 1} & \cdots & a_{i_1 p} & 0 & \cdots & 0 \\ \cdot & \cdots & \cdot & \cdot & \cdots & \cdot \\ a_{i_m 1} & \cdots & a_{i_m p} & 0 & \cdots & 0 \end{array} \right\| ,$$

where

$$1 \leq \begin{matrix} i_1 < i_2 < \cdots < i_m \\ k_1 < k_2 < \cdots < k_m \end{matrix} \leq n.$$

NOTE.

1) For the notation applied in this and the next diagram see the note to formula (3.3) below.

§ 3 EVALUATION OF CERTAIN DETERMINANTS

3.1. In this section the problem will be to evaluate certain determinants of a special form. These determinants will occur repeatedly in the Chapters II and III.

Let $A = \|a_{ij}\|_{i,j=1}^{n}$ be a matrix of order n and let p and r be natural numbers, where $p+r \leqq n$. We consider two sets of indices

$$(1\leqq)\begin{matrix} i_1 < i_2 < \cdots i_{p+r} \\ j_1 < j_2 < \cdots j_{p+r} \end{matrix}(\leqq n)$$

From the set $\{i_1,i_2,\cdots,i_{p+r}\}$ of p+r indices we choose some set

$$\{\mu_1,\mu_2,\cdots,\mu_p\}$$

consisting of p indices, indexed here in arbitrary order, and in the same way a set

$$\{\nu_1,\nu_2,\cdots,\nu_p\}$$

consisting of p indices from the set $\{j_1,j_2,\cdots,j_{p+r}\}$. Let $\alpha_1 < \alpha_2 < \cdots < \alpha_r$ and $\beta_1 < \beta_2 < \cdots < \beta_r$ be the complements of the sets $\{\mu_1,\mu_2,\cdots,\mu_p\}$ and $\{\nu_1,\nu_2,\cdots,\nu_p\}$ in the sets $\{i_1,i_2,\cdots,i_{p+r}\}$ and $\{j_1,j_2,\cdots,j_{p+r}\}$, respectively, and

$$A_r = A\begin{pmatrix} \alpha_1 & \alpha_2 & \cdots & \alpha_r \\ \beta_1 & \beta_2 & \cdots & \beta_r \end{pmatrix}$$

the minor of order r of the matrix A (see § 1.1)

Further we denote through $A^{(\zeta)}$ the matrix, obtained from the matrix A through the substitution of its entries $a_{\mu_1\nu_1}, a_{\mu_2\nu_2},\cdots,a_{\mu_p\nu_p}$ by numbers $\zeta_1,\zeta_2,\cdots,\zeta_p$, respectively, and we consider of the matrices A and $A^{(\zeta)}$ the corresponding minors of order p+r

$$M_p^{(r)} = A\begin{pmatrix} i_1 & i_2 & \cdots & i_{p+r} \\ j_1 & j_2 & \cdots & j_{p+r} \end{pmatrix},$$

$$M_p^{(r)}(\zeta) = A^{(\zeta)}\begin{pmatrix} i_1 & i_2 & \cdots & i_{p+r} \\ j_1 & j_2 & \cdots & j_{p+r} \end{pmatrix}. \qquad (3.1)$$

LEMMA 3.1 *If* r *is the rank of the matrix*

$$A = \|a_{ij}\|_{i,j=1}^{n},$$

and σ_μ *and* σ_ν *are the number of inversions in the sets*

$$\{\mu_1,\mu_2,\cdots,\mu_p,\alpha_1,\alpha_2,\cdots,\alpha_r\}$$

and

$$\{\nu_1,\nu_2,\cdots,\nu_p,\beta_1,\beta_2,\cdots,\beta_r\}$$

respectively, then the determinant $M_p^{(r)}(\zeta)$ *(see (3.1)) is evaluated by the rule* [1)]

$$M_p^{(r)}(\zeta) = (-1)^{\sigma_\mu+\sigma_\nu} A_r \prod_{\omega=1}^{p} (\zeta_\omega - a_{\mu_\omega\nu_\omega}).$$

PROOF. This is particulary clear in the special case where $\mu_\omega = i_\omega$,

$\nu_\omega = j_\omega$ $(\omega = 1,2,\dots,p)$, i.e. [2)]

$$M_p^{(r)}(\zeta) = \left|\begin{array}{cccc|cc} \zeta_1 & a_{i_1j_2} & \dots & a_{i_1j_2} & \dots & a_{i_1j_{p+r}} \\ a_{i_2j_1} & \zeta_2 & \dots & a_{i_2j_p} & \dots & a_{i_2j_{p+r}} \\ \cdot & \cdot & \dots & \cdot & \dots & \cdot \\ a_{i_pj_1} & a_{i_pj_2} & \dots & \zeta_p & \dots & a_{i_pj_{p+r}} \\ \hline \cdot & \cdot & \dots & \cdot & & \\ a_{i_{p+r}j_1} & a_{i_{p+r}j_2} & \dots & a_{i_{p+r}j_p} & & A_r \end{array}\right| . \qquad (3.3)$$

The polynomial $M_p^{(r)}(\zeta) \equiv M_p^{(r)}(\zeta_1,\zeta_2,\dots\zeta_p)$ in the parameters $\zeta_1,\zeta_2,\dots,\zeta_p$ vanishes if one replaces any of these parameters, for example ζ_q, by the element $a_{i_qj_q}$ of the matrix A which place this parameter occupies. Indeed,with $\zeta_q = a_{i_qj_q}$ in the determinant $M_p^{(r)}(\zeta)$ of order p+r, there appear r+1 of the rows (columns) of the original matrix A, which are linearly dependent, as this matrix, by assumption, has rank r. Thus the polynomial $M_p^{(r)}(\zeta)$ is divisible without remainder by the product

$$(\zeta_1 - a_{i_1j_1})(\zeta_2 - a_{i_2j_2})\cdots(\zeta_p - a_{i_pj_p}).$$

It is clear, that the quotient under this division will be the leading coefficient of the polynomial, i.e., the coefficient of the product $\zeta_1\zeta_2\cdots\zeta_p$. But from (3.3) it is evident that the minor A_r is this coefficient, i.e.,

$$M_p^{(r)}(\zeta) = A_r \prod_{\omega=1}^{p} (\zeta_\omega - a_{i_\omega j_\omega}).$$

Since in the present case $\sigma_\mu(\equiv \sigma_i) = 0$, $\sigma_\nu(\equiv \sigma_j) = 0$, formula (3.2) is established for the special case under consideration.

The reasoning we followed is also applicable in the general case with only this difference, that now the coefficient of the product $\zeta_1\zeta_2\cdots\zeta_p$ in the determinant $M_p^{(r)}(\zeta)$ is distinguished from A_r by a factor $\pm$ 1, depending on the location of the entries $\zeta_1,\zeta_2,\dots,\zeta_p$ in the minor $M_p^{(r)}(\zeta)$, i.e., of the elements $a_{\mu_1\nu_1}, a_{\mu_2\nu_2},\dots,a_{\mu_p\nu_p}$ in the minor $M_p^{(r)}$ (see (3.1)). But, as is known from the theory of determinants, this factor is $(-1)^{\sigma_\mu+\sigma_\nu}$. Lemma 3.1 is proved.

The shape of formula (3.2) permits to deduct from Lemma 3.1 this

COROLLARY. *The value of the determinant* $M_p^{(r)}(\zeta)$ *doesn't vary, if arbitrary elements of the matrix A, with exclusion of* $a_{\mu_\omega \nu_\omega}$ $(\omega = 1,2,\cdots,p)$ *and elements which enter in the structure of the minor* A_r*, are changed in such way that the rank of the matrix* A *always remains equal to* r.

3.2. We shall mainly have to apply Lemma 3.1 in two special cases on which we shall go into some detail. The first of these is the case, where (here, it is convenient to substitute the index p by the index k)

$$M_k^{(r)} = \left|\begin{array}{c|cccc} & \cdots & \cdots & \cdots & \cdots \\ A_r & \cdots & \cdots & \cdots & \cdots \\ & \cdots & \cdots & \cdots & \cdots \\ \hline \cdots & \cdots & & & a_{n-k+1,n} \\ & \cdot & & \cdot & \cdot \\ \cdots & \cdot & \cdot & & \cdot \\ & & \cdot & & \cdot \\ \cdots & a_{n,n-k+1} & & \cdots & a_{nn} \end{array}\right|$$

$$M_k^{(r)}(\zeta) = \left|\begin{array}{c|cccc} & \cdots & \cdots & \cdots & \cdots \\ A_r & \cdots & \cdots & \cdots & \cdots \\ & \cdots & \cdots & \cdots & \cdots \\ \hline \cdots & \cdots & & & \zeta_1 \\ & \cdot & & \cdot & \\ \cdots & \cdot & \cdot & & \\ & & \cdot & & \\ \cdots & \zeta_k & & \cdots & a_{nn} \end{array}\right|$$

i.e., when

$$\mu_\omega = n-k+\omega, \quad \nu_\omega = n+1-\omega \quad (\omega = 1,2,\cdots,k).$$

We note, that now (see Lemma 3.1) the complementary indices $\alpha_1 < \alpha_2 < \cdots < \alpha_r$ are smaller than all μ_ω $(\omega = 1,2,\cdots,k)$, the indices $\beta_1 < \beta_2 < \cdots < \beta_r$ smaller all ν_ω $(\omega = 1,2,\cdots,k)$, and the sets $\mu_1 < \cdots < \mu_k$ and $\nu_1 > \cdots > \nu_k$ are monotonous. Hence in the set of indices

$$\mu_1,\mu_2,\cdots,\mu_k,\alpha_1,\alpha_2,\cdots,\alpha_r$$

the numer of inversions is equal to

$$\sigma_\mu = \underbrace{r+r+\cdots+r}_{\text{k times}} = kr,$$

and in the set

$$\nu_1,\nu_2,\cdots,\nu_k,\beta_1,\beta_2,\cdots,\beta_r$$

the number of inversions is equal to

$$\sigma_\nu = (k-1+r)+(k-2+r)+\cdots+(1+r)+r = kr + \frac{k(k-1)}{2} .$$

Thus, in the given case, formula (3.2) takes the form

$$M_k^{(r)}(\zeta) \equiv \left|\begin{array}{c|ccc} A_r & \cdots & \cdots & \cdots \\ & \cdots & \cdots & \cdots \\ & \cdots & \cdots & \cdots \\ \hline \cdots & \cdot\;\cdot\;\cdot & & \zeta_1 \\ \cdots & \cdot\;\cdot\;\cdot & \cdot\quad\cdot & \vdots \\ \cdots & \zeta_k & \cdot\;\cdot\;\cdot & a_{nn} \end{array}\right| =$$

$$= (-1)^{k(k-1)/2} A_r \prod_{\omega=1}^{k} (\zeta_\omega - a_{n-k+\omega,n-\omega+1}) . \qquad (3.4)$$

In particular, for $a_{n-k+\omega,n-\omega+1} \equiv a(\omega = 1,2,\cdots,k)$ and $\zeta_1 = \zeta_2 = \cdots = \zeta_n \equiv \zeta$ (so, namely, it will be in Ch. II) we have

$$M_k^{(r)}(\zeta) = (-1)^{k(k-1)/2} A_r(\zeta-a)^k \qquad (3.5)$$

We make now the following remark, important for the application of formulae (3.4) and (3.5). In the corollary to Lemma 3.1 there were indications on the possibility of variation within specific bounds of some entries of the determinant $M_k^{(r)}(\zeta)$, without changing its value. In the present case this statement can be sharpened, having noted, for example, that

1°. *The determinant* $M_k^{(r)}(\zeta)$ *is not changed if all its entries, standing in the righthand lower corner under the diagonal* $\zeta_1,\cdots,\zeta_k$ *in diagram* (3.4) *are substituted by arbitrary numbers.*

Indeed, we partition the last column of the determinant $M_k^{(r)}(\zeta)$ (see (3.4)) in two parts

$$\begin{pmatrix} 0 \\ \vdots \\ 0 \\ \zeta_1 - a_{n-k+1,n} \\ 0 \\ \vdots \\ 0 \end{pmatrix} + \begin{pmatrix} * \\ \vdots \\ * \\ a_{n-k+1,n} \\ * \\ \vdots \\ * \end{pmatrix} ,$$

where all other entries in the partitioned column (i.e. of the matrix A) are denoted by asterikses, and, corresponding to this, we shall partition the determinant $M_k^{(r)}(\zeta)$ in two terms:

14/ GENERAL THEORY OF MATRICES AND FORMS

$$M_k^{(r)}(\zeta) = \begin{vmatrix} A_r & \begin{matrix} * & * & \cdots & * & 0 \\ \cdot & \cdot & \cdots & \cdot & \cdot \\ * & * & \cdots & * & 0 \end{matrix} \\ \begin{matrix} * & \cdots & * \\ * & \cdots & * \\ \cdot & \cdots & \cdot \\ * & \cdots & * \end{matrix} & \begin{matrix} * & * & \cdots & * & \zeta_1 - a_{n-k+1,n} \\ * & * & \cdots & \zeta_2 & 0 \\ \cdot & \cdot & \cdots & \cdot & \cdot \\ \zeta_k & * & \cdots & * & 0 \end{matrix} \end{vmatrix} +$$

$$+ \left. \begin{vmatrix} A_r & \begin{matrix} * & * & \cdots & * & * \\ \cdot & \cdot & \cdots & \cdot & \cdot \\ * & * & \cdots & * & * \end{matrix} \\ \begin{matrix} * & \cdots & * \\ * & \cdots & * \\ \cdot & \cdots & \cdot \\ * & \cdots & * \end{matrix} & \begin{matrix} * & * & \cdots & * & a_{n-k+1,n} \\ * & * & \cdots & \zeta_2 & * \\ \cdot & \cdot & \cdots & \cdot & \cdot \\ \zeta_1 & * & \cdots & * & * \end{matrix} \end{vmatrix} \right\} r+1$$

Since in the second of these one has $r+1$ rows which are part of the corresponding rows of the matrix A of rank r, this determinant is equal to zero, and this moreover identically relative to the entries of all remaining rows and, in particular, to arbitrary values of the entries, standing in the righthand lower corner, below the diagonal $a_{n-k+1,n}, \zeta_2, \ldots, \zeta_k$. Thus

$$M_k^{(r)}(\zeta) = (-1)^{(r+1)(r+k)} (\zeta_1 - a_{n-k+1,n}) \begin{vmatrix} A_r & \begin{matrix} * & * & \cdots & * \\ \cdot & \cdot & \cdots & \cdot \\ * & * & \cdots & * \end{matrix} \\ \begin{matrix} * & \cdots & * \\ * & \cdots & * \\ \cdot & \cdots & \cdot \\ * & \cdots & * \end{matrix} & \begin{matrix} * & * & \cdots & \zeta_2 \\ * & * & \cdots & * \\ \cdot & \cdot & \cdots & \cdot \\ \underbrace{\zeta_k & * & \cdots & *}_{k-1} \end{matrix} \end{vmatrix} \Bigg\} k-1$$

and repeating the same method another $(k-1)$ times, we are convinced that proposition 1° is correct.

REMARK. Incidentally, we obtain a new, independent proof of formula (3.4). The simple verification that the signs in front of the product do coincide we leave to the reader as an exercise (cf.[4o]).

3.3 Another case of an application of Lemma 3.1 is found if one

considers the minor $M_p^{(r)}$ of the matrix A, which has the form
$M_p^{(r)} \equiv M_{k,\ell}^{(r)} =$

$$= \begin{array}{|ccc|c|ccc|}
\multicolumn{3}{c}{\overbrace{\qquad\qquad}^{k}} & \multicolumn{1}{c}{\overbrace{\qquad}^{r}} & \multicolumn{3}{c}{\overbrace{\qquad\qquad}^{\ell}} \\
a_{11} & \cdots & a_{1k} & \cdot\ \cdots\ \cdot & a_{1,\tau} & \cdots & a_{1n} \\
\cdot & \cdots & \cdot & \cdot\ \cdots\ \cdot & \vdots & & \vdots \\
a_{\ell 1} & \cdots & a_{\ell k} & \cdot\ \cdots\ \cdot & a_{\ell,\tau} & \cdots & a_{\ell n} \\
\hline
\cdot & \cdots & \cdot & & \cdot & \cdots & \cdot \\
\cdot & \cdots & \cdot & A_r & \cdot & \cdots & \cdot \\
\cdot & \cdots & \cdot & & \cdot & \cdots & \cdot \\
\hline
a_{\sigma,1} & \cdots & a_{\sigma,k} & \cdot\ \cdots\ \cdot & a_{\sigma,\tau} & \cdots & a_{\sigma,n} \\
\vdots & \ddots & \vdots & \cdot\ \cdots\ \cdot & \cdot & \cdots & \cdot \\
a_{n1} & \cdots & a_{nk} & \cdot\ \cdots\ \cdot & a_{n,\tau} & \cdots & a_{nn}
\end{array}
\begin{array}{l}
\left.\begin{array}{c}\\ \\ \\ \end{array}\right\} \ell \\
\left.\begin{array}{c}\\ \\ \\ \end{array}\right\} r \\
\left.\begin{array}{c}\\ \\ \\ \end{array}\right\} k
\end{array}
\tag{3.6}$$

(where $\sigma=n-k+1$, $\tau=n-\ell+1$), and the corresponding minor $M_p^{(r)}(\zeta)$ of the matrix $A^{(\zeta)}$. Here (cf. Lemma 3.1)

$$p = k+\ell > 0, k \geqq 0, \ell \geqq 0,$$

$$\mu_1 = n-k+1,\ \mu_2 = n-k+2, \ldots, \mu_k = n,$$

$$\mu_{k+1} = 1,\ \mu_{k+2} = 2, \ldots, \mu_{k+\ell} = \ell;$$

$$\nu_1 = 1,\ \nu_2 = 2, \ldots, \nu_k = k,$$

$$\nu_{k+1} = n-\ell+1,\ \nu_{k+2} = n-\ell+2, \ldots, \nu_{k+1} = n,$$

such that

$$\mu_{k+1} < \mu_{k+2} < \ldots < \mu_{k+\ell} < \alpha_1 < \alpha_2 < \ldots < \alpha_r < \mu_1 < \mu_2 < \ldots \mu_k,$$

$$\nu_1 < \nu_2 < \ldots < \nu_k < \beta_1 < \beta_2 < \ldots < \beta_r < \nu_{k+1} < \nu_{k+2} < \ldots < \nu_{k+\ell} .$$

Hence

$$\sigma_\mu = k(\ell+r),\ \sigma_\nu = \ell r$$

and formula (3.2) yields (cf. [37])
$M_p^{(r)}(\zeta) \equiv M_{k,}^{(r)}(\xi,\eta) =$

$$= \begin{array}{|ccc|c|ccc|}
a_{11} & \cdots & a_{1k} & \cdot\ \cdot\ \cdot & \eta_1 & \cdots & a_{1n} \\
\cdot & \cdots & \cdot & \cdot\ \cdot\ \cdot & \vdots & & \vdots \\
a_{\ell 1} & \cdots & a_{\ell k} & \cdot\ \cdot\ \cdot & a_{\ell,\tau} & \cdots & \eta_\ell \\
\hline
\cdot & \cdots & \cdot & & \cdot & \cdots & \cdot \\
\cdot & \cdots & \cdot & A_r & \cdot & \cdots & \cdot \\
\cdot & \cdot & \cdot & & \cdot & \cdots & \cdot \\
\hline
\xi_1 & \cdots & a_{\sigma,k} & \cdot\ \cdot\ \cdot & a_{\sigma,\tau} & \cdots & a_{\sigma,n} \\
\vdots & \ddots & \vdots & \cdot\ \cdot\ \cdot & \cdot & \cdots & \cdot \\
a_{n1} & \cdots & \xi_k & \cdot\ \cdot\ \cdot & a_{n,\tau} & \cdots & a_{nn}
\end{array} =$$

$$= (-1)^{k\ell+r(k+\ell)} A_r \prod_{\omega=1}^{k} (\xi_\omega - a_{n-k+\omega,\omega}) \prod_{\omega=1}^{\ell} (\eta_\omega - a_{\omega,n-\ell+\omega}) \qquad (3.7)$$

(where $\sigma=n-k+1$, $\tau=n-\ell+1$).

In particular, for $a_{n-k+\omega,\omega} \equiv a, \zeta_\omega \equiv \xi$ $(\omega=1,2,\cdots,k)$; $a_{\omega,n-\ell+\omega} \equiv b$, $\eta_\omega = \eta$ $(\omega=1,2,\cdots,\ell)$ (this case will be met in Ch. III)

$$M^{(r)}_{k,\ell}(\xi,\eta) = (-1)^{k\ell+r(k+\ell)} A_r (\xi-a)^k (\eta-b)^\ell. \qquad (3.8)$$

This formula take an even more simple shape when the matrix A is HERMITIAN and $k=\ell$. Then $b=\bar{a}$, and if, besides, one assumes $\eta=\bar{\xi}$, then we obtain

$$M^{(r)}_{k,k}(\xi,\bar{\xi}) \equiv M^{(r)}_{k}(\xi) = (-1)^k A_r |\xi-a|^{2k}. \qquad (3.9)$$

In an analogous way, in the case of a (complex) symmetric matrix A for $k=\ell$ and $\eta=\xi$ we have instead of (3.8)

$$M^{(r)}_{k,k}(\xi,\xi) \equiv M^{(r)}_{k}(\xi) = (-1)^k A_r (\xi-a)^{2k} \qquad (3.1o)$$

In conclusion we note that, quite in analogy to proposition 1°, one obtains the proposition

2°. *The determinant* $M^{(r)}_{k,\ell}(\xi,\eta)$ *isn't changed if all its elements standing in the lefthand lower and righthand upper corner, respectively below and above the diagonals* $\xi_1,\cdots,\xi_k$ *and* $\eta_1,\cdots,\eta_k$ *in diagram* (3.7) *are interchanged by arbitrary numbers.*

It is clear, that this result expands, in particular, the sphere of application for formulae (3.8)-(3.1o).

EXAMPLES AND EXERCISES

1. Let

$$A = \begin{Vmatrix} -1 & 3 & 0 & 4 & -5 & 2 & 1 \\ 3 & 2 & -2 & 5 & 7 & 0 & -4 \\ 2 & 5 & -2 & 9 & 2 & 2 & -3 \\ 1 & 8 & -2 & 13 & -3 & 4 & -2 \\ -4 & 1 & 2 & -1 & -12 & 2 & 5 \\ -3 & 9 & 0 & 12 & -15 & 6 & 3 \\ 5 & 7 & -4 & 14 & 9 & 2 & -7 \end{Vmatrix}.$$

Here $n=7, r=2$ (all rows are a linear combination of the first two independent rows). Let $(r=2)$, $p=2, i_1=2, i_2=4, i_3=6, i_4=7, j_1=1, j_2=3, j_3=5, j_4=6$, i.e. (see (3.1))

$$M^{(r)}_p = M^{(2)}_2 = A\begin{pmatrix} 2 & 4 & 6 & 7 \\ 1 & 3 & 5 & 6 \end{pmatrix} = \begin{vmatrix} 3 & -2 & 7 & 0 \\ 1 & -2 & -3 & 4 \\ -3 & 0 & -15 & 6 \\ 5 & -4 & 9 & 2 \end{vmatrix}.$$

Let

$$\mu_1 = i_4 = 7,\ \mu_2 = i_1 = 2,\ \nu_1 = j_1 = 1,\ \nu_2 = j_3 = 5,$$

i.e.

$$\alpha_1 = i_2 = 4,\ \alpha_2 = i_3 = 6;\quad \beta_1 = j_2 = 3,\ \beta_2 = j_4 = 6,$$

$$A_r = A_2 = A\begin{pmatrix} \alpha_1 & \alpha_2 \\ \beta_1 & \beta_2 \end{pmatrix} = A\begin{pmatrix} 4 & 6 \\ 3 & 6 \end{pmatrix} = \begin{vmatrix} -2 & 4 \\ 0 & 6 \end{vmatrix} = -12.$$

Now, on one hand

$$M_2^{(2)}(\zeta) = \begin{vmatrix} 3 & -2 & \zeta_2 & 0 \\ 1 & -2 & -3 & 4 \\ -3 & 0 & -15 & 6 \\ \zeta_1 & -4 & 9 & 2 \end{vmatrix} = \begin{vmatrix} 2 & 0 & \zeta_2+3 & -4 \\ 1 & -2 & -3 & 4 \\ -3 & 0 & -15 & 6 \\ \zeta_1-2 & 0 & 15 & -6 \end{vmatrix} =$$

$$= -2\begin{vmatrix} 2 & \zeta_2+3 & -4 \\ -3 & -15 & 6 \\ \zeta_1-2 & 15 & -6 \end{vmatrix} =$$

$$= -2[-(\zeta_2+3)(18-6\zeta_1+12)-4(-45+15\zeta_1-30)] =$$

$$= 2\ [(\zeta_2+3)(30-6\zeta_1)+4(15\zeta_1-75)] =$$

$$= 2\ [-6(\zeta_2+3)(\zeta_1-5)+60(\zeta_1-5)] = -12(\zeta_1-5)(\zeta_2+3-10) =$$

$$= -12(\zeta_1-5)(\zeta_2-7).$$

On the other hand, in the sets

$$\{\mu_1,\mu_2,\alpha_1,\alpha_2\} \equiv \{7,2,4,6\},\ \{\nu_1,\nu_2,\beta_1,\beta_2\} \equiv \{1,5,3,6\}$$

the number of inversions is equal to: $\sigma_\mu = 3$, $\sigma_\nu = 1$. In this way (see formula (3.2))

$$(-1)^{\sigma_\mu+\sigma_\nu} A_r(\zeta_1-a_{71})(\zeta_2-a_{25}) = (-1)^4(-12)(\zeta_1-5)(\zeta_2-7) = -12(\zeta_1-5)(\zeta_2-7)$$

completely in accordance with Lemma 3.1.

2. Evaluate (without developing it!) the determinant

$$\Delta = \begin{vmatrix} -1 & 0 & 4 & \eta_1 & 2 & 1 \\ 3 & -2 & 5 & 7 & \eta_2 & -4 \\ 2 & -2 & 9 & 2 & 2 & \eta_3 \\ 1 & -2 & 13 & -3 & 4 & -2 \\ -4 & 2 & -1 & -12 & 2 & 5 \\ \xi_1 & -4 & 14 & 9 & 2 & -7 \end{vmatrix}$$

Solution: $\Delta = 24(\xi_1-5)(\eta_1+5)\eta_2(\eta_3+3)$.

HINT. Consider the minor $M_{k,\ell}^{(r)} = M_{1,3}^{(2)} = A\begin{pmatrix} 1 & 2 & 3 & 4 & 5 & 7 \\ 1 & 3 & 4 & 5 & 6 & 7 \end{pmatrix}$ of the matrix A of example 1 and use formula (3.7).

3. Find, not carrying out a calculation, the roots of the third degree polynomial

$$P_3(\lambda) = \begin{vmatrix} 3 & 0 & 4 & -5 & 2 & 1 \\ 2 & -2 & 5 & 7 & 0 & -4 \\ 8 & -2 & 13 & -3 & 4 & -2 \\ 1 & 2 & -1 & -12 & 2 & \lambda \\ 9 & 0 & 12 & -15 & \lambda & -8 \\ 7 & 4 & 14 & \lambda & 11 & -3 \end{vmatrix} .$$

Solution. $\lambda_1 = 5,\ \lambda_2 = 6,\ \lambda_3 = 9.$

HINT. Again, starting with the matrix of example 1, use formula (3.4) and proposition 1°.

4. Evaluate (without developing it) the determinant

$$\Delta = \begin{vmatrix} 1 & i & -1 & \bar{\xi} & 0 \\ -i & 1 & i & -1 & \bar{\xi} \\ -1 & -i & 1 & i & -1 \\ \xi & -1 & -i & 1 & i \\ -2 & \xi & -1 & -i & 1 \end{vmatrix} .$$

Solution. $\Delta = |\xi - i|^4$.

HINT. Having selected the corresponding matrix A, use Lemma 3.1 in the shape of formula (3.9) and proposition 2°.

5. Generalize proposition 1° and 2°, having noted that in their conditions the requirement that the elements, which are interchangable with arbitrary numbers, are located strictly on one (but still completely defined) side of the diagonal $\zeta_1,\cdots,\zeta_k$ in diagram (3.4) and of the diagonals $\xi_1,\cdots,\xi_k$ and $\eta_1,\cdots,\eta_\ell$ in the diagram (3.7) is unessential.

NOTES.

1) The minor $M_p^{(r)}$ (see (3.1)) is under the conditions of Lemma 3.1 equal to zero, as its order $p+r>r$, and r is the rank of the matrix A.

2) In diagram (3.3) we use a certain abbrevation for shortness of notation, replacing a well-defined "part" or "block" of the matrix of the determinant $M_p^{(r)}(\zeta)$ by the symbol A_r, not meaning, of course, the number A_r, but the matrix, related with the minor A_r. Such a notation is used several times, alsc in the sequel (in this respect we follow [3]) in all cases, where it cannot cause misunderstanding. Actually, above we used it already in the diagrams of § 2.2.

§ 4. MATRICES AND LINEAR OPERATORS. SPECTRUM.

4.1 We recall that *eigenvalues* or *eigennumbers* (in a different terminology *characteristic numbers)* of the matrix $A = \|a_{ij}\|_{i,j=1}^n$ are called the roots $\lambda_1,\lambda_2,\cdots,\lambda_n$ (here each root is repeated according to its multiplicity) of the *characteristic polynomial*,

$$|A-\lambda E| = \begin{vmatrix} a_{11}-\lambda & a_{12} & \cdots & a_{1n} \\ a_{21} & a_{22}-\lambda & \cdots & a_{2n} \\ \cdot & \cdot & \cdots & \cdot \\ a_{n1} & a_{n2} & \cdots & a_{nn}-\lambda \end{vmatrix} =$$

$$= (-\lambda)^n + (a_{11}+a_{22}+\cdots+a_{nn})(-\lambda)^{n-1} + \cdots + |A| \qquad (4.1)$$

of this matrix. We note (this will be essential to us in the sequel) that because of the abovementioned definition *the eigennumbers of a matrix are a continuous function of its elements* [1].

The eigenvalues of a matrix have a simple geometrical meaning. Let E^n be the complex n-dimensional linear space, and $\{e_1, e_2, \cdots, e_n\}$ a basis for it. To the matrix $A = \|a_{ij}\|_{i,j=1}^n$ and to this basis, as is well-known, one can put in connection a linear operator A, working in the space E^n, defined through its elements on this basis (and thus also on the whole space E^n) by the formula [2]

$$Ae_j = a_{j1}e_1 + a_{j2}e_2 + \cdots + a_{jn}e_n \quad (j=1,2,\cdots,n). \qquad (4.2)$$

Then *the numbers* $\lambda_1, \lambda_2, \cdots, \lambda_n$ *defined above, and they only, are the eigenvalues of the operator* A, i.e., for such $\lambda = \lambda_j (j=1,2,\cdots,n)$ there exists a vector

$$x = \xi_1 e_1 + \xi_2 e_2 + \cdots + \xi_n e_n \qquad (\neq 0)$$

from E^n such that $Ax = \lambda x$.

This statement is obtained immediately, if one notes, that the equation $Ax = \lambda x$ is equivalent to the system of linear homogeneous equations

$$\begin{aligned} (a_{11}-\lambda)\xi_1 + a_{21}\xi_2 + \cdots + a_{n1}\xi_n &= 0, \\ a_{12}\xi_1 + (a_{22}-\lambda)\xi_2 + \cdots + a_{n2}\xi_n &= 0, \\ \cdots\cdots\cdots\cdots\cdots\cdots\cdots\cdots & \\ a_{1n}\xi_1 + a_{2n}\xi_2 + \cdots + (a_{nn}-\lambda)\xi_n &= 0, \end{aligned}$$

which admits a nontrivial solution $x = \{\xi_1, \xi_2, \cdots, \xi_n\}$ if and only if λ is a root of the equation $|A-\lambda E| = 0$. The vector x is called in this case an *eigenvector of the operator* A, corresponding to the eigenvalue λ.

From (4.1) it is clear among other things, that

$$\lambda_1 \lambda_2 \cdots \lambda_n = |A|. \qquad (4.3)$$

4.2. Thus, the totality of all eigenvalues of the matrix A, or its *spectrum* coincides with the *spectrum* $\sigma(A)$ (the set of all eigenvalues) of the linear operator A induced by this matrix on some basis

$\{e_1,e_2,\cdots,e_n\}$ of the space E^n [3]. Thence at once follow two corollaries:

a) *The spectra of all linear operators induced on* E^n *which are given by the matrix* $A = \|a_{ij}\|_{i,j=1}^n$, *through formula* (4.2) *for different choices of the basis* $\{e_1,e_2,\cdots,e_n\}$,*coincide.*

b) *The spectra of all matrices* $\|a_{ij}\|_{i,j=1}^n$ *generated through formula* (4.2) *by one and the same linear operator* $A \in E^n$ *for different choices of the basis* $\{e_1,e_2,\cdots,e_n\}$ *coincide.*

REMARK. Corollary b) is also easily discovered through direct calculation, not resorting to the concept of eigenvectors of a linear operator. Indeed, let on the basis $\{e_1,e_2,\cdots,e_n\}$ the operator A be given by the matrix $A = \|a_{ij}\|_{i,j=1}^n$ (see (4.2)). An arbitrary other basis $\{g_1,g_2,\cdots,g_n\}$ of the space E^n is, as is well-known ([3],p.73), connected with the basis $\{e_1,e_2,\cdots,e_n\}$ through some linear transformation

$$e_k = \sum_{i=1}^n t_{ki}g_i \quad (k = 1,2,\cdots,n) \tag{4.4}$$

with some nonsingular matrix $T = \|t_{ki}\|_{k,i=1}^n$, $(|T| \neq 0)$. On the basis $\{g_1,g_2,\cdots,g_n\}$ the operator A will correspond also to a new matrix $B = \|b_{ij}\|_{i,j=1}^n$ defined by the relations

$$Ag_i = \sum_{j=1}^n b_{ij}g_j \quad (i = 1,2,\cdots,n) \tag{4.5}$$

Then from (4.4) and (4.5) it follows, that

$$Ae_k = \sum_{i=1}^n t_{ki}\, Ag_i = \sum_{i=1}^n t_{ki} \sum_{j=1}^n b_{ij}g_j = \sum_{j=1}^n \Big(\sum_{i=1}^n t_{ki}b_{ij}\Big)g_j \quad (k=1,2,\cdots,n).$$

Meanwhile we have from (4.2) and (4.4)

$$Ae_k = \sum_{j=1}^n a_{kj}e_j = \sum_{j=1}^n a_{kj} \sum_{i=1}^n t_{ji}g_i = \sum_{i=1}^n \Big(\sum_{j=1}^n a_{kj}t_{ji}\Big)g_i .$$

Comparison between the obtained decompositions of the vectors Ae_k over the basis $\{g_1,.g_2,\cdots,g_n\}$ shows, that the matrices A,T and B are connected through the relation

$$AT = TB \quad \text{or} \quad B = T^{-1}AT, \tag{4.6}$$

whence follows that the characteristic polynomials

$$|A-\lambda E| = |T^{-1}(A-\lambda E)T| = |T^{-1}AT-\lambda E| = |B-\lambda E| \tag{4.7}$$

of the matrices A and B (and a forteriori their spectra) coincide.

4.3. If, starting with some basis $\{e_1,e_2,\cdots,e_n\}$ one introduces in E^n the *scalar product*

$$(x,y) = (\xi_1 e_1 + \xi_2 e_2 + \cdots + \xi_n e_n, \eta_1 e_1 + \eta_2 e_2 + \cdots + \eta_n e_n) =$$
$$= \xi_1 \overline{\eta}_1 + \xi_2 \overline{\eta}_2 + \cdots + \xi_n \overline{\eta}_n, \tag{4.8}$$

i.e., if one introduces in E^n the so-called structure of a *unitary* (or Euclidean) space, then in relation to this scalar product the basis $\{e_i, e_2, \cdots, e_n\}$ will have the property

$$(e_1, e_j) = \delta_{ij} \tag{4.9}$$

where δ_{ij} is the Kronecker symbol. Such a basis is called *orthonormal*. Now each Hermitian matrix

$$A = \|a_{ij}\|_{i,j=1}^{n}, \quad a_{ij} = \overline{a}_{ij} \quad (i,j=1,2,\cdots,n)$$

will induce on this basis (i.e. through formula (4.2)) a Hermitian operator A. The latter means, by definition, that for all $x,y \in E^n$

$$(\mathsf{A}x,y) = (x,\mathsf{A}y). \tag{4.1o}$$

The scalar product (4.8) evidently has the properties

$$(x,x) > 0 \quad (x \neq 0), \quad (x_1 + x_2, y) = (x_1,y) + (x_2,y)$$
$$(\alpha x, y) = \alpha\ (x,y), \quad (y,x) = \overline{(x,y)}$$

for all vectors x, x_1, x_2, y from E^n and for all complex numbers α. From these properties and relation (4.1o) follow, in particular, the following propositions.

1° *All eigenvalues* λ *of a Hermitian matrix* A *are real.*

Indeed, for the Hermitian operator A, which corresponds to the matrix A and the basis $\{e_1, e_2, \cdots, e_n\}$ (see (4.9)) through formula (4.2), the equation $\mathsf{A}x = \lambda x$ $(x \neq 0)$ implies that $(\mathsf{A}x,x) = \lambda(x,x)$. But the number $(\mathsf{A}x,x) = (x,\mathsf{A}x) = \overline{(\mathsf{A}x,x)}$ is real and $(x,x) > 0$, so λ is real as well.

2° *Eigenvectors* x,y *of a Hermitian operator* A, *corresponding to different eigenvalues* λ, μ *respectively, are orthogonal i.e.* $(x,y) = 0$.

The statement follows from the evident identity $\lambda(x,y) = (\mathsf{A}x,y) = (x,\mathsf{A}y) = \mu(x,y)$ $(\lambda \neq \mu)$.

Less evident is the next proposition, which is proved in courses on linear algebra.

3° *If* $\lambda_1, \lambda_2, \cdots, \lambda_n$ *are all eigenvalues of the Hermitian matrix* A *(counted with their multiplicities) and* A *is the linear Hermitian operator on the space* E^n *corresponding to the matrix* A *and some orthonormal basis* $\{e_1, e_2, \cdots, e_n\}$, *then in* E^n *there exists some orthonormal basis* $\{f_1, f_2, \cdots, f_n\}$ *consisting of eigenvectors of the operator* A *such that*

$$\mathsf{A}f_i = \lambda_i f_i \quad (i=1,2,\cdots,n).$$

We shall present, for the sake of completeness, a variant of the proof of proposition 3°. At first we consider some eigenvector f_1 of the operator A, belonging to the eigenvalue λ_1 (see Sec. 4.1). Without loss of generality one may assume this vector to be normed i.e., assume $(f_1,f_1) = 1$ (in the opposite case one must take instead of f_1 the eigenvector $\frac{1}{(f_1,f_1)} f_1$, which belongs to the eigenvalue λ_1 as well).

We consider the so-called *orthogonal complement* in E^n to the vector f_1 (exactely formulated, to the onedimensional subspace spanned by the vector f_1). It consists, by definition, of all vectors orthogonal to f_1. As is well-known (see, for example [9],Sec.8o) it will be some (n-1)-dimensional subspace E^{n-1} of the space E^n. Take a vector $x \in E^{n-1}$, i.e., $(x,f_1) = 0$. Then

$$(Ax,f_1) = (x,Af_1) = (x,\lambda_1 f_1) = \lambda_1(x,f_1) = 0,$$

i.e. $Ax \in E^{n-1}$. This fact is expressed through the words: E^{n-1} is an *invariant subspace* of the operator A.

In the subspace E^{n-1} the operator A acts again as a Hermitian operator, and it is clear as well, that all eigenvalues and corresponding eigenvectors of the operator A as operator in E^{n-1} are, respectively, eigenvectors and eigenvalues of A as operator in E^n. Now we choose a new orthonormal basis $\{g_1,g_2,\cdots,g_n\}$ having for its first element the vector $g_1 = f_1$ and with its remaining elements form E^{n-1} (this is always possible - see [3], p.237). Then in representation (4.5) for i = 1 appears $Ag_1 = \lambda_1 g_1$, i.e., $b_{11} = \lambda_1$, $b_{12} = b_{13} = \cdots = b_n = 0$ (we note that, although not essential for us, also $b_{21} = b_{31} = \cdots = b_{n1} = 0$, as the operator A is Hermitian). This means, that the structure of the matrix $B = \|b_{ij}\|_{i,j=1}^{n}$ is

$$B = \left(\begin{array}{c|c} \lambda_1 & 0 \cdots 0 \\ \hline 0 & \\ \vdots & \tilde{B} \\ 0 & \end{array}\right), \qquad (4.11)$$

where $\tilde{B}$ is the matrix which induces (on the basis $\{g_2,g_3,\cdots,g_n\}$) the operator A in the invariant subspace E^{n-1}. But from (4.11) it is obvious, that the characteristic polynomial for the matrix $\tilde{B}$ is obtained from the characteristic polynomial of the matrix B, i.e., (see (4.7)), from $|A-\lambda E|$ by division through the binomial $\lambda_1-\lambda$. Hence the eigenvalues of the operator A in E^{n-1} will be the numbers $\lambda_2,\lambda_3,\cdots,\lambda_n$.

Now, having chosen in E^{n-1} a normalized eigenvector f_2 of the operator

A, belonging to the eigenvalue λ_2, we can repeat the same reasoning having constructed in E^{n-1} a subspace E^{n-2} (of dimension n-2), orthogonal to f_2, invariant with respect to the operator A, and so on.

It is clear, that this procedure will completed through the construction in n steps of the desired orthonormal system of eigenvectors $f_1, f_2, \dots, f_n$ ($Af_i = \lambda_i f_i$), which forms, because of its linear independence ([9],Sec.78,Theorem 1), a basis of the space E^n.

EXAMPLES AND EXERCISES

1. Let a matrix $A = \|a_{ij}\|_{i,j=1}^{n}$, be given. We consider the *adjoint matrix*

$$A^* = \|a^*_{ij}\|_{i,j=1}^{n}$$

where $a^*_{ij} = \overline{a}_{ji}$ (i,j = 1,2,...,n). Then, if $\lambda_1, \lambda_2, \dots, \lambda_n$ is the spectrum of the matrix A (taking into account the multiplicity of the eigennumbers), then $\overline{\lambda}_1, \overline{\lambda}_2, \dots, \overline{\lambda}_n$ form the spectrum of A^*. Symbolically

$$\sigma(A^*) = \overline{\sigma(A)}.$$

2. The matrices A and A^* (see exercise 1) define on a fixed orthonormal bases $\{e_1, e_2, \dots, e_n\}$ of the unitary space E^n the so-called adjoint linear operators A and A^*, respectively, for which

$$(Ax,y) = (x,A^*y) \qquad \text{for all } x,y \in E^n. \tag{4.12}$$

Thus a Hermitian operator A, corresponding to a Hermitian matrix A ($=A^*$) is nothing else than a selfadjoint operator: $A = A^*$.

3. Invert the first statement of exercise 2: if A and A^* are adjoint operators (in the sense of definition (4.12)), then on an arbitrary orthonormal basis correspond to them the adjoint matrices A and A^*, respectively.

4. If $AA^* = A^*A$ then the linear operator A (and also A^*) is called *normal*. Generalize proposition 2° to normal operators.

5. The matrix A of a normal operator A (on an orthonormal basis) is *normal*, i.e., it commutates with its adjoint: $AA^* = A^*A$. The converse of this statement is also true (formulate and prove it!).

6. Generalize proposition 3° to normal matrices.

NOTES

1) Indeed, the coefficients of the characteristic polynomial are, as is obvious from (4.1), entire rational, and therefore continuous functions

of the elements of the matrix. The roots of every polynomial $P_n(\lambda) = \alpha_0\lambda^n + \alpha_1\lambda^{n-1} + \cdots + \alpha_{n-1}\lambda + \alpha_n$ $(\alpha_0 \neq 0)$ dependend continuously on its coefficients. The correct meaning of the latter statement is as follows: if for fixed values $\alpha_0, \alpha_1, \cdots, \alpha_n$ the different roots $\lambda_1, \lambda_2, \cdots, \lambda_r$ of the polynomial $P_n(\lambda)$ have the multiplicities $s_1, s_2, \cdots, s_r$, respectively $(s_1+s_2+\cdots+s_r = n)$, then for arbitrary $\varepsilon > 0$ there exists $\delta > 0$ such that for $|\tilde{\alpha}_i - \alpha_i| < \delta$ $(i=0,1,\cdots,n)$ in an ε-neighbourhood of each of the number λ_k there are exactly s_k (with regard to multiplicity) roots of the polynomial $\tilde{P}_n(\lambda) = \tilde{\alpha}_0\lambda^n + \alpha_1\lambda^{n-1} + \cdots + \alpha_{n-1}\lambda + \tilde{\alpha}_n$ $(k=1,2,\cdots,r)$ - for a proof see, for example [11], § 73.

2) Clearly, the converse holds as well: if a linear operator A on the space E^n is given, then formulae (4.2) relate to it and to a chosen basis $\{e_1, e_2, \cdots, e_n\}$ in a unique way a matrix $A = \|a_{ij}\|_{i,j=1}^n$.

3) Here we leave aside the deeper question of the relation between the mulitplicity of the eigennumber λ as root of the characteristic equation $|A-\lambda E| = 0$ and its so-called eigen- or geometrical multiplicity as eigenvalue of the operator A (see [3], Ch. VII).

§ 5. HERMITIAN AND QUADRATIC FORMS. LAW OF INERTIA. SIGNATURE

5.1 Now we can procede to consider the Hermitian form

$$A(x,x) \equiv \sum_{i,j=1}^{n} a_{ij}\xi_i\bar{\xi}_j, \quad a_{ji} = \bar{a}_{ij} \quad (i,j=1,2,\cdots,n) \tag{5.1}$$

where $\xi_1, \xi_2, \cdots, \xi_n$ are complex parameters and a_{ij} $(i,j=1,2,\cdots,n)$ coefficients. Each such form is, evidently, entirely defined by a (Hermitian) matrix $A = \|a_{ij}\|_{i,j=1}^n$ through its coefficients and conversely. The order n and the rank r of the matrix A are called, respectively, the *order* and the *rank* of the form $A(x,x)$.

One of the important tasks of the theory of Hermitian forms is reduction of the form to a "sum of squares", i.e., to the shape

$$A(x,x) = \sum_{k=1}^{n} \alpha_k |\eta_k|^2, \tag{5.2}$$

where

$$\eta_k \equiv L_k(x) = c_{1k}\xi_1 + c_{2k}\xi_2 + \cdots + c_{nk}\xi_n \quad (k=1,2,\cdots,n). \tag{5.3}$$

are some linear forms[1], and the α_k are real numbers. Usually, one is only interested in those representations in which the linear forms (5.3) are linearly independent. The latter is, as is well-known, equivalent to the nonsingularity of the matrix $C = \|c_{ik}\|_{i,k=1}^n$.

The reduction of the form $A(x,x)$ to a sum of squares by a linear transformation of the type (5.3) can be realised in different ways and, in particular, it follows from a geometrical interpretation of the form (5.1), for example. If, again as in § 4, one considers in the space E^n some basis $\{e_1, e_2, \cdots, e_n\}$, then every vector $x \in E^n$ is represented in the form $x = \xi_1 e_1 + \xi_2 e_2 + \cdots + \xi_n e_n$. The matrix A defines on this basis an operator A which will be Hermitian with respect to the scalar product (4.8). Having compared the representations (4.2), (4.8) and relation (4.9), we convince ourselves that

$$A(x,x) = \sum_{i,j=1}^{n} a_{ij}\xi_i\xi_j = (Ax,x) \tag{5.4}$$

Now we remember, that on the base of proposition 3° of § 4 there corresponds to the eigenvalues $\lambda_1, \lambda_2, \cdots, \lambda_n$ of the matrix A (i.e., of the operator A) a system of eigenvectors $f_1, f_2, \cdots, f_n$ [2] which one can take as new basis of the space E^n where

$$Af_i = \lambda_i f_i ; (f_i, f_j) = \delta_{ij} \quad (i,j=1,2,\cdots,n). \tag{5.5}$$

These two bases of the space E^n are connected by some nonsingular transformation. Consequently

$$e_k = q_{k1}f_1 + q_{k2}f_2 + \cdots + q_{kn}f_n \quad (k=1,2,\cdots,n)$$

where $Q = \|q_{ij}\|_{i,j=1}^{n}$ is some nonsingular matrix ($|Q| \neq 0$). Now for an arbitrary vector $x \in E^n$ have

$$x = \sum_{k=1}^{n} \xi_k e_k = \sum_{k=1}^{n} \xi_k \sum_{j=1}^{n} q_{kj} f_j = \sum_{j=1}^{n} \left(\sum_{k=1}^{n} q_{kj}\xi_k \right) f_j = \sum_{j=1}^{n} \eta_j f_j ,$$

where

$$\eta_j = q_{1j}\xi_1 + q_{2j}\xi_2 + \cdots + q_{nj}\xi_n \quad (j=1,2,\cdots,n) \tag{5.6}$$

are linearly independent linear forms.

Taking into account (5.4) and (5.5) we obtain

$$\left.\begin{aligned} Ax &= \eta_1\lambda_1 f_1 + \eta_2\lambda_2 f_2 + \cdots + \eta_n\lambda_n f_n , \\ A(x,x) &= (Ax,x) = \left(\sum_{i=1}^{n} \eta_i\lambda_i f_i , \sum_{i=1}^{n} \eta_i f_i \right) = \sum_{i=1}^{n} \lambda_i |\eta_i|^2 \end{aligned}\right\} \tag{5.7}$$

i.e., the form A(x,x) is reduced to the sum of n independent squares.

We note that in fact in this sum (5.7) there can be less than n terms, as some of the eigennumbers $\lambda_1, \lambda_2, \cdots, \lambda_n$ can be equal to zero. It is not difficult to clarify, exactly how many terms the sum contains which are different from (identically) zero. Indeed, we have seen (see proposition 3° from § 4) that on the basis $\{f_1, f_2, \cdots, f_n\}$ the linear operator A is represented through the diagonal matrix

$$\Lambda = \begin{Vmatrix} \lambda_1 & & & 0 \\ & \lambda_2 & & \\ & & \ddots & \\ 0 & & & \lambda_n \end{Vmatrix},$$

connected (cf.(4.6)) to the original matrix $A = \|a_{ij}\|_{i,j=1}^{n}$ with some transformation $A = T^{-1}\Lambda T$ by a nonsingular matrix T. But then, as is wellknown ([3],p.27) the ranks of the matrices Λ and A coincide. But the rank r of the matrix A is, evidently, equal to the number of eigenvalues (taking into account their multiplicities) of the matrix A which are different from zero. Now it is clear that an arbitrary other nonsingular linear transformation of the type (5.3) which reduces the form A(x,x) to a sum of squares $\sum_{k=1}^{n} \alpha_k |\eta_k|^2$, and in that way transforms the matrix A to the diagonal form

$$\begin{Vmatrix} \alpha_1 & & & 0 \\ & \alpha_2 & & \\ & & \ddots & \\ 0 & & & \alpha_n \end{Vmatrix},$$

also preserves the rank of the form A, i.e.,

1° *If* r *is the rank of the form* A(x,x) *then in the sum* (5.2) *there are, under the condition that the forms* $\eta_k = L_k(x)$ $(k=1,2,\cdots,n)$ *are linearly independent, always exactly* r *coefficients* α_k $(k=1,2,\cdots,n)$ *different from zero.*

5.2 It is easy to understand that the reduction of the form A(x,x) to a sum of squares can be reached in infinitely many ways (even under the requirement of linear independence of its squares and a forteriori if this restriction is waived). Hence the following result, proved for the first time by Sylvester, is of interest.

THEOREM 5.1 (LAW OF INERTIA). *For an arbitrary way of reducing the form* A(x,x) *(see* (5.1)*) to a sum* (5.2) *of independent squares there are among the coefficients* α_k $(k=1,2,\cdots,n)$ *always the same number* π $(\pi \geq 0)$ *of positive and the same number* ν $(\nu \geq 0)$ *of negative coefficents. Moreover,* $\pi + \nu = r$, *where* r *is the rank of the form* A(x,x).

PROOF. We suppose, using proposition 1°, that two nonsingular transformations of the type (5.3) transform the form

$$A(x,x) = \sum_{i,j=1}^{n} \alpha_{ij}\xi_i\xi_j$$

of rank r, the first one to the shape

$$A(x,x) = \alpha_1|\eta_1|^2 + \alpha_2|\eta_2|^2 + \cdots + \alpha_p|\eta_p|^2 - \alpha_{p+1}|\eta_{p+1}|^2 - \cdots - \alpha_r|\eta_r|^2 \quad (5.8)$$

and the other one to the shape

$$A(x,x) = \beta_1|\zeta_1|^2 + \beta_2|\zeta_2|^2 + \cdots + \beta_q|\zeta_q|^2 - \beta_{q+1}|\zeta_{q+1}|^2 - \cdots - \beta_r|\zeta_r|^2 \quad (5.9)$$

where $\alpha_k > 0$, $\beta_k > 0$, $k=1,2,\cdots,r$ and the linear forms (cf.(5.6)) $\eta_j = L_j(x)$ $(j=1,2,\cdots,n)$ (respectively $\zeta_j = \tilde{L}_j(x)$ $(j=1,2,\cdots,n)$) are linearly independent. Because of this linear independence the parameters $\eta_1,\eta_2,\cdots,\eta_n$ are defined by a unique transform as linear forms in the parameters $\zeta_1,\zeta_2,\cdots,\zeta_n$ and conversely.

Now we prove that p = q. Assuming at first, for example, that $p < q$, we consider the equation which follows from (5.8) and (5.9)

$$\alpha_1|\eta_1|^2 + \cdots + \alpha_p|\eta_p|^2 + \beta_{q+1}|\zeta_{q+1}|^2 + \cdots + \beta_r|\zeta_r|^2 =$$

$$= \beta_1|\zeta_1|^2 + \cdots + \beta_q|\zeta_q|^2 + \alpha_{p+1}|\eta_{p+1}|^2 + \cdots + \alpha_r|\eta_r|^2 \quad (5.10)$$

as an identity with respect to the parameters $\zeta_1,\zeta_2,\cdots,\zeta_n$ (taking into account that $\eta_1,\eta_2,\cdots,\eta_n$ are expressed linearly through these parameters).

We consider the system of linear, homogeneous (relative to $\zeta_1,\zeta_2,\cdots,\zeta_n$) equations

$$\eta_1 = 0,\ \eta_2 = 0,\cdots,\ \eta_p = 0;\ \zeta_{q+1} = 0,\cdots,\ \zeta_n = 0 \quad (5.11)$$

This system contains $p + (n-q) = n - (q-p)$ $(< n)$ equations over n variables $\zeta_1,\zeta_2,\cdots,\zeta_n$, and hence it has a nontrivial solution

$$\zeta_1 = \zeta_1^*,\cdots,\zeta_q = \zeta_q^*,\ \zeta_{q+1} = 0,\cdots,\zeta_n = 0.$$

Inserting these values in the identity (5.1o) and taking (5.11) into account we obtain

$$0 = \beta_1|\zeta_1^*|^2 + \cdots + \beta_q|\zeta_q^*|^2 + \alpha_{p+1}|\eta_{p+1}|^2 + \cdots + \alpha_r|\eta_r|^2,$$

whence $\zeta_1^* = \zeta_2^* = \cdots = \zeta_q^* = 0$, in spite of the assumption.

Theorem 5.1 on the law if inertia is proved.

In view of the law of inertia it is clear, that along with the rank of a Hermitian form important characteristics of it are the number π of the so-calle *positive squares* (i.e. with coefficient $\alpha_k > 0$) and the number ν of the so-called *negative squares* (i.e. with coefficient $\alpha_k < 0$) in the representation (5.2) of the form A(x,x) in the shape of a sum of independent squares, which we shall call a *canonical* [3] representation. These numbers, like the rank of the form, do not change under arbitrary nonsingular transformations (5.3) of the parameters (Theorem 5.1), or, as one says, they are *invariants* under such transformations.

We note, that in fact, we are not dealing here with three invariants (r,π,ν), but merely with two, for example π and ν, since $r = \pi+\nu$. Instead of these two invariants one often considers two other invariants: r and $\sigma = \pi-\nu$. The latter value σ is called the *signature* of the Hermitian form $A(x,x)$. It is clear that the signature σ, like the values r,π and ν, is an integer, but, in contrast to these, it can take negative values as well. From the formulae

$$r = \pi+\nu, \quad \sigma = \pi-\nu, \quad \pi = \frac{1}{2}(r+\sigma), \quad \nu = \frac{1}{2}(r-\sigma)$$

it is clear, that the pairs of numbers (π,ν) and (r,σ) mutually define each other and that *the integers* r *and* σ *always have the same parity.*

From the law of inertia and the arguments of Sec. 5.1 (see (5.7)) follows the propostion

2° *The number* π *of positive squares and the number* ν *of negative squares in an arbitrary canonical representation of the form* $A(x,x)$ *are equal to the number (with respect to the mulitplicity) of positive and negative eigenvalues of the matrix* $A = \|a_{ij}\|_{i,j=1}^{n}$, *respectively.*

The eigenvalues $\lambda_1,\lambda_2,\cdots,\lambda_n$ of the Hermitian matrix A are also called the *eigenvalues* of the corresponding *Hermitian form* $A(x,x)$. In accordance to this the determinant $|A| = \lambda_1\lambda_2\cdots\lambda_n$ (see (4.3)) is called *discriminant of the form* $A(x,x)$. Hence, to nonsingular matrices correspond, by definition, *nondegenerate* (or *regular*) forms $A(x,x)$ with a discriminant which is different from zero ($|A| \neq 0$), and to singular matrices - *degenerate* (or *singular*) forms ($|A| = 0$).

5.3 We show yet one simple, but for the sequel important proposition:

3° *If the Hermitian form* $A(x,x)$ *of order* n *and rank* r (>0) *is represented through whatever method in the shape of a sum of exactly* r *squares:*

$$A(x,x) = \sum_{k=1}^{r} \alpha_k |L_k(x)|^2,$$

then the forms

$$\eta_k = L_k(x) = c_{1k}\xi_1 + c_{2k}\xi_2 + \cdots + c_{nk}\xi_n \quad (k=1,2,\cdots,r) \tag{5.12}$$

are linearly independent, i.e. the given representation is canonical.

Indeed, by assumption, the rank of the form $A(x,x)$, i.e., the rank of the matrix $A = \| a_{ij} \|_{i,j=1}^{n}$ is equal to r. The assertion of the linear independence of the forms (5.12) means that the rank of the (generally speaking, rectangular) matrix $C = \| c_{ki} \|_{k=1,2,\cdots,r}^{i=1,2,\cdots,n}$ is also equal to r, i.e., maximal. Now we note that

$$A(x,x) = \sum_{k=1}^{r} \alpha_k |L_k(x)|^2 = \sum_{k=1}^{r} \alpha_k |c_{k1}\xi_1 + c_{k2}\xi_2 + \cdots + c_{kn}\xi_n|^2 =$$

$$= \sum_{i,j=1}^{n} \left(\sum_{k=1}^{r} \alpha_k c_{ki} \bar{c}_{kj} \right) \xi_i \bar{\xi}_j .$$

Comparison of this identity with the original shape of the form

$$A(x,x) = \sum_{i,j=1}^{n} a_{ij} \xi_i \bar{\xi}_j$$

shows that

$$a_{ij} = \sum_{k=1}^{r} \alpha_k c_{ki} \bar{c}_{kj} = \sum_{k=1}^{r} c_{ik}^{t} \alpha_k \bar{c}_{kj} \quad (i,j=1,2,\cdots,n) \tag{5.13}$$

where $C^t = \| c_{ik}^t \|_{i=1,2,\cdots,n}^{k=1,2,\cdots,r}$ is the transposed matrix with respect to C.

But relation (5.13) is equivalent to the identity

$$A = C^t \begin{pmatrix} \alpha_1 & & & 0 \\ & \alpha_2 & & \\ & & \ddots & \\ 0 & & & \alpha_r \end{pmatrix} \bar{C} , \tag{5.14}$$

where $\bar{C} = \| \bar{c}_{kj} \|_{k=1,2,\cdots,r}^{j=1,2,\cdots,n}$.

Assuming now, that the rank of the matrix C (and therefore also of C^t and $\bar{C}$) is less than r, we would obtain from (5.14) (see[3],p.22) that the rank of the matrix A were less than r, but this would contradict the assumption.

A completely analoguous reasoning shows that the following helpful proposition is also correct:

4° *If the form* $A(x,x) = \sum_{i,j=1}^{n} \alpha_{ij} \xi_i \bar{\xi}_j$ *of order* n *and rank* r *is represented in the shape of a sum of squares*

$$A(x,x) = \sum_{k=1}^{n} \alpha_k |L_k(x)|^2 \qquad (\alpha_k \gtreqless 0)$$

then in this sum not less than r *terms are different from identically zero.*

We note that here the linear forms

$$\eta_k = L_k(x) = c_{k1}\xi_1 + c_{k2}\xi_2 + \cdots + c_{kn}\xi_n$$

are not subject of any restriction (in particular, it is not assumed that they are linearly independent), but the statement again follows straight from the identity

$$A = C^t \begin{pmatrix} \alpha_1 & & & 0 \\ & \alpha_2 & & \\ & & \ddots & \\ 0 & & & \alpha_n \end{pmatrix} \overline{C},$$

where $A = \|a_{ij}\|_{i,j=1}^n$, and $C = \|c_{ki}\|_{k,i=1}^n$.

5.4. The Hermitian form $A(x,x)$ is called *nonnegative*, if $A(x,x) \geqq 0$ for all $x = \{\xi_1, \xi_2, \cdots, \xi_n\}$ and *positive definite* if $A(x,x) > 0$ for all $x \neq 0$ (i.e., $|\xi_1| + |\xi_2| + \cdots + |\xi_n| > 0$).

Here we shall confine ourselves just to those facts concerning these classes - from which the second one is, evidently, contained in the first one - which are absolutely necessary for further intelligence (for more details see, for example, [3]).

THEOREM 5.2. *The Hermitian form* $A(x,x)$ *is nonnegative if and only if all its eigenvalues are nonnegative, and it is positive definite if and only if all its eigenvalues are positive.*

PROOF. If all eigenvalues $\lambda_1, \lambda_2, \cdots, \lambda_n$ of the form $A(x,x)$ are nonnegative, then it follows from representation (5.7) that the form $A(x,x)$ is nonnegative. Moreover, if all $\lambda_k > 0$ $(k=1,2,\cdots,n)$ then it is evident from the same representation that $A(x,x) > 0$ for $x \neq 0$, since under this condition it is impossible that all (linearly independent) forms $\eta_1, \eta_2, \cdots, \eta_n$ become zero simultanuously (see (5.3)).

Conversely, even if only one of the number λ_k is negative, say $\lambda_n < 0$, then, substituting in (5.3) parameters $\xi_1, \xi_2, \cdots, \xi_n$ such that $\eta_1 = \eta_2 = \cdots \eta_{n-1} = 0$ and $\eta_n = 1$ (the latter is possible because of the linear independence of these forms), we obtain from (5.7) that for the $\xi_1, \xi_2, \cdots, \xi_n$ we mentioned

$$A(x,x) = \lambda_n < 0,$$

i.e., the form $A(x,x)$ is not nonnegative.

Finally, if $A(x,x)$ is a positive definite form, then, for that reason, all $\lambda_k \geqq 0$ $(k=1,2,\cdots,n)$. If thereby only one eigenvalues is equal to zero, say $\lambda_n = 0$, then, substituting again $\xi_1, \xi_2, \cdots, \xi_n$ such that $\eta_1 = \eta_2 = \cdots = \eta_{n-1} = 0$ and $\eta_n = 1$ we would obtain the identity $A(x,x) = 0$ for $x \neq 0$, which was impossible.

Theorem 5.2. is proved.

COROLLARY 1. *A nonnegative form* A(x,x) *is positive if and only if it is nondegenerate.*

PROOF. This follows from Theorem 5.2 and the relation (see(4.3))

$$|A| = \lambda_1\lambda_2\cdots\lambda_n.$$

COROLLARY 2. *An arbitrary representation* (5.2) *of a nonnegative form* A(x,x) *in the shape of a sum of independent squares contains no negative squares. The presence in such a canonical representation of exactly* n *positive squares (where* n *is the order of the form) is necessary and sufficient for the positive definitess of the form.*

This statement is obtained if one combines representation (5.2), Theorem 5.2 and the law of inertia.

5.5. In conclusion of the present section we consider the case where $A = \|a_{ij}\|^n_{i,j=1}$ is a real symmetric matrix: $a_{ij} = a_{ji}$ $(i,j=1,2,\cdots,n)$ It is natural to consider in this case instead of the Hermitian form $\sum_{i,j=1}^{n} a_{ij}\xi_i\overline{\xi}_j$ a *quadratic form* 4)

$$A(x,x) = \sum_{i,j=1}^{n} a_{ij}\xi_i\xi_j$$

where the ξ_i $(i=1,2,\cdots,n)$ are real parameters. In this situation, as one easily sees, all propositions, stated in the present section, remain valid, with the difference that for the geometrical interpretation one must now consider the real Euclidian space E^n in which the scalar product (4.8) of the vectors $x = \xi_1e_1 + \xi_2e_2 + \cdots + \xi_ne_n$ and $y = \eta_1e_1 + \eta_2e_2 + \cdots + \eta_ne_n$ is defined by the formula

$$(x,y) = \xi_1\eta_1 + \xi_2\eta_2 + \cdots + \xi_n\eta_n,$$

and representation (5.2) is written in the form

$$A(x,x) = \sum_{k=1}^{n} a_k\eta_k^2$$

where

$$\eta_k = L_k(x) = c_{k1}\xi_1 + c_{k2}\xi_2 + c_{kn}\xi_n \qquad (k=1,2,\cdots,n)$$

are real linear forms.

We note that in the sequel (see §§ 6-8, below) all results, even if we formulate them for Hermitian forms, remain valid for quadratic forms as well; we shall not especially memorate this by that time.

EXAMPLES AND EXERCISES

1. Find the rank r and the signature σ of the Hermitian form

$$A(x,x) = 3|\xi_1|^2 + \xi_1\overline{\xi}_2 + \overline{\xi}_1\xi_2 + 2i\xi_1\overline{\xi}_3 - 2i\overline{\xi}_1\xi_3 .$$

Here

$$A = \begin{Vmatrix} 3 & 1 & 2i \\ 1 & 0 & 0 \\ -2i & 0 & 0 \end{Vmatrix}, r = 2, \ |A-\lambda E| = \begin{vmatrix} 3-\lambda & 1 & 2i \\ 1 & -\lambda & 0 \\ -2i & 0 & -\lambda \end{vmatrix} =$$

$$= -2i(2i\lambda) - \lambda(\lambda^2-3\lambda-1) = -\lambda^3 + 3\lambda^2 + 5\lambda .$$

The eigenvalues are : $\lambda_1 = \frac{1}{2}(3+\sqrt{29}) > 0$, $\lambda_2 = \frac{1}{2}(3-\sqrt{29}) < 0$, $\lambda_3 = 0$. Thus (see proposition 2°) $\pi=1$, $\nu=1$, so the signature $\sigma=0$.

2. We consider a so-called *Hankel* (real) quadratic form (to such forms § 12 below is dedicated)

$$\sum_{j,k=0}^{n-1} s_{j+k}\xi_j\xi_k \tag{5.15}$$

of order n with the matrix

$$H_{n-1} \equiv \| s_{j+k} \|_{j,k=0}^{n-1} = \begin{Vmatrix} s_0 & s_1 & \cdots & s_{n-1} \\ s_1 & s_2 & \cdots & s_n \\ \cdot & \cdot & \cdots & \cdot \\ s_{n-1} & s_n & \cdots & s_{2n-2} \end{Vmatrix} .$$

If, in particular, the numbers $s_0, s_1, \dots, s_{2n-2}$ form an arithmetic sequence

$$s_0 = a, \ s_1 = a+d, \ s_2 = a+2d, \dots, \ s_{2n-2} = a+(2n-2)d,$$

then the rank r of the form (5.15) for d=0 is, evidently, equal to 1, if $a \neq 0$ (but what value has the signature in this case?) and 0 if a=0. For $d \neq 0$ the situation is different. Prove, that in this case for arbitrary a (and for arbitrary $n \geq 2$) the rank of the form (5.15) is r=2 and its signature $\sigma=0$, i.e., $\pi=\nu=1$.

3. Is the representation of the Hermitian form A(x,x) of order 3 in the shape of the sum of squares

$$A(x,x) = |\xi_1 + \xi_2|^2 - |2\xi_1 - \xi_3|^2 + |2\xi_2 + \xi_3|^2$$

canonical? Which rank and signature has this form?

Solution. No, r=2, $\sigma=0$.

4. For what values of the real parameter a is the quadratic form

$$A(x,x) = (a^2+1)\ \xi_1^2 + 2(a-1)\xi_1\xi_2$$

nonnegative? What are rank and signature of the form A(x,x) for these a and for all other values of the parameter a?

Solution. a=1, $r=\sigma=1$; for $a \neq 1$: r=2, $\sigma=0$

NOTES

1) In the notations $A(x,x)$ and $L_k(x)$ the symbol x represents, as in § 4, a set of n numbers (vector): $x = (\xi_1, \xi_2, \cdots, \xi_n)$. Relation (5.2) is understood as an identity with respect to the parameters $\xi_1, \xi_2, \cdots, \xi_n$.

2) We note that in proposition 3° from § 4 there was proved only the existence, but by no means the uniqueness of such a system.

3) Besides, sometimes more special representations, on which we shall not dwell here, are called canonical representations.

4) Sometimes under a quadratic form is understood an expression $\sum_{i,j=1}^{n} a_{ij}\xi_i\xi_j$, where it is not required that the coefficients a_{ij} and the parameters ξ_i are real. However, we shall always assume that these conditions are satisfied.

§ 6. TRUNCATED FORMS

6.1 Along with a given Hermitian form

$$A(x,x) = \sum_{i,j=1}^{n} a_{ij}\xi_i\overline{\xi}_j$$

of order n with discriminant $|A|$ we must often consider the so-called *truncated* forms

$$A_k(x,x) = \sum_{i,j=1}^{k} a_{ij}\xi_i\overline{\xi}_j$$

of order k $(k=1,2,\cdots,n)$ (induced by the given form) with discriminant

$$\Delta_k = \det \|a_{ij}\|_{i,j=1}^{k} \qquad (k=1,2,\cdots,n-1).$$

It is natural to put $A_n(x,x) \equiv A(x,x)$, $\Delta_n \equiv |A|$. Moreover, it will be convenient to assume

$$\Delta_0 \equiv 1.$$

The numbers $\Delta_0, \Delta_1, \Delta_2, \cdots, \Delta_n$ are called the *successive principal minors of the form* $A(x,x)$.

In the sequel the comparison between "adjacent" truncated forms $A_{k+1}(x,x)$ and $A_k(x,x)$ $(k=1,2,\cdots,n-1)$ will play an especially important role.

LEMMA 6.1 (THE BASIC IDENTITY). *The form*

$$A_{k+1}(x,x) = \sum_{i,j=1}^{k+1} a_{ij}\xi_i\overline{\xi}_j$$

of order k+1 *is connected to the truncated form*

$$A_k(x,x) = \sum_{i,j=1}^{n} a_{ij}\xi_i\xi_j$$

of order k (k=1,2,···,n-1) *through the identity*

$$A_{k+1}(x,x) = A_k(x,x) + $$
$$+ \frac{1}{2}\left|a_{1,k+1}\xi_1 + \cdots + a_{k,k+1}\xi_k + (\frac{1}{2}a_{k+1,k+1}+1)\xi_{k+1}\right|^2 -$$
$$- \frac{1}{2}\left|a_{1,k+1}\xi_1 + \cdots + a_{k,k+1}\xi_k + (\frac{1}{2}a_{k+1,k+1}-1)\xi_{k+1}\right|^2 . \qquad (6.1)$$

PROOF. This is reduced to a direct calculation:

$$\frac{1}{2}\left|a_{1,k+1}\xi_1 + \cdots + a_{k,k+1}\xi_k + (\frac{1}{2}a_{k+1,k+1}+1)\xi_{k+1}\right|^2 -$$
$$- \frac{1}{2}\left|a_{1,k+1}\xi_1 + \cdots + a_{k,k+1}\xi_k + (\frac{1}{2}a_{k+1,k+1}-1)\xi_{k+1}\right|^2 =$$
$$= \frac{1}{2}\left[a_{1,k+1}\xi_1 + \cdots + a_{k,k+1}\xi_k + (\frac{1}{2}a_{k+1,k+1}+1)\xi_{k+1}\right] \times$$
$$\times \left[a_{k+1,1}\overline{\xi}_1 + \cdots + a_{k+1,k}\overline{\xi}_k + (\frac{1}{2}a_{k+1,k+1}+1)\overline{\xi}_{k+1}\right] -$$
$$- \frac{1}{2}\left[a_{1,k+1}\xi_1 + \cdots + a_{k,k+1}\xi_k + (\frac{1}{2}a_{k+1,k+1}-1)\xi_{k+1}\right] \times$$
$$\times \left[a_{k+1,1}\overline{\xi}_1 + \cdots + a_{k+1,k}\overline{\xi}_k + (\frac{1}{2}a_{k+1,k+1}-1)\overline{\xi}_{k+1}\right] =$$
$$= \sum_{i=1}^{k} a_{i,k+1}\xi_i\overline{\xi}_{i+k} + \sum_{i=1}^{k} a_{k+1,i}\xi_{k+1}\overline{\xi}_i + a_{k+1,k+1}|\xi_{k+1}|^2 .$$

Having added to this sum the form

$$A_k(x,x) = \sum_{i,j=1}^{k} a_{ij}\xi_i\overline{\xi}_k$$

we obtain

$$\sum_{i,j=1}^{k+1} a_{ij}\xi_i\overline{\xi}_j = A_{k+1}(x,x).$$

COROLLARY. *If the rank of the forms* $A_{k+1}(x,x)$ *and* $A_k(x,x)$ *are equal to* r_{k+1} *and* r_k *respectively, then* [1)]

$$0 \leq r_{k+1} - r_k \leq 2. \qquad (6.2)$$

Indeed, from comparison of the matrices A_{k+1} and A_k of the considered forms it is clear, that $r_{k+1} \geq r_k$, but from identity (6.1) and propositions 1° and 4° of § 5 it follows that

$$r_{k+1} - r_k \leq 2.$$

6.2 Unlike the absolutely elementary algebraical Lemma 6.1, the next Lemma has an analytical nature and it relies on facts from the analysis.

Lemma 6.2.[2] *If under continuous variation of the coefficients of*[3] *the form*

$$A(x,x) = \sum_{i,j=1}^{k} a_{ij}\xi_i\bar{\xi}_j,$$

its rank r *remains invariant, then the signature* σ *doesn't change either.*

PROOF. The rank r of the form A(x,x) is equal to the number of its eigenvalues which are different from zero, taking into account their multiplicities (§ 2, proposition 2°). Let for some fixed value of the coefficients of the form among its eigenvalues there be π positive and ν negative ($r = \pi+\nu$, $\sigma = \pi-\nu$) and the remaining $d(=n-r)$ equal to zero. As the eigenvalues of the form depend continuously on its coefficients (Sec. 4.1), then for a sufficiently small variation of these the eigenvalues which are different from zero retain their own sign, but none of those which are equal to zero becomes different from zero, as this would cause an increase in the rank of the form, which would contradict the conditions of the lemma.

Thus in a small neighbourhood of an arbitrary set of coefficients the signature of the form remains constant. Hence, since a segment is connected, the signature remains constant if the coefficients of the form are continuous functions on a segment $[t_o,T]$ and the rank doesn't vary on $[t_o,T]$ (cf. note [3]).

6.3 Returning to the truncated forms, we can now, using the Lemmas 6.1 and 6.2, provide a more precise description of the character of the variation in the signature at the transition to the form $A_{k+1}(x,x)$ from the truncated form $A_k(x,x)$ and conversely.

The answer to these questions is given in the next three theorems.

THEOREM 6.1 *If in relation* (6.2) *equality does hold, i.e., if* $r_{k+1} = r_k + 2$, *then the signatures of the forms* $A_{k+1}(x,x)$ *and* $A_k(x,x)$ *coincide*: $\sigma_{k+1} = \sigma_k$.

PROOF. We assume that on the righthand side of the basic identity the form $A_k(x,x)$ is represented in the shape of a sum of r_k independent squares (§ 5, proposition 1°). Then (6.1) turns into a representation of the form A_{k+1} in the shape of a sum of $r_k + 2$ squares. But as $r_k + 2 = r_{k+1}$, by assumption, these squares are, according to proposition 3° of § 4, linearly independent. Returning again to identity (6.1), we see that the form $A_{k+1}(x,x)$ has gained in comparison to $A_k(x,x)$ one positive and one negative square, whence, according to the inertia theorem, also follows the identity $\sigma_{k+1} = \sigma_k$.

THEOREM 6.2. *If the ranks of the forms* $A_{k+1}(x,x)$ *and* $A_k(x,x)$ *are equal, i.e.,* $r_{k+1} = r_k$, *then their signatures are equal as well:* $\sigma_{k+1} = \sigma_k$.

PROOF. We represent the form $A_{k+1}(x,x)$ in the shape of a sum

$$A_{k+1}(x,x) = \sum_{j=1}^{r_k} \alpha_j |L_j(x)|^2 \tag{6.3}$$

of r_k independent squares (§ 5, proposition 1°), and then we insert in this identity $\xi_{k+1} = 0$. Then the form $A_{k+1}(x,x)$ on the left turns (see, for example, (6.1)) into $A_k(x,x)$, and on the righthand side of (6.3) none of the squares is annihilated, as in the opposite case the form $A_k(x,x)$ of rank r_k would be represented in the shape of a sum of less than r_k squares, which is not possible (§ 5, proposition 4°).

Thus we have obtained a representation of $A_k(x,x)$ in the shape of a sum of r_k independent (§ 5, proposition 3°) squares with the same coefficients $\alpha_j (j=1,2,\cdots,r_k)$ as in (6.3). Hence the equality $\sigma_{k+1} = \sigma_k$ holds

In the theorems 6.1 und 6.2 the two "extreme" cases in inequality (6.2) have been considered. The remaining "intermediate" case, where $r_{k+1} = r_k + 1$ is settled by

THEOREM 6.3. *If the rank* r_{k+1} *of the form* $A_{k+1}(x,x)$ *exceeds the rank of the truncated form* $A_k(x,x)$ *by one unity, i.e.,* $r_{k+1} = r_k + 1$, *then for the corresponding signatures* σ_{k+1} *and* σ_k *holds* $|\sigma_{k+1} - \sigma_k| = 1$, *i.e., either* $\sigma_{k+1} = \sigma_k + 1$ *or* $\sigma_{k+1} = \sigma_k - 1$.

PROOF.[4)] The proof relies on the continuity argument of Lemma 6.2. Let $r = r_k$ be the rank of the form $A_k(x,x)$. As this form depends only on the parameters $\xi_1, \xi_2, \cdots, \xi_k$ there exist linearly independent forms $L_j(x)$ $(j=1,2,\cdots,r)$ in these parameters such that

$$A_k(x,x) = \sum_{j=1}^{r} \alpha_j |L_j(x)|^2$$

with $\alpha_j \neq 0$ $(j=1,2,\cdots,r)$. Define forms P,N in the parameters $\xi_1, \xi_2, \cdots, \xi_{k+1}$

$$P(x) = a_{1,k+1}\xi_1 + \cdots + a_{k,k+1}\xi_k + (\tfrac{1}{2} a_{k+1,k+1} + 1)\xi_{k+1},$$

$$N(x) = a_{1,k+1}\xi_1 + \cdots + a_{k,k+1}\xi_k + (\tfrac{1}{2} a_{k+1,k+1} - 1)\xi_{k+1}.$$

First, we deal with the case where either P(x) or N(x) is a linear combination of $L_1(x),\cdots,L_r(x)$. If, for example, N is a linear combination of $L_1,\cdots,L_r$, then N depends on $\xi_1,\cdots,\xi_k$ only (i.e., $\frac{1}{2}a_{k+1,k+1}-1=0$). We note that for each t the Hermitian form

$$\tilde{B}^{(t)}(x,x) = A_k(x,x) - \frac{t}{2}|N(x)|^2$$

is of rank less or equal r [5)]; on the other hand, as the matrix associated with $\tilde{B}^{(t)}(x,x)$ is an extension of the matrix A_k, the rank of $\tilde{B}^{(t)}(x,x)$ must be r. As $\tilde{B}^{(o)}(x,x) = A_k(x,x)$, it follows from Lemma 6.2 that the signature of $\tilde{B}^{(1)}(x,x)$ is equal to the signature of $A_k(x,x)$, that is, equal to σ_k. We write

$$\tilde{B}^{(1)}(x,x) = \sum_{i=1}^{r} \beta_i|\tilde{L}_i(x)|^2,$$

where $\tilde{L}_1(x),\cdots,\tilde{L}_r(x)$ are linearly independent linear forms in the parameters $\xi_1,\cdots,\xi_k$. Then $\tilde{L}_1(x),\cdots,\tilde{L}_r(x)$, P(x) are also linearly independent, as $\frac{1}{2}a_{k+1,k+1}+1 \neq 0$, so P(x) does depend on ξ_{k+1}. Note that

$$A_{k+1}(x,x) = \tilde{B}^{(1)}(x,x) + \frac{1}{2}\,P(x)\;^2 = \sum_{i=1}^{r}\beta_i|\tilde{L}_i(x)|^2 + \frac{1}{2}|P(x)|^2$$

so the signature σ_{k+1} of $A_{k+1}(x,x)$ must be equal to σ_k+1. In the same way one proves that $\sigma_{k+1} = \sigma_k - 1$ if P is a linear combination of $L_1,\cdots,L_r$.

Next we assume that neither P nor N is a linear combination of $L_1,\cdots,L_r$. For $0 \leq t \leq 1$ we define the Hermitian form $B^{(t)}(x,x)$ by

$$B^{(t)}(x,x) = \sum_{j=1}^{r}\alpha_j|L_j(x)|^2 + (1-t)|P(x)|^2 - t|N(x)|^2. \tag{6.4}$$

Note that $B^{(1/2)}(x,x) = A_{k+1}(x,x)$, and that both $B^{(o)}(x,x)$ and $B^{(1)}(x,x)$ must have rank $r_k+1 = r+1$. Below we shall prove that there exists exactly one $0 < t_o < 1$, such that the rank of $B^{(t)}(x,x)$ is equal to $r+1$ for $0 \leq t \leq 1$, $t \neq t_o$, and the rank of $B^{(t_o)}(x,x)$ is less or equal $r = r_k$. Let $t_o > \frac{1}{2}$. Then the rank of $B^{(t)}(x,x)$ is r+1 for $0 \leq t \leq \frac{1}{2}$, and from Lemma 6.2 follows that the signature σ_{k+1} of the form $A_{k+1}(x,x) = B^{(1/2)}(x,x)$ is equal to the signature of $B^{(o)}(x,x)$. From formula (6.4) it is clear that the latter is equal to σ_k+1, as for $t=0$ formula (6.4) defines $B^{(o)}(x,x)$ as sum or $r+1$ independent squares. So $\sigma_{k+1} = \sigma_k+1$. In an analoguous way one proves that σ_{k+1} is equal to the signature of $B^{(1)}(x,x)$ if $t_o < 1/2$, and in this case if follows from (6.4) for $t=1$ that the signature of $B^{(1)}(x,x)$, and hence σ_{k+1}, is equal to σ_k-1.

It remains to prove that the rank of $B^{(t)}(x,x)$ is equal to r_k+1 for all but one $0 \leq t \leq 1$. Clearly, the rank of $B^{(t)}(x,x)$ must be less than

$r+1$ for at least one $t = t_o$, because otherwise it would follow from Lemma 6.2 that the signatures of $B^{(o)}(x,x)$ and $B^{(1)}(x,x)$ which are $\sigma_k + 1$ and $\sigma_k - 1$, respectively, would coincide. We use the method of the proofs of the propositions 3° and 4° of § 5, and set

$$L_j(x) = c_{j,1}\xi_1 + \cdots + c_{j,k}\xi_j + c_{j,k+1}\xi_{k+1} \qquad (j=1,2,\cdots,r),$$

$$P(x) = c_{r+1,1}\xi_1 + \cdots + c_{r+1,k+1}\xi_{k+1}, \quad N(x) = c_{r+2,1}\xi_1 + \cdots + c_{r+2,k+1}\xi_{k+1}.$$

Note that $c_{j,k+1} = 0 \quad (j=1,2,\cdots,r)$. Let $C = \|c_{ji}\|_{j=1,2,\cdots,r+2}^{i=1,2,\cdots,k+1}$, $\tilde{C} = \|c_{ji}\|_{j=1,2,\cdots,r+1}^{i=1,2,\cdots,k+1}$. Note that

$$B^{(t)} = C^t \begin{pmatrix} \alpha_1 & & & \\ & \ddots & & 0 \\ & & \alpha_r & \\ 0 & & & (1-t) \\ & & & & t \end{pmatrix} \overline{C} \tag{6.4.a}$$

As the forms $L_1, L_2, \cdots, L_r$, P,N are linearly dependent, the rank of C is $r+1$, and there exist $\mu_1, \cdots, \mu_r, \mu_{r+1}$ such that $c_{r+2,i} = \sum_{j=1}^{r+1} \mu_j c_{ji}$

Let $\tilde{C}_o$ be the $(r+2)\times(k+1)$-matrix obtained by adding a row of zeros to $\tilde{C}$. Then we have

$$C = \begin{pmatrix} 1 & & & \\ & \ddots & & 0 \\ & & \cdot & \\ 0 & & 1 & \\ \mu_1 & \cdots & \mu_{r+1} & 1 \end{pmatrix} \tilde{C}_o = M\tilde{C}_o$$

Multiplying the diagonal matrix in (6.4.a) on the lefthand side with M^t, and on the righthand side with the complex conjugate $\overline{M}$, and crossing out the (superfluous) last row and column in this product we obtain

$$B^{(t)} = \tilde{C}^t \begin{pmatrix} \alpha_1 - t\mu_1\overline{\mu}_1 & -t\mu_1\overline{\mu}_2 & \cdots & \cdot & -t\mu_1\overline{\mu}_{r+1} \\ -t\mu_2\overline{\mu}_1 & \alpha_2 - t\mu_2\overline{\mu}_2 & \cdots & \cdot & \cdot \\ \vdots & \vdots & & \vdots & \vdots \\ -t\mu_r\overline{\mu}_1 & \cdot & & \alpha_r - t\mu_r\overline{\mu}_r & -t\mu_r\overline{\mu}_{r+1} \\ -t\mu_{r+1}\overline{\mu}_1 & -t\mu_{r+1}\overline{\mu}_2 & \cdots & -t\mu_{r+1}\overline{\mu}_1 & (1-t)-t\mu_{r+1}\overline{\mu}_{r+1} \end{pmatrix} \overline{\tilde{C}} =$$

$= \tilde{C}^t K^{(t)} \overline{\tilde{C}}$. As both $\tilde{C}^t$ and $\overline{\tilde{C}}$ have maximal rank $r+1$, one sees that the rank of $B^{(t)}$ is equal to $r+1$ if and only if

$$\Delta^t = \det K^{(t)} = \det \begin{pmatrix} \alpha_1 - t\mu_1\bar{\mu}_1 & \cdots & -t\mu_1\bar{\mu}_{r+1} \\ \vdots & & \vdots \\ -t\mu_{r+1}\bar{\mu}_1 & \cdots & (1-t) - t\mu_{r+1}\bar{\mu}_{r+1} \end{pmatrix}$$

is nonzero. Evaluating Δ^t (using $\mu_{r+1} \neq 0$) yields

$$\Delta^t = (\alpha_1 \cdot \alpha_2 \cdots \alpha_r \cdot ((1-t) - t|\mu_{r+1}|^2)) + t(t-1) \sum_{j=1}^{r} \alpha_1 \cdots \alpha_{j-1} |\mu_j|^2 \alpha_{j+1} \cdots \alpha_r .$$

Note that $\Delta^o = \alpha_1 \cdot \alpha_2 \cdots \alpha_r \cdot$, $\Delta^1 = -\alpha_1 \cdots \alpha_r |\mu_{r+1}|^2 = -|\mu_{r+1}|^2 \Delta^o$. As Δ^o and Δ^1 have opposite signs, and Δ^t is polynomial in t of degree 2, it follows that there is exactly one t_o between 0 and 1 with $\Delta^{t_o} = 0$. This completes the proof.

REMARK. The Theorems 6.1-6.3 can be considered as special cases of common facts from the (variational) theory of eigenvalues of linear bundles of Hermitian forms (see, for example, [3], Ch. X, §§ 7,9). However on one hand, in none of the accounts of this theory known to us did we find the ready-formulated Theorems 6.1-6.3, which are necessary in order to obtain the basic result of the Chapters II and III, and on the other hand, it seems attractive to introduce the direct proofs of these Theorems, using a minimum of tools, without references to the theory of bundles.

6.4. In conclusion we introduce, following [3], yet one useful proposition concerning truncated forms.

1° *If the successive principal minor of order* r *of the form* A(x,x) *of rank* r *is different from zero* ($\Delta_r \neq 0$), *then the truncated form*

$$A_r(x,x) = \sum_{i,j=1}^{r} \alpha_{ij} \xi_i \bar{\xi}_j$$

has the same rank and the same signature as the complete form A(x,x).

Indeed, the statement on the rank is trivial, as $(0 \neq)\ \Delta_r$ is the discriminant of the form $A_r(x,x)$. If $r = n$, then the assertion on the signature is trivial.

Now let $r < n$, and

$$A(x,x) = \sum_{k=1}^{r} \alpha_r |L_k(x)|^2 \tag{6.5}$$

be a representation of the form A(x,x) in the shape of a sum of independent squares.

In (6.5) we insert

$$\xi_n = \xi_{n-1} = \cdots = \xi_{r+1} = 0.$$

Then on the lefthand side the form $A(x,x)$ turns into $A_r(x,x)$, and on the righthand side one obtains a representation of the form $A_r(x,x)$ in the shape of a sum of r squares. But as the rank of the form $A_r(x,x)$ is equal to r $(\Delta_r \neq 0)$, it follows from proposition 3° of § 5 that these squares are linearly independent. But since (§ 5, proposition 1°) $\alpha_k \neq 0$ $(k=1,2,\cdots,r)$ the signature of $A_r(x,x)$ is also that of $A(x,x)$.

EXAMPLES AND EXERCISES

1. We consider the form (cf. example 1 of § 5):

$$A_3(x,x) \equiv A(x,x) = 3|\xi_1|^2 + \xi_1\overline{\xi}_2 + \overline{\xi}_1\xi_2 + 2i\xi_1\overline{\xi}_3 - 2i\overline{\xi}_1\xi_3$$

and we shall represent it with help of identity (6.1). Here (setting $\xi_3 = 0$) we have

$$A_2(x,x) = 3|\xi_1|^2 + \xi_1\overline{\xi}_2 + \overline{\xi}_1\xi_2,$$

and as $a_{13} = 2i$, $a_{23} = 0$, $a_{33} = 0$, then (see(6.1))

$$A_3(x,x) = A_2(x,x) + \frac{1}{2}|2i\xi_1 + \xi_3|^2 - \frac{1}{2}|2i\xi_1 - \xi_3|^2.$$

Is the last representation canonical? What are the ranks r_1, r_2, r_3 and the signatures $\sigma_1, \sigma_2, \sigma_3$ of the forms $A_1(x,x), A_2(x,x), A_3(x,x)$ respectively? Compare the results with the Theorems 6.1, 6.2 and 6.3.

2. We consider the Hermitian form

$$\sum_{p,q=0}^{n-1} c_{p-q}\xi_p\overline{\xi}_q \qquad (c_{-p} = \overline{c}_p,\ p = 0,1,\cdots,n-1) \tag{6.6}$$

of order n (such a form is called a *Toeplitz* form, § 16 is dedicated to these forms). Let $n = 4$ and let the matrix of the form (6.6) have the shape

$$\begin{Vmatrix} c_0 & c_{-1} & c_{-2} & c_{-3} \\ c_1 & c_0 & c_{-1} & c_{-2} \\ c_2 & c_1 & c_0 & c_{-1} \\ c_3 & c_2 & c_1 & c_0 \end{Vmatrix} = \begin{Vmatrix} 0 & 0 & 0 & -i \\ 0 & 0 & 0 & 0 \\ 0 & 0 & 0 & 0 \\ i & 0 & 0 & 0 \end{Vmatrix},$$

i.e., the form (6.6) is reduced to

$$A_4(x,x) = i\xi_0\overline{\xi}_3 + i\overline{\xi}_0\xi_3$$

(take notice of the unusual way in which the parameters are indexed; however, we have already seen this in exercise 2 of § 5).

What is the signature σ_4 of this form and how does it change under transistion to the (Toeplitz) form

$$A_5(x,x) = i\xi_0\overline{\xi}_3 + i\overline{\xi}_0\xi_3 - i\xi_1\overline{\xi}_4 + i\overline{\xi}_1\xi_4 + \overline{\zeta}\xi_0\overline{\xi}_4 + \zeta\overline{\xi}_0\xi_4$$

with the matrix

$$\left\|\begin{matrix} 0 & 0 & 0 & -i & \overline{\zeta} \\ 0 & 0 & 0 & 0 & -i \\ 0 & 0 & 0 & 0 & 0 \\ i & 0 & 0 & 0 & 0 \\ \zeta & i & 0 & 0 & 0 \end{matrix}\right\| ,$$

where ζ is an arbitrary complex number? How do the rank r_5 and the signature σ_5 of the form $A_5(x,x)$ depend on the parameter ζ?

3. Is it possible to determine the signature σ_4 of the (Hankel - see § 5, exercise 2) quadratic form

$$A_4(x,x) = \xi_0^2 - \xi_1^2 + \xi_2^2 + \xi_3^2 - 2\xi_0\xi_2 + 2\xi_1\xi_3$$

with the matrix

$$A_4 = \left\|\begin{matrix} 1 & 0 & -1 & 0 \\ 0 & -1 & 0 & 1 \\ -1 & 0 & 1 & 0 \\ 0 & 1 & 0 & 1 \end{matrix}\right\| ,$$

by considering a somehow more simple quadratic form?

HINT. Apply proposition 1°.

NOTES

1) The stated corollary represent in itself a very special case of a general proposition: if an arbitrary (even rectangular) matrix is bordered by a row and a column the rank of the new matrix is either the original rank or it exceeds it with not more than two units. And this, in turn, follows from the evident fact, that under the addition of one arbitrary row (or column) to an arbitrary matrix the rank of this matrix cannot increase by more than one unit.

2) Taken from [3], p.28o. But the proof mentioned in [3] is to our mind presented in an insufficiently convincing way.

3) The exact meaning of this condition is the following: the coefficients $a_{ij}(=\overline{a_{ji}})$ $(i,j=1,2,\cdots,n)$ are continuous functions of a real parameter t, which runs trough some segment $[t_o,T]$. From the proof of the lemma thc reader will see, how one can generalize this condition and at the same time Lemma 6.2. Indeed, the proof of Lemma 6.2 shows that the signature is locally constant on X, if the coefficients of the form are continuous functions on the topological space X, and the rank of the form is the same for each value of $t \in X$. If, X is, in addition connected, then the signature will be constant on X (in coordination with the

author the present text of note [3], just as the proof of Lemma 6.2, was somewhat modified compared with the original-*translator*).

[4] The present proof differs from the proof in the original, as the original argument was not entirely convincing [Note of the translator].

[5] The method of constructing "intermediate" forms we use in this proof is called a homotopy, and the forms $\tilde{B}^{(0)}(x,x)$ and $\tilde{B}^{(1)}(x,x)$ are called homotopic.

§ 7. THE SYLVESTER FORMULA AND THE REPRESENTATION OF A HERMITIAN FORM AS A SUM OF SQUARES BY THE METHOD OF JACOBI.

7.1 We return to the Hermitian form

$$A(x,x) = \sum_{i,j=1}^{n} a_{ij}\xi_i\overline{\xi}_j \equiv A_n(x,x)$$

with discriminant $|A| \equiv \Delta_n$ and to the truncated forms $A_k(x,x)$ with discriminants

$$\Delta_k = \det \|a_{ij}\|_{i,j=1}^{k} \quad (k=1,2,\cdots,n-1),\ \Delta_o \equiv 1.$$

For the simplification of subsequent notations it is appropriate to expand somewhat the use of the symbol $A\begin{pmatrix} i_1,i_2,\cdots,i_p \\ j_1,j_2,\cdots,j_p \end{pmatrix}$, introduced in in the beginning (see § 1.1), denoting in this way furtheron any determinant of arbitrary order $p(\lesseqgtr n)$, consisting of the rows of the matrix A with index $i_1,i_2,\cdots,i_p$ and the columns with index $j_1,j_2,\cdots,j_p$. Here in neither set the indices are nessesarily arranged in increasing order, and recurrences are possible (and for $p > n$, clearly, inevitable).

THEOREM 7.1 (SYLVESTER FORMULA). *If for some* $r(1 \leq r \leq n)$ *we have* $\Delta_r \neq 0$ *then*

$$A(x,x) = -\frac{1}{\Delta_r}\begin{vmatrix} a_{11} & a_{12} & \cdots & a_{1r} & \overline{A_1(x)} \\ a_{21} & a_{22} & \cdots & a_{2r} & \overline{A_2(x)} \\ \cdot & \cdot & \cdots & \cdot & \cdot \\ a_{r1} & a_{r2} & \cdots & a_{rr} & \overline{A_r(x)} \\ A_1(x) & A_2(x) & \cdots & A_r(x) & 0 \end{vmatrix} +$$

$$+\frac{1}{\Delta_r}\sum_{i,j=1}^{n} A\begin{pmatrix} 1 & 2 & \cdots & r & i \\ 1 & 2 & \cdots & r & j \end{pmatrix}\xi_i\overline{\xi}_j \qquad (7.1)$$

where the $A_j(x) = \sum_{i=1}^{n} a_{ij}\xi_j$ $(j=1,2,\cdots,n)$ *are linear forms.*

PROOF. For the ascertainment of the Sylvester formula (7.1) we multiply both sides with Δ_r and we bring the first term on the righthand side to the left. On the lefthand side we obtain

$$A(x,x)\Delta_r + \begin{vmatrix} a_{11} & a_{12} & \cdots & a_{1r} & \overline{A_1(x)} \\ a_{21} & a_{22} & \cdots & a_{2r} & \overline{A_2(x)} \\ \cdot & \cdot & \cdots & \cdot & \cdot \\ a_{r1} & a_{r2} & \cdots & a_{rr} & \overline{A_r(x)} \\ A_1(x) & A_2(x) & \cdots & A_r(x) & 0 \end{vmatrix} =$$

$$= \begin{vmatrix} a_{11} & a_{12} & \cdots & a_{1r} & \overline{A_1(x)} \\ a_{21} & a_{22} & \cdots & a_{2r} & \overline{A_2(x)} \\ \cdot & \cdot & \cdots & \cdot & \cdot \\ a_{r1} & a_{r2} & \cdots & a_{rr} & \overline{A_r(x)} \\ 0 & 0 & \cdots & 0 & A(x,x) \end{vmatrix} + \begin{vmatrix} a_{11} & \cdots & a_{1r} & \overline{A_1(x)} \\ a_{21} & \cdots & a_{2r} & \overline{A_2(x)} \\ \cdot & \cdots & \cdot & \cdot \\ a_{r1} & \cdots & a_{rr} & \overline{A_r(x)} \\ A_1(x) & \cdots & A_r(x) & 0 \end{vmatrix} =$$

$$= \begin{vmatrix} a_{11} & a_{12} & \cdots & a_{1r} & \overline{A_1(x)} \\ a_{21} & a_{23} & \cdots & a_{2r} & \overline{A_2(x)} \\ \cdot & \cdot & \cdots & \cdot & \cdot \\ a_{r1} & a_{r2} & \cdots & a_{rr} & \overline{A_r(x)} \\ A_1(x) & A_2(x) & \cdots & A_r(x) & A(x,x) \end{vmatrix} =$$

$$= \begin{vmatrix} a_{11} & a_{12} & \cdots & a_{1r} & \sum_{j=1}^{n} a_{1j}\overline{\xi}_j \\ a_{21} & a_{22} & \cdots & a_{2r} & \sum_{j=1}^{n} a_{2j}\overline{\xi}_j \\ \cdot & \cdot & \cdots & \cdot & \cdot \\ a_{r1} & a_{r2} & \cdots & a_{rr} & \sum_{j=1}^{n} a_{rj}\overline{\xi}_j \\ \sum_{i=1}^{n} a_{i1}\xi_i & \sum_{i=1}^{n} a_{i2}\xi_i & \cdots & \sum_{i=1}^{n} a_{ir}\xi_i & \sum_{i,j=1}^{n} a_{ij}\xi_i\overline{\xi}_j \end{vmatrix} .$$

Applying to the last determinant the addition theorem, we partition it in n^2 terms, each of which has the form

$$\begin{vmatrix} a_{11} & a_{12} & \cdots & a_{1r} & a_{1j} \\ a_{21} & a_{22} & \cdots & a_{2r} & a_{2j} \\ \cdot & \cdot & \cdots & \cdot & \cdot \\ a_{r1} & a_{r2} & \cdots & a_{rr} & a_{rj} \\ a_{i1} & a_{i2} & \cdots & a_{ir} & a_{ij} \end{vmatrix} \xi_i\overline{\xi}_j \quad . \qquad (i,j=1,2,\cdots,n).$$

The sum of these expressions over all i and j gives

$$\sum_{i,j=1}^{n} A\begin{pmatrix} 1 & 2 & \ldots & r & i \\ 1 & 2 & \ldots & r & j \end{pmatrix} \xi_i\overline{\xi}_j \qquad (7.2)$$

which was to be proved.

We note, that for $r = n$ all terms of the sum (7.2) are equal to zero and for $r < n$ all terms in it vanish for which at least one of the indices i,j does not exceed r.

COROLLARY. Let the rank of the form be equal to r and $\Delta_r \neq 0$. Then

$$A(x,x) = -\frac{1}{\Delta_r}\begin{vmatrix} a_{11} & a_{12} & \cdots & a_{1r} & \overline{A_1(x)} \\ a_{21} & a_{22} & \cdots & a_{2r} & \overline{A_2(x)} \\ \cdot & \cdot & \cdots & \cdot & \cdot \\ a_{r1} & a_{r2} & \cdots & a_{rr} & \overline{A_r(x)} \\ A_1(x) & A_2(x) & \cdots & A_r(x) & 0 \end{vmatrix} \tag{7.3}$$

In the literature this relation is known as the *Kronecker identity*.

7.2 If the rank of the Hermitian form

$$A(x,x) = \sum_{i,j=1}^{n} a_{ij}\xi_i\overline{\xi}_j \tag{7.4}$$

is equal to r $(1 \leqq r \leqq n)$ then this form (see § 5, propostion 1°) can be represented in a sum of r independent squares. In some cases such a representation can be realized in a very simple way through standard formulae. In particular, we shall consider the Jacobi method.

THEOREM 7.2. *Let the rank of the Hermitian form* (7.4) *be equal to* r *and* $\Delta_1 \neq 0$, $\Delta_2 \neq 0$, $\cdots$, $\Delta_r \neq 0$. *We denote* $X_1(x) = A_1(x)$,

$$X_k(x) = \begin{vmatrix} a_{11} & a_{12} & \cdots & a_{1,k-1} & a_{1k} \\ a_{21} & a_{22} & \cdots & a_{2,k-1} & a_{2k} \\ \cdot & \cdot & \cdots & \cdot & \cdot \\ a_{k-1,1} & a_{k-1,2} & \cdots & a_{k-1,k-1} & a_{k-1,k} \\ A_1(x) & A_2(x) & \cdots & A_{k-1}(x) & A_k(x) \end{vmatrix} \qquad (k=1,2,\cdots,r).$$

Then

$$A(x,x) = \sum_{k=1}^{r} \frac{|X_k(x)|^2}{\Delta_{k-1}\Delta_k} \tag{7.5}$$

PROOF. Starting with the pattern the Kronecker identity provides, we define the Hermitian forms

$$B_k(x,x) = -\frac{1}{\Delta_k}\begin{vmatrix} a_{11} & a_{12} & \cdots & a_{1k} & \overline{A_1(x)} \\ a_{21} & a_{22} & \cdots & a_{2k} & \overline{A_2(x)} \\ \cdot & \cdot & \cdots & \cdot & \cdot \\ a_{k1} & a_{k2} & \cdots & a_{kk} & \overline{A_k(x)} \\ A_1(x) & A_2(x) & \cdots & A_k(x) & 0 \end{vmatrix} \qquad (k=1,2,\cdots,r)$$

We multiply both sides of each of these equations with $(-\Delta_k)$ respectively, and then we apply to the determinant on the righthand side the Sylvester identity in the shape of (2.5) (see § 2), where $k+1$ plays the role of n in the given case. We obtain

$$-B_k(x,x)\Delta_k\Delta_{k-1} = \Delta_k(-B_{k-1}(x,x)\Delta_{k-1}) - X_k(x)\overline{X_k(x)} \qquad (k=2,3,\cdots,r).$$

We rewrite this sequence of identities in the form

$$B_2(x,x) = B_1(x,x) + \frac{|X_2(x)|^2}{\Delta_1\Delta_2},$$

$$B_3(x,x) = B_2(x,x) + \frac{|X_3(x)|^3}{\Delta_2\Delta_3},$$

$$\cdots\cdots\cdots\cdots\cdots\cdots$$

$$B_r(x,x) = B_{r-1}(x,x) + \frac{|X_r(x)|^2}{\Delta_{r-1}\Delta_r}.$$

Supplement it further with the directly verifiable identity

$$B_1(x,x) = \frac{|X_1(x)|^2}{\Delta_0\Delta_1}$$

and add all the extracted identities termwise. Then, after simplification, we obtain

$$B_r(x,x) = \sum_{k=1}^{r} \frac{|X_k(x)|^2}{\Delta_{k-1}\Delta_k}.$$

It remains to note that, because of the Kronecker identity (7.3),

$$B_r(x,x) = A(x,x).$$

REMARK. On the strength of proposition 3° In § 5 the linear forms $X_1(x), X_2(x), \cdots, X_r(x)$ are linearly independent; it is not difficult to convince oneself of this directly.

EXAMPLES AND EXERCISES

1. Let the Hermitian form

$$A(x,x) = 2|\xi_1|^2 + (1+i)\xi_1\overline{\xi}_2 + (1-i)\overline{\xi}_1\xi_2 + |\xi_3|^2 + 2\xi_1\overline{\xi}_3 + 2\overline{\xi}_1\xi_3$$

with the matrix

$$A = \begin{Vmatrix} 2 & 1+i & 2 \\ 1-i & 0 & 0 \\ 2 & 0 & 1 \end{Vmatrix}$$

be given. Represent $A(x,x)$ through the Sylvester formula (7.1) for $r=2$ (here $\Delta_2 = -|1+i|^2 = -2 \neq 0$).

2. For the form

$$A(x,x) = 2|\xi_1|^2 + (1+i)\xi_1\bar{\xi}_2 + (1-i)\bar{\xi}_1\xi_3 + |\xi_3|^2 + 2\xi_1\bar{\xi}_3 + 2\bar{\xi}_1\xi_3 + \xi_2\bar{\xi}_3 + \bar{\xi}_2\xi_3$$

with the matrix

$$A = \left\| \begin{matrix} 2 & 1+i & 2 \\ 1-i & 0 & 1 \\ 2 & 1 & 1 \end{matrix} \right\|$$

which is a slight modification of the form in exercise 1, the identity

$$A(x,x) \equiv \frac{1}{2} \begin{vmatrix} 2 & 1+i & 2\bar{\xi}_1 + (1+i)\bar{\xi}_2 + 2\bar{\xi}_3 \\ 1-i & 0 & (1-i)\bar{\xi}_1 + \bar{\xi}_3 \\ 2\xi_1 + (1-i)\xi_2 + 2\xi_3 & (1+i)\xi_1 + \xi_3 & 0 \end{vmatrix}.$$

holds. How can one check it without developing the determinant on the righthand side?

3. Prove, that for the real (Hankel) quadratic form

$$A(x,x) = \sum_{j+k=0}^{n-1} s_{j+k}\xi_j\xi_k$$

of order $n(\geqq 2)$ with the matrix

$$\left\| \begin{matrix} s_0 & s_1 & \dots & s_{n-1} \\ s_1 & s_2 & \dots & s_n \\ \cdot & \cdot & \cdot\ \cdot\ \cdot & \cdot \\ s_{n-1} & s_n & \dots & s_{2n-2} \end{matrix} \right\| = \left\| \begin{matrix} a & a+d & \dots & a+(n-1)d \\ a+d & a+2d & \dots & a+nd \\ \cdot & \cdot & \cdot\ \cdot\ \cdot & \cdot \\ a+(n-1)d & a+nd & \dots & a+(2n-2)d \end{matrix} \right\|$$

the canonical representation

$$A(x,x) = \frac{\left[\sum_{j=0}^{n-1} (a+jd)\xi_j\right]^2}{a} - \frac{\left[a\sum_{j=0}^{n-1} \{a+(j+1)d\}\xi_j - (a+d)\sum_{j=0}^{n-1} (a+jd)\xi_j\right]^2}{ad^2}$$

is correct for $a \neq 0$ and $d \neq 0$. In particular, for $n = 2$,

$$A(x,x) = a\xi_0^2 + 2(a+d)\xi_0\xi_1 + (a+2d)\xi_1^2 = \frac{1}{a}[a\xi_0 + (a+d)\xi_1]^2 - \frac{d^2}{a}\xi_1^2.$$

HINT. Use the result of ex. 2 in § 5 and the Jacobi method (Thm. 7.2.).

§ 8 THE SIGNATURE RULE OF JACOBI AND ITS GENERALIZATIONS.

8.1. In various propositions in the theory of Hermitian and quadratic forms the problem often arises to determine the signature $\sigma = \pi - \nu$ of the form $A(x,x)$ (see Sec. 5.2) without representing it as a sum of independent squares. In this section rules are proved, which in some cases allow to determine the numbers π and ν if the rank r and the successive principal minors $(1\equiv)\ \Delta_0, \Delta_1, \cdots, \Delta_r$ of the form $A(x,x)$ are known.

First of all we shall formulate a direct consequence of Theorem 7.2.

THEOREM 8.1 (SIGNATURE RULE OF JACOBI). *If the rank of the Hermitian form*

$$A(x,x) = \sum_{i,j=1}^{n} a_{ij}\xi_i\overline{\xi}_j$$

is equal to r *and the successive principal minors* $\Delta_1, \Delta_2, \cdots, \Delta_{r-1}, \Delta_r$ *are different from zero, then*

$$\pi = P(1,\Delta_1,\Delta_2,\cdots,\Delta_r), \quad \nu = V(1,\Delta_1,\Delta_2,\cdots,\Delta_r) \tag{8.1}$$

where the symbols P *and* V *denote, respectively, the number of sign permanences and the number of sign changes in the set standing between the parantheses behind these symbols.*

PROOF. This is derived immediately from formula (7.5) and the remark on Theorem 7.2 [1)]

8.2. For all its attractiveness the rule of Jacobi has the evident defect, that it relies on the highly restrictive conditions

$$\Delta_1 \neq 0,\ \Delta_2 \neq 0, \cdots, \Delta_{r-1} \neq 0,\ \Delta_r \neq 0. \tag{8.2}$$

The violation of only one of these already deprives not only formula (7.5), from which we have derived the rule of Jacobi, of its meaning (this obstacle, as is evident from the note [1)], can be avoided) but also expression (8.1). Therefore, still in the last century, the question arose whether it is possible to preserve rule (8.1) also in those cases where some of the minors $\Delta_1, \Delta_2, \cdots, \Delta_r$ are equal to zero. The exact statement of such a question is:

If the signs plus or minus are known of those from the minors

$$(1\equiv)\ \Delta_0, \Delta_1, \Delta_2, \cdots, \Delta_{r-1}, \Delta_r \tag{8.3}$$

which are different from zero, can one under this condition prescribe signs plus or minus for the other minors of (8.3) *(i.e., for those equal to zero) in such a way, that identity* (8.1) *remains valid?*

In order to show the nontriviality of this question, we shall start with a negative result, which is usually cited in textbooks (see [3]).

EXAMPLE 8.1. Let a and b be real numbers, $a \neq b$, $ab \neq 0$; we consider the quadratic form

$$A(x,x) = a\xi_1^2 + a\xi_2^2 + b\xi_3^2 + 2a(\xi_1\xi_2 + \xi_2\xi_3 + \xi_3\xi_1) \tag{8.4}$$

of order $n = 3$. Its matrix

$$A = \begin{Vmatrix} a & a & a \\ a & a & a \\ a & a & b \end{Vmatrix}$$

has rank 2, whereas

$$\Delta_o = 1,\ \Delta_1 = a(\neq 0),\ \Delta_2 = \begin{vmatrix} a & a \\ a & a \end{vmatrix} = 0 \qquad (\text{i.e., } \Delta_r = 0).$$

It is easy to transform the form (8.4) to a sum of independent squares, writing it in the form

$$A(x,x) = a(\xi_1 + \xi_2 + \xi_3)^2 + (b-a)\xi_3^2 \tag{8.5}$$

We fix, for example, $a > 0$. Thus in the set Δ_o, Δ_1, Δ_2 are fixed the determinants Δ_o, Δ_1 and (a forteriori) their signs. Meanwhile , choosing $b > a$ we see (from (8.5)) that the signature σ of the form A will be equal to two ($\pi = 2$, $\nu = 0$). If $b < a$, then $\sigma = 0$ ($\pi = 1$, $\nu = 1$).

In this way, if $\Delta_r = 0$, then even if all other minors $\Delta_o, \Delta_1, \cdots, \Delta_{r-1}$ are different from zero, their signs cannot, in general,[2] define the signature of the form $A(x,x)$. Therefore, searching a generalization of Jacobi's rule at the expense of relaxation of conditions (8.2), the last of these ($\Delta_r \neq 0$) always remains in force.

8.3. Already in the second half of the XIX-th century at first S. Gundelfinger [26] and later G. Frobenius [19] succeded to generalize Jacobi's rule to the case where in (8.3) there are isolated zeros, i.e., for example $\Delta_{k-1} \neq 0$, $\Delta_k = 0$, $\Delta_{k-1} \neq 0$ (S. Gundelfinger) or isolated pairs of zeros: $\Delta_{k-1} \neq 0$, $\Delta_k = \Delta_{k-1} = 0$, $\Delta_{k+2} \neq 0$ (G. Frobenius). We present the rules which correspond to these cases, deriving them from the Theorems 6.1 - 6.3 (cf. [38]).

THEOREM 8.2 (GUNDESFINGERS RULE)[3] *Let in the set*

$$(1 \equiv)\Delta_o, \Delta_1, \Delta_2, \cdots, \Delta_{r-1}, \Delta_r \tag{8.6}$$

of successive principal minors of the Hermitian form

$$A(x,x) = \sum_{i,j=1}^{n} a_{ij}\xi_i\overline{\xi}_j \tag{8.7}$$

of rank r $(1 \leq r \leq n)$ *one of the minors be equal to zero, but the two adjacent minors different from zero:*

$$\Delta_{k-1} \neq 0,\ \Delta_k = 0,\ \Delta_{k+1} \neq 0 . \tag{8.8}$$

Then $\Delta_{k-1}\Delta_{k+1} < 0$ *and the signature rule of Jacobi remains valid, whatever sign (plus or minus) is written for the (vanishing) minor* Δ_k.

PROOF. We consider the truncated forms

$$A_{k-1}(x,x) = \sum_{i,j=1}^{k-1} a_{ij}\xi_i\overline{\xi}_j,\quad A_k(x,x) = \sum_{i,j=1}^{k} a_{ij}\xi_i\overline{\xi}_j,$$

$$A_{k+1}(x,x) = \sum_{i,j=1}^{k+1} a_{ij}\xi_i\overline{\xi}_j$$

with discriminants Δ_{k-1}, Δ_k and Δ_{k+1}, respectively. The rank of the (nondegenerate) form $A_{k-1}(x,x)$ is equal to k-1; such is also the rank of the degenerate ($\Delta_k = 0$) form $A_k(x,x)$[4]. The rank of the "extended" (and again nondegenerate) form $A_{k+1}(x,x)$ is $k+1$. So Theorem 6.1 allows to assert, that the signatures of $A_{k-1}(x,x)$ and $A_{k+1}(x,x)$ coincide, and as two units makes their ranks equal, the form $A_{k+1}(x,x)$ has exactly one positive square (i.e., one positive eigenvalue) and exactly one negative square (i.e., one negative eigenvalue) more than the form $A_{k-1}(x,x)$. Since the discriminants Δ_{k-1} and Δ_{k+1}, respectively, are equal (see(4.3)) to the products of all eigenvalues of the truncated forms, the signs of these discriminants are opposite (in the composition of the factors forming Δ_{k+1}, there enters one "extra" negative eigenvalue): $\Delta_{k-1}\Delta_{k+1} < 0$.

It remains to note that, having written for the minor Δ_k an arbitrary sign, we obtain in the set

$$(1\equiv)\quad \Delta_0, \Delta_1, \cdots, \Delta_{k-1}, \Delta_k, \Delta_{k+1}$$

exactly one change of sign and one permanence of sign more than in the set

$$(1\equiv)\quad \Delta_0, \Delta_1, \cdots, \Delta_{k-1},$$

i.e., for the form $A_{k+1}(x,x)$ Jacobi's rule remains valid. This rule is also correct for the whole form $A(x,x) = A_n(x,x)$ if all further "extended" forms $A_{k+2}(x,x), \cdots, A_r(x,x)$ are nondegenerate. Then at each transition from $A_{m+1}(x,x)$ to $A_{m+2}(x,x)$ $(m \geq k)$ there emerges either just one positive eigenvalue (if $\Delta_{m+1}\Delta_{m+2} > 0$) or just one negative eigenvalue (if $\Delta_{m+1}\Delta_{m+2} < 0$).

From the preceding reasoning it is clear that Gundelfingers rule can be applied also in the case, where in (8.6) there are several "isolated" zeros, i.e., if situation (8.8) is repeated for several values of the index k.

THEOREM 8.3 (FROBENIUS' RULE). *Let in the set* (8.6) *for the form* (8.7) *two successive minors be equal to zero, but the two neighbouring minors of this pair be different from zero:*

$$\Delta_{k-1} \neq 0,\ \Delta_k = \Delta_{k+1} = 0,\ \Delta_{k+2} \neq 0.$$

If one writes for the minors Δ_k and Δ_{k+1} (arbitrary) identical signs when $\Delta_{k-1}\Delta_{k+2} < 0$ and (arbitrary) opposite signs when $\Delta_{k-1}\Delta_{k+2} > 0$, then *the signature rule of Jacobi remains valid.*

PROOF. Again we consider the truncated forms $A_{k-1}(x,x)$, $A_k(x,x)$, $A_{k+1}(x,x)$ and $A_{k+2}(x,x)$ with the discriminants Δ_{k-1}, Δ_k, Δ_{k+1} and Δ_{k+2}, respectively. The rank of the form $A_{k-1}(x,x)$ is equal to k-1 ($\Delta_{k-1} \neq 0$). That is also the rank of the degenerate form $A_k(x,x)$, and hence the signature of this form is also the same (Theorem 6.2). The rank of the form $A_{k+2}(x,x)$ is equal to k+2 ($\Delta_{k+2} \neq 0$), and hence the rank of the form $A_{k+1}(x,x)$ is not less than k (according to Lemma 6.1). At the same time it doesn't surpass k, as $\Delta_{k+1} = 0$.

So the rank of the form $A_{k+1}(x,x)$ is equal to k. Hence (Theorem 6.1) the signatures of the forms $A_{k+1}(x,x)$ and $A_{k+2}(x,x)$ are the same. In comparison to $A_k(x,x)$, and, of course, also to $A_{k-1}(x,x)$, the form $A_{k+1}(x,x)$ has gained either one positive or one negative square (Theorem 6.3). Therefore the form $A_{k+2}(x,x)$, in comparison to $A_{k-1}(x,x)$ has gained either two positive and one negative square (if $\Delta_{k-1}\Delta_{k+2} < 0$) or two negative and one positive square (if $\Delta_{k-1}\Delta_{k+2} > 0$). It is easy to see that, having written for the minors Δ_k and Δ_{k+1} the signs plus or minus according to the rule indicated in the formulation of the theorem, we obtain in the set

$$\Delta_{k-1},\ \Delta_k,\ \Delta_{k+1},\ \Delta_{k+2}$$

in the former of the mentioned cases two sign permanences and one sign change and in the latter case one sign permanence and two sign changes. It remains to repeat the same conclusive reasoning that was also in the proof of Gundelfingers rule - Theorem 8.2.

REMARK. From the formulation and the proof of Theorems 8.2 and 8.3 it is clear, that they, like the initial rule of Jacobi (see(8.1)) which these theorems extends, remain valid also for (real) quadratic forms

$$A(x,x) = \sum_{i,j=1}^{n} a_{ij}\xi_i\xi_j \qquad (a_{ij} = a_{ji},\ i,j = 1,2,\cdots,i).$$

8.4 The results of Gundelfinger and Frobenius suggest the opportunity of further generalizations of the signature rule of Jacobi. However, in the general case (i.e., for arbitrary Hermitian and quadratic forms), a disappointment, as in Sec. 8.2., awaits us here. This is shown by the following example ([3],p.275).

Example 8.2. We consider the quadratic form

$$A(x,x) = 2a_{14}\xi_1\xi_4 + a_{22}\xi_2^2 + a_{33}\xi_3^2 ,$$

where $a_{14}(=a_{41})$, a_{22}, a_{33} are real coefficients which are different from zero and A is a quadratic form of order four with the matrix

$$A = \begin{Vmatrix} 0 & 0 & 0 & a_{14} \\ 0 & a_{22} & 0 & 0 \\ 0 & 0 & a_{33} & 0 \\ a_{41} & 0 & 0 & 0 \end{Vmatrix} .$$

Here $\Delta_o = 1$, $\Delta_1 = 0$, $\Delta_2 = 0$, $\Delta_3 = 0$, $\Delta_4 = -a_{14}^2 a_{22} a_{33} \neq 0$. The rank r is equal to 4, the condition $\Delta_r(=\Delta_4) \neq 0$ is satisfied. In contrast to the requirements in the Theorem 8.1 - 8.3, however, there are three successive determinants equal to zero at the same time: $\Delta_1 = \Delta_2 = \Delta_3 = 0$. We shall show that the signs of the remaining nonzero minors Δ_o and Δ_4 and even, in general, these minors themselves, in no way determine the signature of the form A(x,x).

Indeed, one easely represents this form as sum of independent squares:

$$A(x,x) = a_{22}\xi_2^2 + a_{33}\xi_3^2 + \frac{1}{2} a_{14}(\xi_1 + \xi_4)^2 - \frac{1}{2} a_{14}(\xi_1 - \xi_4)^2 .$$

From this representation it is clear, that for $a_{22} > 0$, $a_{33} > 0$ the signature $\sigma = +2$, but for $a_{22} < 0$, $a_{33} < 0$ the signature $\sigma = -2$. Meanwhile, in both cases $\Delta_o \equiv 1 > 0$ and $\Delta_4 = -a_{14}^2 a_{22} a_{33} < 0$.

In view of Example 8.2 two classes of quadratic and Hermitian forms are of particular interest, to which §§ 12 and 16 are devoted, respectively: the so-called Hankel- and Toeplitz forms (we have met these already in the form of examples and exercises in §§ 5-7). As we shall see below (in the Chapters II and III, respectively) it is possible to make a maximal extension of Jacobi's rule for these forms. It turns out that for the forms of these classes $\Delta_1, \Delta_2, \cdots, \Delta_r$ always (even if they are all equal to zero at the same time!) completely define the signature of the corresponding form.

EXAMPLES AND EXERCISES

1. Given is the quadratic form of order four

$$A(x,x) = + \xi_1^2 - \xi_2^2 + 2\xi_3^2 - \xi_4^2 + 2\xi_1\xi_3 + 2\xi_2\xi_4 + 2\xi_3\xi_4$$

Determine its signature with help of Jacobi' rule (Theorem 8.1).

Solution. $\sigma = 0$

2. We consider the Hermitian form of order three

$$A(x,x) = (-1+2i)\xi_1\overline{\xi}_2 + (-1-2i)\overline{\xi}_1\xi_2 + \frac{1}{2}|\xi_2|^2 - (3+i)\xi_2\overline{\xi}_3 - (3-i)\overline{\xi}_2\xi_3 - |\xi_3|^2$$

with the matrix

$$A = \left\| \begin{array}{ccc} 0 & -1+2i & 0 \\ -1-2i & 1/2 & -3-i \\ 0 & -3+i & -1 \end{array} \right\| .$$

Here

$$\Delta_0 \equiv 1,\ \Delta_1 = 0,\ \Delta_2 = -|-1+2i|^2 = -5,\ \Delta_3 = |-1+2i|^2 = 5.$$

Hence the form $A(x,x)$ has rank r equal to 3 (is a nondegenerate form). Since the group of successive principal minors $\Delta_0 = 1$, $\Delta_1 = 0$, $\Delta_2 = -5$ contains the isolated zero Δ_1, one has (completely in accordance to Gundesfingers rule) that $\Delta_0\Delta_2 < 0$, and, having written an arbitrary sign for Δ_1 we have

$$P(\Delta_0,\Delta_1,\Delta_2,\Delta_3) = 1, \qquad V(\Delta_0,\Delta_1,\Delta_2,\Delta_3) = 2$$

i.e. (by Theorem 8.2), $\pi = 1$, $\nu = 2$, $\sigma = \pi-\nu = -1$.

Check this calculation by a direct representation of the form $A(x,x)$ in canonical way.

3. Determine the signature σ of the Hermitian form

$$A(x,x) = 2|\xi_1|^2 + |\xi_3|^2 + \xi_1\overline{\xi}_3 + \overline{\xi}_1\xi_3 + \xi_1\overline{\xi}_4 + \overline{\xi}_1\xi_4 + \xi_2\overline{\xi}_4 + \overline{\xi}_2\xi_4 + \xi_3\overline{\xi}_4 + \overline{\xi}_3\xi_4 .$$

Solution. $\sigma = 2$

HINT. Use Frobenius' rule (Theorem 8.3).

Notes

1) With help of Theorem 6.3 one easily obtains another proof of the sign rule of Jacobi. Indeed, if ($\Delta_0 \equiv 1$), $\Delta_1 \neq 0$, $\Delta_2 \neq 0$, $\cdots$, $\Delta_r \neq 0$, where r is the rank of the form $A(x,x)$, then for $k \leqq r$ the transition from the form $A_{k-1}(x,x)$ to $A_k(x,x)$ is accompanied with an increase in the rank of one, and hence, because of Theorem 6.3, the form $A_k(x,x)$ aquires in comparison to $A_{k-1}(x,x)$ either a positive square (i.e., according to proposition 2° of § 5, it gains one additional positive eigenvalue), and then, following (4.3), $\Delta_{k-1}\Delta_k > 0$, or a negative square, and then $\Delta_{k-1}\Delta_k < 0$. Thence follows the rule of Jacobi as well.

2) For the reason of this stipulation see below at the end of Sec. 8.4.

3) For the case of forms of order three, this rule was already known to Gauss [5].

4) Hence the signatures of the forms $A_{k-1}(x,x)$ and $A_k(x,x)$ coincide - see Theorem 6.2.

Chapter II

HANKEL MATRICES AND FORMS

§ 9 HANKEL MATRICES. SINGULAR EXTENSIONS

9.1 A *Hankel matrix* of order $n(=1,2,\cdots)$ is the name for a matrix of the shape $H_{n-1} = \| s_{i+j} \|_{i,j=0}^{n-1}$ where the s_k are arbitrary complex numbers $(k=0,1,2,\cdots,2n-2)$. One can write more explicitly:

$$H_{n-1} = \begin{Vmatrix} s_0 & s_1 & s_2 & \cdots & s_{n-2} & s_{n-1} \\ s_1 & s_2 & s_3 & \cdots & s_{n-1} & s_n \\ s_2 & s_3 & s_4 & \cdots & s_n & s_{n+1} \\ \cdot & \cdot & \cdot & \cdots & \cdot & \cdot \\ s_{n-2} & s_{n-1} & s_n & \cdots & s_{2n-4} & s_{2n-3} \\ s_{n-1} & s_n & s_{n+1} & \cdots & s_{2n-3} & s_{2n-2} \end{Vmatrix} .$$

From the structure of H_{n-1} it is clear, in particular, that a Hankel matrix is always symmetric. (Moreover, its auxiliary diagonal (sometimes called second diagonal) and also all diagonals parallel to it consist of equal (the same for each diagonal) elements). Hence the matrix H_{n-1} is Hermitian if and only if it is real. Sometimes we shall also consider infinite Hankel matrices $H_\infty = \| s_{i+j} \|_{i,j=0}^{\infty}$. But at first we shall study finite matrices H_{n-1} and their so-called extensions.

9.2. An *extension* of the Hankel matrix H_{n-1} is called each Hankel matrix

$$H_{n-1+\nu} = \| s_{i+j} \|_{i,j=0}^{n-1+\nu} \qquad (\nu = 1,2,\cdots),$$

of which the left upper corner ("block") consists of the given matrix $H_{n-1} = \| s_{i+j} \|_{i,j=1}^{n-1}$. In diagram:

$$H_{n-1+\nu} = \left\Vert \begin{array}{ccc|ccc} & H_{n-1} & & s_n & \cdots & s_{n-1+\nu} \\ & & & \cdot & \cdots & \\ & & & \cdot & \cdots & \\ \hline s_n & & \cdots & s_{2n} & \cdots & \cdot \\ \cdot & & & & \ddots & \cdot \\ \cdot & & & & & \cdot \\ s_{n-1+\nu} & & \cdots & \cdots & & s_{2n-2+2\nu} \end{array} \right\Vert$$

A *singular extension* of the matrix H_{n-1} is called such an extension of it, that the rank of the extension coincides with the rank of the matrix H_{n-1}.

Below will be explained a method for studying Hankel matrices by means of the construction of singular extensions of some of its "blocks" and of the comparison of the given matrices with these singular extensions. For the development of the indicated method, however, one must first of all clarify whether for the given matrix H_{n-1} there exist always singular extensions (if only of order n+1) and what kind of "supply" of them. This problem is most simply solved for nonsingular Hankel matrices H_{n-1} $(|H_{n-1}| \neq 0)$.

For shortness of notation we introduce the symbol [1)]

$$D_k \equiv \det \| s_{i+j} \|_{i,j=0}^{k} \qquad (k = 0,1,\cdots),$$

such that D_k is a determinant (a *successive principal minor*) of order k+1. Moreover, we put

$$D_{-1} \equiv 1.$$

THEOREM 9.1 (FIRST EXTENSION THEOREM). *If* H_{n-1} *is a nonsingular Hankel matrix* $(D_{n-1} \neq 0)$, *then it has infinitely many singular extensions of order* n+1.

PROOF. The problem is to determine extensions H_n of order n+1 and rank n. We consider the function of two variables

$$D_n(x,y) = \begin{vmatrix} s_0 & s_1 & \cdots & s_{n-1} & s_n \\ s_1 & s_2 & \cdots & s_n & s_{n+1} \\ \cdot & \cdot & \cdots & \cdot & \cdot \\ s_{n-1} & s_n & \cdots & s_{2n-2} & x \\ s_n & s_{n+1} & \cdots & x & y \end{vmatrix} .$$

The problem, evidently, is reduced to solving the algebraic equation

$$D_n(x,y) = 0, \tag{9.1}$$

and after that it suffices to substitute $s_{2n-1} = x$, $s_{2n} = y$ and the desired extension will be obtained.

After evaluation of the determinant $D_n(x,y)$ the equation (9.1) takes the form (a,b and c are some coefficients)

$$D_{n-1}y + ax^2 + bx + c = 0, \tag{9.2}$$

and from this equation follows (since $D_{n-1} \neq 0$) that the theorem is correct. Moreover, we see that for an arbitrary choice for $x = s_{2n-1}$ one finds a unique $y(=s_{2n})$, which together with x defines a singular extension of the matrix H_{n-1}, and all these extensions are characterized by equation (9.1).

REMARK. In the special case were the matrix H_{n-1} is real, one is interested in its real singular extensions. Since in this case all the coefficients in (9.1) are real, the equation (9.2) defines in the (x,y)-plane a parabola for $a \neq 0$ and a straight line for $a = 0$. From the structure of the determinant $D_n(x,y)$ one sees, among other things, easily that $a = -D_{n-2}$.

9.3 The solution of the problem on the singular extensions of singular Hankel matrices H_{n-1} ($D_{n-1} = 0$), which is more important to us, is somewhat complicated.

THEOREM 9.2 (SECOND EXTENSION THEOREM). *Let* H_{n-1} *be a singular Hankel matrix and its rank* ρ [2] ($<n$). *If the principal minor* $D_{\rho-1} \neq 0$, *then there exists a unique pair of numbers* s_{2n-1}, s_{2n}, *defining a singular extension* H_n *of order* $n+1$ *of the matrix* H_{n-1}.

PROOF. In the case $\rho = 0$ the theorem is evidently valid: $s_{2n-1} = s_{2n} = 0$. Now let $\rho > 0$. We write down in detail the matrix H_{n-1}:

$$H_{n-1} = \left\| \begin{array}{cccc|ccc} s_0 & s_1 & \cdots & s_{\rho-1} & \cdots & s_{n-2} & s_{n-1} \\ s_1 & s_2 & \cdots & s_\rho & \cdots & s_{n-1} & s_n \\ s_2 & s_3 & \cdots & s_{\rho+1} & \cdots & s_n & s_{n+1} \\ \cdot & \cdot & \cdots & \cdot & \cdots & \cdot & \cdot \\ s_{\rho-1} & s_\rho & \cdots & s_{2\rho-2} & \cdots & s_{n+\rho-3} & s_{n+\rho-2} \\ \hline s_\rho & \boxed{s_{\rho+1}} & \cdots & s_{2\rho-1} & \cdots & s_{n+\rho-2} & s_{n+\rho-1} \\ s_{\rho+1} & s_{\rho+2} & \cdots & s_{2\rho} & \cdots & s_{n+\rho-1} & s_{n+\rho} \\ \cdot & \cdot & \cdots & \cdot & \cdots & \cdot & \cdot \\ s_{n-1} & s_n & \cdots & s_{n+\rho-2} & \cdots & s_{2n-3} & s_{2n-2} \end{array} \right\| . \tag{9.3}$$

Since $D_{\rho-1} \neq 0$, the first ρ rows of the matrix H_{n-1} are linearly independent, and the remaining ones are (as the rank of H_{n-1} is equal to ρ) a linear combination of these. In particular, we write the $(\rho+1)$st row as linear combination of the first ρ rows, obtaining ($\nu=\rho,\rho+1,\cdots,n+\rho-1$)

$$s_\nu = \sum_{j=0}^{\rho-1} \alpha_j s_{\nu-j-1} . \tag{9.4}$$

We prove that the same formula also holds for $\nu=n+\rho,n+\rho+1,\cdots,2n-2$.

We turn to the $(\rho+2)$nd row of the matrix H_{n-1}. All its elements, except the last ($s_{n+\rho}$), stand also in the previous row (with a shift to the right of one position, as marked in the diagram (9.3)), and hence for these entries formula (9.4) is true. Now we shall verify its validity for the element $s_{n+\rho}$ as well. To this end we multiply the second row of the matrix H_{n-1} with $\alpha_{\rho-1}$, the third with $\alpha_{\rho-2}$, and so on, finally the $(\rho+1)$st row with α_o, we add the resulting rows termwise and we subtract the result from the $(\rho+2)$nd row. If one has considered the formula (9.4), then it is clear, that after such a transformation (which, clearly, does not change the rank) the matrix H_{n-1} turns into

$$\tilde{H}_{n-1} = \left\| \begin{array}{cc|cccc} \multicolumn{2}{c|}{} & * & \cdots & * & * \\ \multicolumn{2}{c|}{H_{\rho-1}} & \cdot & \cdots & \cdot & \cdot \\ \multicolumn{2}{c|}{} & * & \cdots & * & * \\ \hline s_\rho & s_{\rho+1} & \cdot & \cdots & s_{n+\rho-2} & s_{n+\rho-1} \\ 0 & 0 & \cdot & \cdots & 0 & t \\ * & * & \cdot & \cdots & * & * \\ \cdot & \cdot & \cdot & \cdots & \cdot & \cdot \\ * & * & \cdot & \cdots & * & * \end{array} \right\| ,$$

where $H_{\rho-1}$ is the "block" of the matrix H_{n-1} for which $\det H_{\rho-1} = D_{\rho-1} \neq 0$; here by asterisks are marked the elements of the matrix which are not subject to the transformation and have no further significance, and

$$t = s_{n+\rho} - \alpha_o s_{n+\rho-1} - \alpha_1 s_{n+\rho-2} - \cdots - \alpha_{\rho-1} s_n .$$

The rank of the matrix $\tilde{H}_{n-1}$ is equal to ρ. Hence its minor

$$\det \left\| \begin{array}{cccc|c} & & & & * \\ & & & & \cdot \\ & H_{\rho-1} & & & \cdot \\ & & & & \cdot \\ \hline 0 & 0 & \cdots & 0 & * \\ & & & & t \end{array} \right\| = \left| \begin{array}{cccc|c} & & & & * \\ & & & & \cdot \\ & D_{\rho-1} & & & \cdot \\ & & & & * \\ \hline 0 & 0 & \cdots & 0 & t \end{array} \right| = D_{\rho-1} \cdot t$$

is equal to zero, as is each minor of order $\rho+1$. But as $D_{\rho-1} \neq 0$, t must be zero, and thus formula (9.4) is verified for $\nu = n+\rho$. It is clear, that we can extrapolate formula (9.4) also to $\nu = n+\rho+1, \cdots, 2n-2$ by repeating this step.

Now, if there exists a singular extension H_n of the matrix H_{n-1}, then the same reasoning could be extended, i.e., to obtain formula (9.4) for $\nu = 2n-1, 2n$ as well:

$$\left. \begin{aligned} s_{2n-1} &= \alpha_o s_{2n-2} + \alpha_1 s_{2n-3} + \cdots + \alpha_{\rho-1} s_{2n-\rho-1}, \\ s_{2n} &= \alpha_o s_{2n-1} + \alpha_1 s_{2n-2} + \cdots + \alpha_{\rho-1} s_{2n-\rho} \ . \end{aligned} \right\} \qquad (9.5)$$

By this it is proved, that the desired extension, if it exists, is defined uniquely by formula (9.5).

It remains to prove the converse reasoning; namely: define numbers

s_{2n-1}, s_{2n} through formula (9.5) and verify that they define a singular extension H_n of the matrix H_{n-1}. But formulae (9.5), together with formula (9.4), established above for $\nu=\rho,\rho+1,\cdots,2n-2$, show that also in the extended matrix H_n each row is a linear combination of the preceding ρ rows, i.e., in the end they are all a linear combination of the first ρ (linearly independent!) rows. So the rank of the matrix H_n is equal to ρ.

The Theorem is proved.

COROLLARY. *Under the conditions of Theorem* 9.2, *the formula* (9.4) *for* $\nu=2n-1,2n;2n+1;2n+2;\cdots$ *defines recursively an infinite sequence of pairs of numbers* $s_{2n-1},s_{2n};s_{2n+1},s_{2n+2};\cdots$, *which provide singular extension* $H_n,H_{n+1},\cdots$ *of the matrix* H_{n-1}.

Thus an infinite sequence of elements $s_0,s_1,s_2,\cdots,s_{2n-2},s_{2n-1},\cdots$ is defined, i.e., an infinite Hankel matrix H_∞, which, loosely speaking, can be considered as a singular extension of the matrix H_{n-1}, if one ascribes to the matrix H_∞ the rank ρ [3)].

EXAMPLES AND EXERCISES

1. For the Hankel matrix (of order $n=3$)

$$H_2=\left\|\begin{array}{ccc} 1 & 0 & 1-i \\ 0 & 1-i & -3 \\ 1-i & -3 & 2+i \end{array}\right\|$$

we consider the extensions

$$H_3=\left\|\begin{array}{ccc|c} 1 & 0 & 1-i & -3 \\ 0 & 1-i & -3 & 2+i \\ 1-i & -3 & 2+i & 0 \\ \hline -3 & 2+i & 0 & 1 \end{array}\right\|, \quad H_4=\left\|\begin{array}{ccc|c|c} 1 & 0 & 1-i & -3 & 2+i \\ 0 & 1-i & -3 & 2+i & 0 \\ 1-i & -3 & 2+i & 0 & 1 \\ \hline -3 & 2+i & 0 & 1 & i \\ \hline 2+i & 0 & 1 & i & -5 \end{array}\right\|$$

$$\tilde{H}_3=\left\|\begin{array}{ccc|c} 1 & 0 & 1-i & -3 \\ 0 & 1-i & -3 & 2+i \\ 1-i & -3 & 2+i & 0 \\ \hline -3 & 2+i & 0 & 0 \end{array}\right\|, \quad \tilde{H}_4=\left\|\begin{array}{ccc|c|c} 1 & 0 & 1-i & -3 & 2+i \\ 0 & 1-i & -3 & 2+i & 0 \\ 1-i & -3 & 2+i & 0 & 0 \\ \hline -3 & 2+i & 0 & 0 & 0 \\ \hline 2+i & 0 & 0 & 0 & 0 \end{array}\right\|$$

$$\tilde{H}_5 = \left\| \begin{array}{ccc|c|c|c} 1 & 0 & 1-i & -3 & 2+i & 0 \\ 0 & 1-i & -3 & 2+i & 0 & 0 \\ 1-i & -3 & 2+i & 0 & 0 & 0 \\ \hline -3 & 2+i & 0 & 0 & 0 & 0 \\ \hline 2+i & 0 & 0 & 0 & 0 & 0 \\ \hline 0 & 0 & 0 & 0 & 0 & 0 \end{array} \right\|$$

$$\tilde{\tilde{H}}_3 = \left\| \begin{array}{ccc|c} 1 & 0 & 1-i & -3 \\ 0 & 1-i & -3 & 2+i \\ 1-i & -3 & 2+i & 0 \\ \hline -3 & 2+i & 0 & \frac{14}{17} + \frac{46}{17} i \end{array} \right\|$$

It is easy to see, that H_4 is also an extension of the Hankel matrix H_3, that the matrices $\tilde{H}_4$ and $\tilde{H}_5$ are extensions of $\tilde{H}_3$ and that the matrix $\tilde{H}_5$ is an extension of the matrix $\tilde{H}_4$.

2. Verify, that in Example 1, det $H_2 = -4+i \neq 0$, i.e., the rank ρ of the matrix H_2 is equal to 3. Convince yourself, that $D_3 \equiv \det H_3 \neq 0$, i.e., that H_3 is not a singular extension of H_2. At the same time $\tilde{\tilde{D}}_3 = \det \tilde{\tilde{H}}_3 = 0$, i.e., rank $\tilde{\tilde{H}}_3$ is equal to 3 and $\tilde{\tilde{H}}_3$ is a singular extension of H_2.

3. Evaluate in Example 1 the determinant det $\tilde{H}_3 = \tilde{D}_3$ and convince yourself that $\tilde{D}_3 \neq 0$ (just as $D_3 \neq 0$); compare these results with the identity $\tilde{\tilde{D}}_3 = 0$ and Theorem 9.1.

4. The ranks of the matrices H_3 and $\tilde{H}_3$ of Example 1 coincide and are equal to 4. Hence the ranks of their extension H_4 and $\tilde{H}_4$ (respectively) are not less than 4, so these are not singular extension of H_2. Are they singular extension of H_3 and $\tilde{H}_3$, respectively?

What kind of extension is the matrix $\tilde{H}_5$ of H_2, of $\tilde{H}_3$, of $\tilde{H}_4$?

5. We consider the real Hankel matrix of order two

$$H_1 = \left\| \begin{array}{cc} 0 & 1 \\ 1 & 0 \end{array} \right\|$$

and its extensions

$$H_2 = \left\| \begin{array}{cc|c} 0 & 1 & 0 \\ 1 & 0 & 1 \\ \hline 0 & 1 & 0 \end{array} \right\|, \qquad H_3 = \left\| \begin{array}{cc|c|c} 0 & 1 & 0 & 1 \\ 1 & 0 & 1 & 0 \\ \hline 0 & 1 & 0 & 1 \\ \hline 1 & 0 & 1 & 0 \end{array} \right\|,$$

$$H_4 = \left\| \begin{array}{ccccc} 0 & 1 & 0 & 1 & 0 \\ 1 & 0 & 1 & 0 & 1 \\ 0 & 1 & 0 & 1 & 0 \\ 1 & 0 & 1 & 0 & 1 \\ 0 & 1 & 0 & 1 & 0 \end{array} \right\| , \cdots .$$

$$\tilde{H}_2 = \left\| \begin{array}{ccc} 0 & 1 & 0 \\ 1 & 0 & 0 \\ 0 & 0 & 0 \end{array} \right\| , \qquad \tilde{H}_3 = \left\| \begin{array}{cccc} 0 & 1 & 0 & 0 \\ 1 & 0 & 0 & 0 \\ 0 & 0 & 0 & 0 \\ 0 & 0 & 0 & 0 \end{array} \right\| ,$$

$$\tilde{H}_4 = \left\| \begin{array}{ccccc} 0 & 1 & 0 & 0 & 0 \\ 1 & 0 & 0 & 0 & 0 \\ 0 & 0 & 0 & 0 & 0 \\ 0 & 0 & 0 & 0 & 0 \\ 0 & 0 & 0 & 0 & 0 \end{array} \right\| , \cdots ;$$

$$\overset{\approx}{H}_2 = \left\| \begin{array}{ccc} 0 & 1 & 0 \\ 1 & 0 & 0 \\ 0 & 0 & 1 \end{array} \right\| , \qquad \overset{\approx}{H}_3 = \left\| \begin{array}{cccc} 0 & 1 & 0 & 0 \\ 1 & 0 & 0 & 1 \\ 0 & 0 & 1 & 0 \\ 0 & 1 & 0 & 0 \end{array} \right\| .$$

It is clear, that the matrices $H_2, H_3, H_4, \cdots$, as well as $\tilde{H}_2, \tilde{H}_3, \tilde{H}_4, \cdots$ are singular extensions of the matrix H_1 (their rank, like that of the matrix H_2, is equal 2). Rank $\overset{\approx}{H}_2$ is equal 3, so $\overset{\approx}{H}_2$ is not a singular extension of H_1. The rank of the matrix $\overset{\approx}{H}_3$ is also equal to 3 (verify this!) so $\overset{\approx}{H}_3$ is a singular extension of $\overset{\approx}{H}_2$ (but not of H_1!)

The matrices $H_3, H_4, \cdots$ are also singular extensions of H_2, and as such (but by no means as singular extensions of H_1!) they are uniquely defined (Theorem 9.2): here $n=3; s_5=1, s_6=0; s_7=1, s_8=0; \cdots$.

An analoguous situation holds for the matrix $\tilde{H}_2$ and its singular extensions $\tilde{H}_3, \tilde{H}_4, \cdots$: here again $n=3$, but $s_5=0, s_6=0; s_7=0, s_8=0; \cdots$. Compare these examples with the result, derived below in exercise 7.

6. Show, that the coefficients α_j $(j=0,1,\cdots,\rho-1)$ appearing in relation (9.4) are given by the formulae

$$\alpha_0 = -\frac{\beta_{\rho-1}}{\beta_\rho}, \quad \alpha_1 = -\frac{\beta_{\rho-2}}{\beta_\rho}, \cdots, \alpha_{\rho-1} = -\frac{\beta_0}{\beta_\rho} ,$$

where $\beta_0, \beta_1, \cdots, \beta_{\rho-1}$ and β_ρ $(=D_{\rho-1} \neq 0)$ are the cofactors of the elements $s_\rho, s_{\rho+1}, \cdots, s_{2\rho-1}$ and $s_{2\rho}$ in the last row of the determinant D_ρ $(=0)$, respectively.

7. In the case of a real Hankel matrix H_{n-1} which satisfies the conditions of Theorem 9.2, its singular extensions are also real

HINT. Use the result of exercise 6.

8. For the Hankel matrix

$$H_{n-1} = \left\| \begin{matrix} s_0 & s_1 & \cdots & s_{\rho-1} & s_\rho & \cdots & s_{n-1} \\ s_1 & s_2 & \cdots & s_\rho & s_{\rho+1} & \cdots & s_n \\ \cdot & \cdot & \cdots & \cdot & \cdot & \cdots & \cdot \\ s_{\rho-1} & s_\rho & \cdots & s_{2\rho-2} & s_{2\rho-1} & \cdots & s_{n+\rho-2} \\ s_\rho & s_{\rho+1} & \cdots & s_{2\rho-1} & s_{2\rho} & \cdots & s_{n+\rho-1} \\ \cdot & \cdot & \cdots & \cdot & \cdot & \cdots & \cdot \\ s_{n-1} & s_n & \cdots & s_{n+\rho-2} & s_{n+\rho-1} & \cdots & s_{2n-2} \end{matrix} \right\|$$

of rank ρ we consider all minors of order ρ and with the shape

$$\Delta_\rho^{(\alpha)} \equiv \begin{vmatrix} s_\alpha & s_{\alpha+1} & \cdots & s_{\alpha+\rho-1} \\ s_{\alpha+1} & s_{\alpha+2} & \cdots & s_{\alpha+\rho} \\ \cdot & \cdot & \cdots & \cdot \\ s_{\alpha+\rho-1} & s_{\alpha+\rho} & \cdots & s_{\alpha+2\rho-2} \end{vmatrix} \qquad (\alpha = 0,1,\cdots,2n-2\rho). \tag{9.6}$$

To these minors for $n \geqq 2$, $\rho < n$ clearly belong

$$(D_{\rho-1}=)\,\Delta_\rho^{(o)} = \begin{vmatrix} s_0 & s_1 & \cdots & s_{\rho-1} \\ s_1 & s_2 & \cdots & s_\rho \\ \cdot & \cdot & \cdots & \cdot \\ s_{\rho-1} & s_\rho & \cdots & s_{2\rho-2} \end{vmatrix}$$

and

$$\Delta_\rho^{(1)} = \begin{vmatrix} s_1 & s_2 & \cdots & s_\rho \\ s_2 & s_3 & \cdots & s_{\rho+1} \\ \cdot & \cdot & \cdots & \cdot \\ s_\rho & s_{\rho+1} & \cdots & s_{2\rho-1} \end{vmatrix}.$$

Prove the recurrence formula

$$\Delta_\rho^{(\alpha)}\Delta_\rho^{(o)} = \Delta_\rho^{(\alpha-1)}\Delta_\rho^{(1)} \qquad (\alpha = 1,2,\cdots,2n-2\rho) \tag{9.7}$$

HINT. For $\Delta_\rho^{(o)} \neq 0$ we use formula (9.4) and the result of exercise 6. For $\Delta_\rho^{(o)} = 0$ and $\Delta_\rho^{(1)} = 0$, formula (9.7) is trivial. If suffices to prove, that *the case, where* $\Delta_\rho^{(o)} = 0$ *but* $\Delta_\rho^{(1)} \neq 0$, *does not exist,* using to this end, for example, the Sylvester identity (2.6). For a generalization of the last result, see below in exercise 1o.

9. Under the condition $\Delta_\rho^{(o)} \neq 0$ follows from (9.7), that

$$\Delta_\rho^{(\alpha)} = \left(\frac{\Delta_\rho^{(1)}}{\Delta_\rho^{(0)}}\right)^\alpha \Delta_\rho^{(0)} \qquad (\alpha = 0,1,\cdots,2n-2\rho). \tag{9.8}$$

If $\Delta_\rho^{(1)} \neq 0$, then (see the hint to exercise 8) also $\Delta_\rho^{(0)} \neq 0$, and we see from (9.8) that all $\Delta_\rho^{(\alpha)} \neq 0$ $(\alpha \geqq 1)$.

Show with examples that for the last conclusion the condition $\Delta_\rho^{(0)} \neq 0$ is essential.

1o. (T.Ya. Azizov) If $\Delta_\rho^{(0)} = 0$, then

$$\Delta_\rho^{(\alpha)} = 0 \qquad (\alpha \leqq n-\rho).$$

Show with examples that for $\alpha > n-\rho$ the inequality $\Delta_\rho^{(\alpha)} \neq 0$ may already hold.

HINT. Throw out the first column and the last row from the matrix H_{n-1} and study in the remaining matrix (again a Hankel one) the minors $\Delta_\rho^{(\alpha)}$ $(\alpha \leqq n-\rho)$, using again the fact that for $\Delta_\rho^{(0)} = 0$ also $\Delta_\rho^{(1)} = 0$.

11. If in the Hankel matrix $H_{n-1} = \|s_{i+j}\|_{i,j=0}^{n-1}$ the first h rows are linearly independent, but the first h+1 rows are linearly dependent, then $D_{h-1} \neq 0$ [3].

HINT. Write for the elements of the (h+1)st row the analogon of formulae (9.4) and convince yourself by means of this, that in the strip consisting of the first h rows all minors of order h are equal to zero if $D_{h-1} = 0$.

NOTES.

1) Here (and in the sequel - in the full length of the Chapters II and III) we deliberately change, in connection with the special way of indexing the elements of Hankel- and Toeplitz matrices, the notation, used in Chapter I, for successive principal minors: now D_{k-1} $(k=0,1,\cdots,n)$ denotes the minor Δ_k of order k (see Sec. 6.1).

2) Instead of the symbol r, used in Chapter I, we shall write ρ for the rank of Hankel (and in Chapter III also of Toeplitz) matrices, reserving the letter r for other purposes (see §§ 1o and 11 below).

3) All its minors of order $\rho+1$ are equal to zero, as each of these is part of a singular extension $H_{n-1+\nu}$ $(\nu > 0)$ of H_{n-1}.

§ 1o. THE (r,k)-CHARACTERISTIC OF A HANKEL MATRIX.

1o.1 Theorem 9.2 opens up a way for defining an integral characteristic for a Hankel matrix. In the sequel this characteristic will play the role of a very helpful instrument for the investigation of Hankel matrices.

Let $H_{n-1} = \|s_{i+j}\|_{i,j=0}^{n-1}$ be an arbitrary Hankel matrix of order n (>0) and rank ρ $(0 \leqq \rho \leqq n)$ and let

$$(1\equiv)\ D_{-1}, D_0, D_1, \cdots, D_{r-1}, D_r, \cdots, D_{n-1} \tag{1o.1}$$

be all its successive principal minors. Let in the set (1o.1) the last minor (reading from left to right) which is different from zero be the minor D_{r-1}. Thus is defined an *integral constant* r $(0 \leqq r \leqq \rho)$:

$$D_{r-1} \neq 0,\ D_{\nu-1} = 0\ (\nu > r). \tag{1o.2}$$

Clearly, for $r = n(=\rho)$ the second of these relations isn't important.

Now we introduce another *integral constant* k in the following way: for $r = \rho$ we set $k = 0$. We note, that the identity $r = \rho$, in particular, always holds for $\rho = 0$ and $\rho = n$.

If, however, $r < \rho$, then we consider the "truncated" matrix

$$H_r = \left\| \begin{matrix} s_0 & s_1 & \cdots & s_{r-1} & s_r \\ s_1 & s_2 & \cdots & s_r & s_{r+1} \\ \cdot & \cdot & \cdots & \cdot & \cdot \\ s_{r-1} & s_r & \cdots & s_{2r-2} & s_{2r-1} \\ s_r & s_{r+1} & \cdots & s_{2r-1} & s_{2r} \end{matrix} \right\| .$$

Its determinant is equal to zero (because of (1o.2)), but its rank is equal to r, as $D_{r-1} \neq 0$. Thus the matrix H_r satisfies the conditions of Theorem 9.2. Hence (see the Corollary to Theorem 9.2) there exists a uniquely defined infinite sequence of numbers

$$s'_{2r+1}, s'_{2r+2};\ s'_{2r+3}, s'_{2r+4};\ \cdots , \tag{1o.3}$$

giving singular extensions $H'_{r+1}, H'_{r+2}, \cdots$ of the matrix H_r.

Parallel to (1o.3) we consider the finite set

$$s_{2r+1}, s_{2r+2};\ s_{2r+3}, s_{2r+4};\ \cdots;\ s_{2n-3}, s_{2n-2} \tag{1o.4}$$

of elements of the original matrix H_{n-1}. We note, that this set is non-empty, as $r < n-1$ (since $r < \rho < n$).

Now we compare the set (1o.4) with the sequence (1o.3). If $s_\nu = s'_\nu$ $(\nu = 2r+1, \cdots, 2n-2)$ then it would follow that $\rho = r$, in contradiction to our assumption. Hence there exist a uniquely defined natural number k such that

$$s_{2r+1} = s'_{2r+1},\ s_{2r+2} = s'_{2r+2},\ \cdots ,\ s_{2n-k-2} = s'_{2n-k-2},\ s_{2n-k-1} \neq s'_{2n-k-1} \tag{1o.5}$$

Thus, in the given situation

$$0 < k \leqq 2n - 2r - 2, \tag{1o.6}$$

and here equality on the right is realized if and only if already $s_{2r+1} \neq s'_{2r+1}$.

Here it is useful to make the meaning of the constant k understood by means of the diagram

$H_{n-1} =$

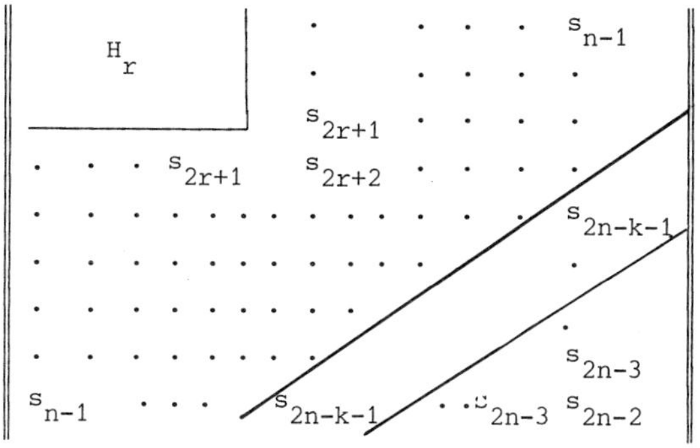

It is clear, that in this diagram the number k indicates the "distance", i.e., the number of diagonals,[1] from the righthand lower corner of the matrix H_{n-1} to the location of the first (moving from the left to the right) "wrong" diagonal after the "block" H_r, i.e., the diagonal in which the elements (for the first time!) do not coincide with those which define a singular extension of H_r. In this interpretation the case $k = 0$ $(r = \rho)$ is not excluded - the "wrong" diagnonal is simply absent. The abovedefined pair of integers (r,k) shall be called the (r,k)-*characteristic* [2] or simply *characteristic* of the Hankel matrix H_{n-1}.

1o.2 As will be explained in the sequel, the parity of the constant k plays a very important role in the (r,k)-characteristic of a Hankel matrix H_{n-1}. At first we consider the case of even k : $k = 2m > 0$ (for $k = 0$ the argument which follows becomes empty). The "truncated" matrix H_{n-m-1} has the shape

$$H_{n-m-1} = \begin{Vmatrix} s_0 & s_1 & \cdots & s_{n-m-1} \\ s_1 & s_2 & \cdots & s_{n-m} \\ \cdot & \cdot & \cdots & \cdot \\ s_{n-m-1} & s_{n-m} & \cdots & s_{2n-2m-2} \end{Vmatrix} \tag{1o.7}$$

i.e., (as $2n-2m-2 = 2n-k-2 < 2n-k-1$) it doesn't yet contain the "wrong" diagonal, and hence it is a singular extension of the matrix H_r[3]: the rank of the matrix H_{n-m-1} is equal to r and its (r,k)-characteristic has the shape (r,0). For the extension H_{n-m} of the matrix H_{n-m-1}, i.e., for the matrix

$$H_{n-m} = \left\| \begin{matrix} s_0 & s_1 & \cdots & s_{n-m-1} & s_{n-m} \\ s_1 & s_2 & \cdots & s_{n-m} & s_{n-m+1} \\ \cdot & \cdot & \cdots & \cdot & \cdot \\ s_{n-m-1} & s_{n-m} & \cdots & s_{2n-2m-2} & s_{2n-k-1} \\ s_{n-m} & s_{n-m+1} & \cdots & s_{2n-k-1} & s_{2n-k} \end{matrix} \right\|$$

this characteristic will be (r,2) already. It is clear, that at each further step of extension, i.e., at transition to the matrices H_{n-m+1}, H_{n-m+2} and so on (as long as the size of the matrix H_{n-1} permits it), in the characteristic only the second component will change, and also at each time it will increase by two units.

The situation turns out to be somewhat different for odd $k = 2m-1\,(>0)$. The matrix H_{n-m-1} (see(1o.7)) again has the rank r (since $2n-2m-2 = 2n-k-3 < 2n-k-1$) and the characteristic (r,0). As to the matrix H_{n-m}, this now has the shape

$$H_{n-m} = \left\| \begin{matrix} s_0 & s_1 & \cdots & s_{n-m-1} & s_{n-m} \\ s_1 & s_2 & \cdots & s_{n-m} & s_{n-m+1} \\ \cdot & \cdot & \cdots & \cdot & \cdot \\ s_{n-m} & s_{n-m+1} & \cdots & s_{2n-2m-1} & s_{2n-k-1} \end{matrix} \right\| \qquad (1o.8)$$

i.e., its (r,k)-characteristic turns out to be (r,1). At the transition to further extensions H_{n-m+1} H_{n-m+2}, ... this characteristic will change into (r,3), (r,5) and so on, since again at the addition of one more row and column the "wrong" diagonal, consisting of the element s_{2n-k-1}, will be moved away from the righthand lower corner each time by two additional positions.

Summing up this, one can state that is proved

THEOREM 1o.1 *Let in the* (r,k)*-characteristic of the Hankel matrix* H_{n-1} *the number* $k > 0$. *We denote* $m = \left[\frac{k+1}{2}\right]$ (*here* [a] *is the entire part of* a). *Then the rank of the matrix* H_{n-m-1} *is equal to* r *and its characteristic has the shape* (r,0). *For the extension* H_{n-m} *of the matrix* H_{n-m-1}, *depending on the evenness or oddness of* k, *the characteristic has the shape* (r,2) *for* $k = 2m$ *or* (r,1) *for* $k = 2m-1$. *For all* *further extensions* $H_{n-m+\nu}$ $(0<\nu\leq m-1)$ *the characteristic for even or odd* k *has, respectively, the shape* $(r,\ 2+2\nu)$ *or* $(r,\ 1+2\nu)$.

EXAMPLES AND EXERCISES.

1. We consider the Hankel matrix (see example 1 in § 9)

$$\tilde{\tilde{H}}_3 = \begin{Vmatrix} 1 & 0 & 1-i & -3 \\ 0 & 1-i & -3 & 2+i \\ 1-i & -3 & 2+i & 0 \\ -3 & 2+i & 0 & \frac{14}{16}+\frac{46}{17}i \end{Vmatrix}.$$

Here

$$\tilde{\tilde{D}}_2 = \begin{Vmatrix} 1 & 0 & 1-i \\ 0 & 1-i & -3 \\ 1-i & -3 & 2+i \end{Vmatrix} = -4+i \neq 0, \text{ but } \tilde{\tilde{D}}_3 = \det \tilde{\tilde{H}}_3 = 0,$$

so $r=3$. As, evidently, for $\tilde{\tilde{H}}_3$ the rank ρ is also equal to 3, one has $k=0$, i.e., the (r,k)-characteristic of the matrix $\tilde{\tilde{H}}_3$ has the shape $(3,0)$.

2. Verify, that for the Hankel matrix

$$\hat{H}_4 = \begin{Vmatrix} 0 & 1 & 0 & 1 & 0 \\ 1 & 0 & 1 & 0 & 1 \\ 0 & 1 & 0 & 1 & 0 \\ 1 & 0 & 1 & 0 & 0 \\ 0 & 1 & 0 & 0 & 0 \end{Vmatrix}$$

the (r,k)-characteristic has the shape $(2,2)$.

HINT. Compare $\hat{H}_4$ with the matrices H_1, H_2, H_3 and H_4 of Example 5 in § 9 and use the conclusions of this example.

3. Find the (r,k)-characteristic of the Hankel matrix

$$H_3 = \begin{Vmatrix} 0 & 4 & 0 & 1 \\ 4 & 0 & 1 & 0 \\ 0 & 1 & 0 & 1/4 \\ 1 & 0 & 1/4 & -6 \end{Vmatrix}$$

Solution. $(2,1)$.

HINT. In order to determine the constant k one can use formulae (9.4) and the results of exercise 6 in § 9.

4. Find the (r,k)-characteristic of the Hankel matrix

$$H_5 = \begin{Vmatrix} 1 & 0 & 1 & 0 & 1 & 0 \\ 0 & 1 & 0 & 1 & 0 & 1 \\ 1 & 0 & 1 & 0 & 1 & 0 \\ 0 & 1 & 0 & 1 & 0 & 2i \\ 1 & 0 & 1 & 0 & 2i & 1-4i \\ 0 & 1 & 0 & 2i & 1-4i & 3 \end{Vmatrix}$$

Solution. $(2,3)$.

HINT. For the calculation of the constant k use the result of exercise 7 in § 9.

5. Construct a Hankel matrix with (r,k)-characteristic (0,5). Find the shape of all Hankel matrices of order six with (r,k)-characteristic (0,5). Idem that of a Hankel matrix of order n (≥ 1) and with (r,k)-characteristic (0,m), where $0 \leq m < n$.

6. Let in the (r,k)-characteristic of the matrix $H_{n-1} = \|s_{i+j}\|_{i,j=0}^{n-1}$ the component r satisfy the condition $1 \leq r < n$. Then the (r+1)-st row of the matrix is a linear combination of its first r rows [3].

HINT. Use relation (1o.2) and apply the result of exercise 11 in § 9.

7. (Frobenius [19], see also [3], Ch. X, § 1o, Lemma 2 and Theorem 23). Under the condition of exercise 6 we consider the bordered minors (cf. § 2)

$$b_{\mu\nu} = \begin{vmatrix} \begin{matrix} & \\ & D_{r-1} \\ & \end{matrix} & \begin{matrix} s_{r+\nu} \\ \vdots \\ s_{2r+\nu-1} \end{matrix} \\ s_{r+\mu} \; \cdots \; s_{2r+\mu-1} & s_{2r+\mu+\nu} \end{vmatrix} \qquad (\mu,\nu = 0,1,\cdots,n-r-1)$$

and the numbers

$$t_{\mu\nu} = \frac{b_{\mu\nu}}{D_{r-1}} \quad (\mu,\nu = 0,1,\cdots,n-r-1).$$

Then the matrix $T_{n-r-1} = \|t_{\mu\nu}\|_{\mu,\nu=0}^{n-r-1}$ is a Hankel matrix and all elements situated on its auxiliary diagonal and above it are equal to zero, i.e., $t_{\mu\nu} \equiv t_{\mu+\nu}$ $(\mu,\nu = 0,1,\cdots,n-r-1)$; $t_0 = t_1 = \cdots = t_{n-r-2} = t_{n-r-1} = 0$:

$$T_{n-r-1} = \left\| \begin{matrix} 0 & 0 & \cdots & 0 & 0 \\ 0 & 0 & \cdots & 0 & t_{n-r} \\ 0 & 0 & \cdots & t_{n-r} & t_{n-r+1} \\ \cdot & \cdot & \cdots & \cdot & \cdot \\ 0 & 0 & \cdots & t_{2n-2r-4} & t_{2n-2r-3} \\ 0 & t_{n-r} & \cdots & t_{2n-2r-3} & t_{2n-2r-2} \end{matrix} \right\|$$

HINT. Consider the truncated matrix $T_{p-1} = \|t_{\mu\nu}\|_{\mu,\nu=0}^{p-1}$ $(p=1,2,\cdots, n-r)$, apply induction to p and use the Sylvester identity (S) (§ 2); also, use the result of exercise 6.

REMARK. In the original memoir of Frobenius [19] the result, mentioned above in exercise 7, is adjoined (more precisely, is preceded) by a whole row of propositions, which represent an independent interest. We shall adduce these in the following exercises.

8. Let

$$H_{n-1} = \|s_{i+j}\|_{i,j=0}^{n-1}$$

be a Hankel matrix, let $p \leqq n$ and

$$x_0, x_1, \cdots, x_p; \quad y_0, y_1, \cdots, y_p; \; z$$

arbitrary numbers. Then the identity

$$\begin{vmatrix} s_0 & \cdots & s_{p-1} & y_0 \\ \cdot & \cdots & \cdot & \cdot \\ s_{p-1} & \cdots & s_{2p-2} & y_{p-1} \\ x_1 & \cdots & x_p & z \end{vmatrix} - \begin{vmatrix} s_0 & \cdots & s_{p-1} & y_1 \\ \cdot & \cdots & \cdot & \cdot \\ s_{p-1} & \cdots & s_{2p-2} & y_p \\ x_0 & \cdots & x_{p-1} & z \end{vmatrix} =$$

$$\begin{vmatrix} s_0 & \cdots & s_{p2} & x_0 & y_0 \\ \cdot & \cdots & \cdot & \cdot & \cdot \\ s_{p-1} & \cdots & s_{2p-3} & x_{p-1} & y_{p-1} \\ s_p & \cdots & s_{2p-2} & x_p & y_p \end{vmatrix} \tag{1o.9}$$

holds.

HINT. Use the fact that the difference between the righthand and lefthand side of formula (1o.9) is a linear form in the parameters $x_0, x_1, \cdots, x_p, z$ in which the coefficients for x_0, x_p and z are equal to zero, so that in fact, it depends only on the p-1 parameters $x_1, \cdots, x_{p-1}$; at the same time, substituting for these parameters $s_{\nu+1}, \cdots, s_{\nu+p-1}$, and $x_0 = s_\nu$, $x_p = s_{p+\nu}$, $z = y_{\nu+1}$ $(\nu = 0, 1, \cdots, p-2)$ all three determinants in (1o.9) vanish.

REMARK. Identity (1o.9) is a special case of the more general proposition (see [19]):

For the minors of order p+2 $(\leqq n)$ *of each symmetric matrix* $A = \|a_{ij}\|_{i,j=0}^{n-1}$ *holds the identity*

$$A\begin{pmatrix} 0 & 1 & \cdots & p-2 & p-1 & p+1 \\ 1 & 2 & \cdots & p-1 & p & p+2 \end{pmatrix} - A\begin{pmatrix} 1 & 2 & \cdots & p-1 & p & p+1 \\ 0 & 1 & \cdots & p-2 & p-1 & p+2 \end{pmatrix} =$$

$$= A\begin{pmatrix} 0 & 1 & \cdots & p-2 & p-1 & p \\ 1 & 2 & \cdots & p-1 & p+1 & p+2 \end{pmatrix} \tag{1o.1o}$$

9. To what matrix A should one apply identity (1o.1o) in order to obtain (1o.9) from it?

1o. Let in the Hankel Matrix $H_{n-1} = \|s_{i+j}\|_{i,j=0}^{n-1}$ the minor D_{p-1} be different from zero, and let the bordered minors with the shape

$$b_{o\nu} = \begin{vmatrix} D_{p-1} & & s_{p+\nu} \\ & & \vdots \\ & & s_{2p+\nu-1} \\ s_p \;\cdots\; s_{2p-1} & \cdots & s_{2p+\nu} \end{vmatrix} = 0 \qquad (\nu = 0,1,\cdots,m \leq n-p-1).$$

Then $b_{\mu\nu} = 0$ for $\mu + \nu < m$ and

$$b_{\mu\nu} = b_{\mu+\nu} \qquad (\mu,\nu = 0,1,\cdots,m).$$

HINT. For $p > 0$, apply (1o.9), having chosen $x_o,\cdots,x_p$, $y_o,\cdots,y_p;z$, and make use of exercise 4 in § 2, on the strength of which the rank of the matrix

$$\begin{Vmatrix} a_o & \cdots & a_{p-1} & a_p & \cdots & a_{p+m-1} \\ \cdot & \cdots & \cdot & \cdot & \cdots & \cdot \\ a_{p-1} & \cdots & a_{2p-2} & a_{2p-1} & \cdots & a_{2p+m-2} \\ a_p & \cdots & a_{2p-1} & a_{2p} & \cdots & a_{2p+m-1} \end{Vmatrix}$$

is equal to p; for $p = 0$ the statement is trivial.

11. (Kronecker [48]). We consider in the Hankel matrix $H_{n-1} = \|s_{i+j}\|_{i,j=0}^{n-1}$ the minor D_{p-1} and the bordered minors

$$b_{\mu\nu} = \begin{vmatrix} D_{p-1} & & s_{p+\nu} \\ & & \vdots \\ & & s_{2p+\nu-1} \\ s_{p+\mu} \;\cdots\; s_{2p+\mu-1} & \cdots & s_{2p+\mu+\nu} \end{vmatrix} \qquad (\mu,\nu = 0,1,\cdots,n-p-1)$$

and we let $D_{p-1} \neq 0$.

Then, if

$$b_{oo} = b_{o1} = \cdots = b_{om} = 0 \qquad (m \leq n-p-1)$$

then

$$D_p = D_{p+1} = \cdots = D_{p+m} = 0$$

(i.e., in the (r,k)-characteristic of the truncated matrix H_{p+m} is the component r equal to p). Conversely if

$$D_p = D_{p+1} = \cdots = D_{p+m} = 0,$$

then $b_{oo} = b_{o1} = \cdots = b_{om} = 0$.

HINT. Use the Sylvester identity (S) of § 2 and the result of exercise 1o.

12. Obtain an affirmation of exercise 7 from the results of the exercises 11 and 1o.

NOTES.

1) In Sec. 11.1 it will be shown, that always $r+k \leqq n$, i.e., a forteriori $k \leqq n$; hence one can speak here of rows or columns instead of diagonals. This fact is reflected in the presented diagram.

2) We note, that the important role of the constant r was already revealed by Frobenius [19]. The other constant k as introduced here was defined at first in [39] (see on this point the Note to Theorem 11.2. below).

3) We recall that from (1o.6) follows the inequality $2n-2m-2 \geqq 2r$

§ 11. THEOREMS ON THE RANK.

11.1 Our immediate aim is to clarify, what are the connections between the (r,k)-characteristic of a Hankel matrix and its rank

LEMMA 11.1 *If* H_{n-1} *is a Hankel matrix with a given* (r,k)-*characteristic then* $r+k \leqq n$ *and the minor* $\tilde{D}_{r+k-1}$ *of order* $r+k$, *consisting of the first* r *lines (rows and columns) of the matrix* H_{n-1} *and the final* k *of its lines, is different from zero.*

PROOF. Although for $r=k=0$ both statements of the Lemma become meaningless we may also in this case formally given a meaning to it, putting, by definition

$$\tilde{D}_{-1} = D_{-1} = 1 \neq 0.$$

Now let $r+k > 0$. We note, that for $k=0$ both statemants of the Lemma are trivial, since by the definition of the constant r (see Sec. 1o.1) we have $r \leqq n$ and $\tilde{D}_{r-1} = D_{r-1} \neq 0$.

In the case $k > 0$ we assume at first, that the inequality $r+k \leqq n$ is already proved. Then

$$\tilde{D}_{r+k-1} = \left|\begin{array}{ccc|cccc}
 & & & * & * & \dots, & * \\
 & D_{r-1} & & \cdot & \cdot & \dots & \cdot \\
 & & & * & * & \dots & * \\
\hline
* & \dots & * & \dots & \cdot & & s_{2n-k-1} \\
\cdot & \cdot & \cdot & \dots & & & \\
\cdot & \cdot & \cdot & \dots & & & \cdot \\
\cdot & \cdot & \cdot & \dots & s_{2n-k-1} & \dots & \\
* & \dots & * & s_{2n-k-1} & & & \dots
\end{array}\right| ,$$

where all other elements of the final k lines of the matrix H_{n-1} (which don't interest us now) are denoted by asterisks. In order to evaluate the determinant $\tilde{D}_{r+k-1}$ we use the results of § 3. Namely, we consider the matrix H'_{n-1}, the singular extension of the matrix H_r (with rank r) and we construct the minor $M_k^{(r)}$ of the matrix H'_{n-1} ($\equiv A$) according to the diagram (3.1) where the minor D_{r-1} plays the role of A_r:

$$M_k^{(r)} = \left|\begin{array}{c|cccc}
D_{r-1} & \cdot & \dots & \dots & \\
 & \cdot & \dots & \dots & \\
\hline
\dots & \cdot & \dots & s'_{2n-k-1} & \\
\dots & \cdot & & \cdot & \\
 & & & \cdot & \\
\dots & s'_{2n-k-1} & \dots & s'_{2n-2} &
\end{array}\right| \qquad (11.1)$$

(here, for $r=0$, the lefthand upper corner, i.e., D_{r-1}, is nonexistent). Here by the symbols s'_ν ($\nu = 2n-k-1,\cdots,2n-2$) are denoted those elements of the sequence (1o.3), defining all singular extensions of the matrix H_r, which occur in the structure of the minor $M_k^{(r)}$ (we note, that because of formulae (1o.5) all other elements of the minor $M_k^{(r)}$ belong to the original matrix H_{n-1}).

Now we replace in the determinant $M_k^{(r)}$ (see diagram (11.1)) the diagonal consisting of the elements s'_{2n-k-1}, by the corresponding diagonal of the original matrix H_{n-1}, i.e., by the elements s_{2n-k-1}. Taking into account proposition 1° of § 3, we do not change the value of the new determinant $M_k^{(r)}(s_{2n-k-1})$, obtained in this way, (see (3.4)) if we interchange all its other elements, i.e., the elements situated in its righthand lower corner beneath the diagonal s'_{2n-k-1}, also by the corresponding elements of the original matrix H_{n-1}. But in the latter case we obtain (see the conditions of the Lemma) the minor $\tilde{D}_{r+k-1}$, so that, taking into account Lemma 3.1 in the shape of (3.5), we have

$$\tilde{D}_{r+k-1} = M_k^{(r)}(s_{2n-k-1}) = (-1)^{\frac{k(k-1)}{2}} D_{r-1}(s_{2n-k-1}-s'_{2n-k-1})^k . \qquad (11.2)$$

It suffices to recall, that $D_{r-1} \neq 0$ and that by definition (1o.5)

$s_{2n-k-1} - s'_{2n-k-1} \neq 0$, and the inequality $\tilde{D}_{r+k-1} \neq 0$ is obtained.

From this follows yet the conclusion, that for $k > 0$ the identity $r + k = n$ is impossible, since it would imply the relation

$$D_{n-1} = D_{r+k-1} = \tilde{D}_{r+k-1} \neq 0,$$

i.e., $p = n = r$, which contradicts the condition $k > 0$ (see the definition of the constant k in Sec. 10.1).

For the completion of the proof of Lemma 11.1., we only have to exclude the possibility of the inequality $r + k > n$ for $k > 0$. To this end we recall that (see (10.6))

$$0 < k \leqq 2n - 2r - 2.$$

We truncate the matrix H_{n-1}, by taking away the final $r + k - n$ of its lines.[1)] We obtain a Hankel matrix $H_{\tilde{n}-1}$ of order

$$\tilde{n} = n-(r+k-n) = 2n - r - k \geqq 2n - r-(2n-2r-2) = 2 + r.$$

Hence it follows, that $H_{\tilde{n}-1}$ is, in any case, an extension of the matrix H_r (and a forteriori of the matrix H_{r-1}) and in the $(\tilde{r},\tilde{k})$-characteristic of the matrix $H_{\tilde{n}-1}$ the component $\tilde{r}$ is equal to r, and as $r + k - n < n$ one has

$$\tilde{k} = k - 2(r+k-n) = 2n - 2r - k \geqq 2 > 0.$$

But then $\tilde{r} + \tilde{k} = 2n - r - k = \tilde{n}$ and this is, as was shown above, impossible.

11.2. Lemma 11.1 permits, in combination with the results of § 10, tc establish the following fact, which is fundamental for the entire theory we consider.

THEOREM 11.1 (FUNDAMENTAL THEOREM ON THE RANK). *If* H_{n-1} *is an arbitrary Hankel matrix with a given* (r,k)*-characteristic, and* ρ *is the rank of the matrix, then*

$$\rho = r + k.$$

PROOF. For $r = \rho$ the statement of the Theorem is trivial, since in this case $k = 0$, by definition.

Let $r < \rho$, i.e., (see (10.6)), $k > 0$. We introduce, as also in Theorem 10.1, the quantity $m = \left[\frac{k+1}{2}\right]$. Because of Theorem 10.1 the "truncated" matrix H_{n-m-1} has rank r, but the matrix H_{n-m} has a rank which already surpasses r. But because of the Corollary of Lemma 6.1 this rank can be equal to $r + 1$ or $r + 2$.

For even $k(=2m)$, again because of Theorem 10.1, the characteristic of the matrix H_{n-m} has the shape $(r,2)$. But then it follows from Lemma 11.1 that the matrix H_{n-m} contains a nonzero minor of order $r + 2$, and therefore its rank is equal to $r + 2$. Each further step of extension, i.e., transistion from H_{n-m} to $H_{n-m-1}, H_{n-m+2}, \cdots, H_{n-m+\nu}, \cdots$, will, by Theorem 1o.1, give matrices with characteristic $(r,4),(r,6),\cdots,$ $(r,2+2\nu)\cdots$, and the rank of these will be equal to $r+4, r+6,\cdots,$ $r+2+2\nu,\cdots$, respectively. The process is finished through the complete reconstruction of the matrix $H_{n-1} = H_{n-m+(m-1)}$, the rank of which is calculated by this rule as

$$\rho = r+2+2(m-1) = r+2m = r+k.$$

If, however, k is odd $(k = 2m-1)$, then the characteristic of the matrix H_{n-m} has the shape $(r,1)$. Because of Lemma 11.1 the rank of the matrix H_{n-m} is not less than $r+1$. But it is exactely equal to $r+1$, as

$$H_{n-m} = \left\| \begin{array}{c|c} H_{n-m-1} & \vdots \\ \hline \cdots & s_{2n-k-1} \end{array} \right\| ,$$

and the rejection of the last row reduces it to rectangular matrix not containing the "wrong" element s_{2n-k-1} (see Sec. 1o.2), but inculuding the block H_{n-m-1} (and a forteriori H_r). Hence the rank of the rectangular matrix is equal to r, and so the rank of the matrix H_{n-m} doesn't surpass $r+1$.

Again it remains to apply the same method of constructing extensions $H_{n-m+1}, H_{n-m+2}, \cdots, H_{n-m+\nu}, \cdots$, with characteristics $(r,3),(r,5),\cdots,$ $(r,1+2\nu),\cdots$. Because of Lemma 11.1 the ranks of these extensions will be equal to $r+3, r+5,\cdots, r+1+2\nu,\cdots$, respectively. Therefore, the rank of the matrix $H_{n-1} = H_{m-1+(m-1)}$ is equal to

$$\rho = r+1+2(m-1) = r+2m-1 = r+k.$$

Theorem 11.1 is proved.

11.3 Theorem 11.1 on the rank yields on its own an entire row of corollaries. First of all is directly obtained from it

THEOREM 11.2 (FROBENIUS). *If* H_{n-1} *is a Hankel matrix of rank* ρ, *and the number* r *is defined according to* (1o.2), *then the minor* $\tilde{D}_{\rho-1}$ *(of order* ρ*) of this matrix, consisting of the first* r *and the final* $\rho - r$ *of its lines, is nonzero.*

PROOF. This is evident, since $\rho - r = k$, and Theorem 11.2 simply paraphrases Lemma 11.1.[2)]

Theorem 11.2 was the starting point in the theory of Hankel matrices constructed by Frobenius (see the memoir by Frobenius [19], or, for example, the presentation of this theory in [3][3)]). In our construction it is, on the contrary, a "by-product".

We note that already in this theorem by Frobenius was contained, basically, an algorithm for the determination of the rank of a Hankel matrix,[4] namely

THEOREM 11.3. *Let* H_{n-1} *be a Hankel matrix and let the number* r *be given by relation* (1o.2). *If* $r = n$, *then the rank of the matrix* H_{n-1} *is equal to* n: $\rho = n$. *Now, if* $r < n$, *then, having adjoined to the minor* D_{r-1} ($\neq 0$) *(and in the case* $r = 0$ *simply having taken) at first the final row and column of the matrix* H_{n-1} *we form the minor* $\tilde{D}_r$ *of order* $r+1$; *next we adjoin to* D_{r-1} *the last two rows and the last two columns of the matrix* H_{n-1}, *and we form the minor* $\tilde{D}_{r+1}$ *of order* $r+2$, *and so on, as long as the relative sizes of the matrix* H_{n-1} *and the minor* D_{r-1} *permit this.*

We consider the maximal (with respect to the order) one of the minors $\tilde{D}_{r-1+\nu}$ ($\nu = 0,1,2,\cdots$) *which is different from zero* ($\tilde{D}_{r-1} \equiv D_{r-1}$) *Then*

$$\max_{\tilde{D}_{r-1+\nu} \neq 0} \nu = k$$

and $r + k = \rho$ *is the rank of the matrix* H_{n-1}.

Another almost immediate consequence of Theorem 11.1 is

THEROREM 11.4. *If in the* (r,k)*-characteristic of the Hankel matrix* H_{n-1} *of rank* ρ *the number* $k > 0$, *then an arbitrary extension* H_n *of the matrix* H_{n-1} *(i.e., an extension of order* $n+1$, *defined by an arbitrary pair of numbers* s_{2n-1}, s_{2n}*) has rank* $\tilde{\rho} = \rho + 2$.

PROOF. If the matrix H_n is nonsingular ($D_n = \det H_n \neq 0$), then $\tilde{\rho} = n+1$. Further, since $k > 0$, $D_{n-1} = 0$ (see Sec. 1o.1) and hence $\rho \leqq n-1$. But $\tilde{\rho} - \rho \leqq 2$ (see the Corollary of Lemma 6.1), so that $\rho = n-1$ and $\tilde{\rho} = \rho+2$. Now, if $D_n = 0$, then in the $(\tilde{r},\tilde{k})$-characteristic of the matrix H_n we have $\tilde{r} = r$ and $\tilde{k} = k+2$, i.e., $\tilde{\rho} = \tilde{r} + \tilde{k} = r + k + 2 = \rho + 2$.

From Theorem 11.4, in turn, one obtains a general criterium for the existence of singular extensions (see § 9) of arbitrary Hankel matrices, namely

THEOREM 11.5. *The Hankel matrix* H_{n-1} *of rank* ρ *allows singular extensions if and only if* $D_{\rho-1} \neq 0$.

PROOF. The sufficiency of the condition was already proved in the Theorems 9.1 and 9.2. The necessity follows from Theorem 11.4, as for $D_{\rho-1} = 0$ in the (r,k)-characteristic of the Hankel matrix H_{n-1} necessarily $k > 0$ (as $r < \rho$), and hence there are no singular extensions of

H_{n-1}: an arbitrary expansion H_n of order $n+1$ already has always rank $\rho+2$.

From this is, in particular, obtained a well-known theorem of Kronecker on infinite Hankel matrices. Just as for finite matrices we agree to ascribe to an infinite Hankel matrix the finite rank ρ ($\rho \geqq 0$ is an integer), if all its minors of order $\rho+1$ are equal to zero, but among the minors of order ρ there are some different from zero. Essentially, we have already used this definition in § 9 (see the Corollary to Theorem 9.2).

THEOREM 11.6 (KRONECKER).[5)] *If H_∞ is an infinite Hankel matrix of finite rank ρ, then its minor $D_{\rho-1}$ is different from zero.*

PROOF. We consider the truncated (finite) matrices H_{n-1} of order $n(=1,2,\cdots)$ obtained from H_∞. By assumption, the matrix H_{n-1} will have the rank ρ for sufficiently large n. If it would turn out that $D_{\rho-1} = 0$, then because of Theorem 11.4 the matrix H_n would already have the rank $\rho+2$, which is impossible.

The theorem is proved.

In view of Theorem 11.1 one can now, among other things, also take a look at Theorem 1o.1. For convenience of future references we reformulate it in the form of the following theorem.

THEOREM 11.7 (ON THE JUMP IN THE RANK BY EXTENSION). *If in the* (r,k)-*characteristic of the matrix H_{n-1} the number $k>0$, and $m = \left[\frac{k+1}{2}\right]$, then the truncated matrix H_{n-m-1} has rank r. The matrix H_{n-m} has the rank $r+2$ for $k=2m$, and the rank $r+1$ for $k=2m-1$. At each subsequent step of extension of the matrix H_{n-m} (until the complete reconstruction of the matrix H_{n-1}) the rank increases by two units.*

11.4 With Theorem 11.7 (or Theorem 1o.1) is closely connected the question on the determination of the general shape of a Hankel matrix H_{n-1} by its (r,k)-characteristic with $r \geqq 0$ and $k>0$.

Let be given entire numbers r, k and n which satisfy the conditions

$$r \geqq 0, \quad k > 0, \quad n > r+k. \tag{11.3}$$

We take an arbitrary nonsingular Hankel matrix H_{r-1} of order r (for $r=0$ this step is left out). Further, in agreement to Theorem 9.1., we construct an arbitrary singular extension H_r of the matrix H_{r-1} (all such extensions are described by equation (9.1) for $n=r$). For $r=0$ we put $H_0 = (0)$.

We denote, as before, $m \equiv \left[\frac{k+1}{2}\right]$. From (11.3) it follows that

$n \geq r+k+1$, and hence

$$n-m-1 \geq r+k-m = r+k-\left[\frac{k+1}{2}\right] \geq r.$$

We shall construct a Hankel matrix H_{n-m-1}. If $n-m-1 = r$, then this matrix $H_{n-m-1} = H_r$ was already chosen above. If $n-m-1 > r$, then the matrices $H_{r+1}, H_{r+2}, \cdots, H_{n-m-1}$ are the singular extensions of the matrix H_r. In this capacity they are defined in a unique way by means of Theorem 9.2, as the matrix H_r satisfies the conditions of this theorem. Theorem 9.2 guarantees the existence of a unique pair of numbers (we shall denote these with $s'_{2(n-m)-1}$, $s'_{2(n-m)}$) giving a singular extension (we denote it with H'_{m-m}) of the matrix H_{n-m-1} (and of H_r).

If $k = 2m$, then, having chosen any $s_{2(n-m)-1}$ ($\neq s'_{2(n-m)-1}$) and an arbitrary $s_{2(n-m)}$, we obtain a matrix H_{n-m} of rank $r+2$. Indeed, if $D_{n-m} \neq 0$, then the rank of the matrix H_{n-m} is equal to its order $n-m+1$, and our claim follows from the Corollary to Lemma 6.1 (since $D_{n-m-1} = 0$ and, therefore, the rank of the matrix H_{n-m-1} of order $n-m$ is, by construction, equal to r, which satisfies the inequality $r \leq n-m-1$). If, however $D_{n-m} = 0$ as well, then the statement follows directly from Theorem 1o.1 (or Theorem 11.7), applied to the matrix H_{n-m} (instead of H_{n-1}).

By exactly the same reasoning it is shown, that for arbitrary choices of the further elements $s_{2(n-m)+1}, \cdots, s_{2n-3}, s_{2n-2}$, the ranks of the matrices $H_{n-m+1}, \cdots, H_{n-2}, H_{n-1}$ will be equal to $r+4, \cdots, r+2(m-1)$, $r+2m (=r+k)$, respectively.

In the case of odd $k = 2m-1$ we put $s_{2(n-m)-1} = s'_{2(n-m)-1}$ and for $s_{2(n-m)}$ we take any number different from $s'_{2(n-m)}$. Then the rank of the matrix H_{n-m} is equal to $r+1$ (only the last diagonal, consisting of the single element $s_{2(n-m)}$ is "wrong"). At further extensions of the matrix H_{n-m} by arbitrary elements [6] $s_{2(n-m)+1}, \cdots, s_{2n-3}, s_{2n-2}$ the ranks of the matrices $H_{n-m+1}, \cdots, H_{n-2}, H_{n-1}$ will be equal to $r+3, \cdots, r+2m-3$, $r+2m-1 (=r+k)$, respectively. This is obtained again from Theorem 11.7 and the Corollary to Lemma 6.1.

Since for an arbitrary Hankel matrix H_{n-1} of order n and rank ρ with given (r,k)-characteristic $(r+k = \rho)$ the conditions (11.3) are, as is easily seen, always realized for $\rho > r$, the construction, followed in the given situation, yields the complete description (the general shape) of all Hankel matrices for which in the (r,k)-characteristic the number $k > 0$, i.e., $\rho > r$.

Thus we have proved

THEOREM 11.8. *The general shape of a Hankel matrix* $H_{n-1} = \| s_{i+j} \|_{i,j=0}^{n-1}$

of given order n $(\geqq 2)$ *and with given* (r,k)*-characteristic* $r \geqq 0$, $k > 0$, $n > r+k$ *is determined in the following way:*

1) *to* $r \geqq 1$ *is assigned an arbitrary nonsingular Hankel matrix* H_{r-1} *of order* r *(for* $r = 0$ *this step drops out);*

2) *as* H_r *is taken an arbitrary singular extension of the matrix* H_{r-1} *(for* $r = 0$ *we put* $H_0 = (0)$*);*

3) *if one sets* $m \equiv \left[\frac{k+1}{2}\right]$, *then* $n - m - 1 \geqq r$. *For* $n - m - 1 > r$ *matrices* $H_{r+1}, H_{r+2}, \ldots, H_{n-m-1}$ *are defined in a unique way as singular extensions of the matrix* H_r *(for* $n - m - 1 = r$ *the matrix* H_{n-m-1} *is already defined);*

4) *further is defined in a unique way the pair of numbers* $s'_{2(n-m)-1}$, $s'_{2(n-m)}$ *giving the singular extension* H'_{n-m} *(of order* $n - m + 1$*) of the matrix* H_r.

5) *for* $k = 2m$ *the element* $s'_{2(n-m)-1}$ *in* H'_{n-m} *is replaced by an arbitrary but different number* $s_{2(n-m)-1}$, *and* $s'_{2(n-m)}$ *may be replaced by an arbitrary number* $s_{2(n-m)}$; *in the case* $k = 2m-1$ *the element* $s'_{2(n-m)-1}$ $(\equiv s_{2(n-m)-1})$ *is retained but* $s'_{2(n-m)}$ *is replaced by an arbitrary number* $s_{2(n-m)}$ *which is different from it; in both cases we obtain a new matrix* H_{n-m} *of order* $n - m + 1$.

6) *for* $k = 1$ *we have* $m = 1$, *and the construction of the matrix is completed; for* $k > 1$ *the remaining elements* $s_{2(n-m)+1}, s_{2(n-m)+2}, \ldots, s_{2n-2}$ *are chosen arbitrarily.*

EXAMPLES AND EXERCISES.

1. For the Hankel matrix (see exercise 2 in § 10)

$$H_4 = \left\| \begin{matrix} 0 & 1 & 0 & 1 & 0 \\ 1 & 0 & 1 & 0 & 1 \\ 0 & 1 & 0 & 1 & 0 \\ 1 & 0 & 1 & 0 & 0 \\ 0 & 1 & 0 & 0 & 0 \end{matrix} \right\|$$

with (r,k)-characteristic $(2,2)$ the minors D_{r-1} and $\tilde{D}_{r+k-1}$ are, respectively, equal to

$$D_{r-1} = D_1 = \begin{vmatrix} 0 & 1 \\ 1 & 0 \end{vmatrix} = -1, \quad \tilde{D}_{r+k-1} = \tilde{D}_3 = \begin{vmatrix} 0 & 1 & 1 & 0 \\ 1 & 0 & 0 & 1 \\ 1 & 0 & 0 & 0 \\ 0 & 1 & 0 & 0 \end{vmatrix} = 1 \quad (\neq 0)$$

(Lemma 11.1)

Here $\rho = r+k = 4$.

2. For the matrix (cf. exercise 3 in § 10)

$$H_3 = \left\| \begin{array}{cccc} 0 & 4 & 0 & 1 \\ 4 & 0 & 1 & 0 \\ 0 & 1 & 0 & 1/4 \\ 1 & 0 & 1/4 & -6 \end{array} \right\|$$

with (r,k)-characteristic (2,1) we find

$$D_{r-1} = D_1 = \begin{vmatrix} 0 & 4 \\ 4 & 0 \end{vmatrix} = -16, \quad \tilde{D}_{r+k-1} = \tilde{D}_2 = \begin{vmatrix} 0 & 4 & 1 \\ 4 & 0 & 0 \\ 1 & 0 & -6 \end{vmatrix} = 96 \ (\neq 0)$$

(Lemma 11.1)

Here $\rho = r+k = 3$.

3. We consider the Hankel matrix of order six

$$H_5 = \left\| \begin{array}{cccccc} 1 & 0 & 0 & 1 & 0 & 0 \\ 0 & 0 & 1 & 0 & 0 & 1 \\ 0 & 1 & 0 & 0 & 1 & 0 \\ 1 & 0 & 0 & 1 & 0 & 0 \\ 0 & 0 & 1 & 0 & 0 & 1 \\ 0 & 1 & 0 & 0 & 1 & 1 \end{array} \right\| .$$

We see immediatiely, that $D_0 = 1$, $D_1 = 0$, $D_2 = -1$, $D_3 = D_4 = D_5 = 0$. Thus $r = 3$, $D_{r-1} = D_2$ $(= -1) \neq 0$. The constant k (and with it the rank $\rho = r+k$) and the minor $\tilde{D}_{r+k-1}$ $(\neq 0)$ we shall find simultanuously, by applying the rule of Theorem 11.3. At first we border the minor

$$D_2 = \begin{vmatrix} 1 & 0 & 0 \\ 0 & 0 & 1 \\ 0 & 1 & 0 \end{vmatrix}$$

with the final row and the final column of the matrix H_5:

$$\tilde{D}_{2+1} = \tilde{D}_3 = \left| \begin{array}{ccc|c} 1 & 0 & 0 & 0 \\ 0 & 0 & 1 & 1 \\ 0 & 1 & 0 & 0 \\ \hline 0 & 1 & 0 & 1 \end{array} \right| = -1 \ (\neq 0) .$$

Next we use for the bordering of D_2 the last two rows and the last two columns of the matrix H_5:

$$\tilde{D}_{2+2} = \tilde{D}_4 = \left| \begin{array}{ccc|cc} 1 & 0 & 0 & 0 & 0 \\ 0 & 1 & 1 & 0 & 1 \\ 0 & 1 & 0 & 1 & 0 \\ \hline 0 & 0 & 1 & 0 & 1 \\ 0 & 1 & 0 & 1 & 1 \end{array} \right| = 0$$

As $\tilde{D}_{2+3} = \tilde{D}_5 = D_5 = 0$, one has, according to Theorem 11.3

$$k = \max_{\tilde{D}_{2+\nu} \neq 0} \nu = 1, \ \rho = r+k = 4, \ \tilde{D}_{r+k-1} = \tilde{D}_3 = -1.$$

4. For the matrix H_5 of example 3, for which the (r,k)-characteristic has the shape (3,1), trace the jumps in the rank at the construction

of extension of the "block" $H_{r-1} = H_2$:

$$H_2 = \begin{Vmatrix} 1 & 0 & 0 \\ 0 & 0 & 1 \\ 0 & 1 & 0 \end{Vmatrix}, \quad H_3 = \begin{Vmatrix} 1 & 0 & 0 & 1 \\ 0 & 0 & 1 & 0 \\ 0 & 1 & 0 & 0 \\ 1 & 0 & 0 & 1 \end{Vmatrix}, \quad H_4 = \begin{Vmatrix} 1 & 0 & 0 & 1 & 0 \\ 0 & 0 & 1 & 0 & 0 \\ 0 & 1 & 0 & 0 & 1 \\ 1 & 0 & 0 & 1 & 0 \\ 0 & 0 & 1 & 0 & 0 \end{Vmatrix}$$

$$H_5 = \begin{Vmatrix} 1 & 0 & 0 & 1 & 0 & 0 \\ 0 & 0 & 1 & 0 & 0 & 1 \\ 0 & 1 & 0 & 0 & 1 & 0 \\ 1 & 0 & 0 & 1 & 0 & 0 \\ 0 & 0 & 1 & 0 & 0 & 1 \\ 0 & 1 & 0 & 0 & 1 & 1 \end{Vmatrix}$$

Compare the result with Theorem 11.7.

5. Prove the following (cf. [4o], Theorem 4):

If for the Hankel matrix H_{n-1} *the rank is* ρ *and in the* (r,k)*-characteristic the number* $k > 0$ $(r < \rho < n)$, *and the defect of the matrix is equal* d $(=n-\rho>0)$, *then after* d *steps of extension of this matrix* [7)] *by arbitrary pairs of numbers*

$$s_{2n-1}, s_{2n}; s_{2n+1}, s_{2n+2}, \ldots, s_{2n+2(d-1)-1}, s_{2n+2(d-1)}$$

we obtain (for the first time) a nonsingular matrix H_{n-1+d}.

HINT. Apply the Theorems 11.4 and 1o.1.

6. We shall illustrate exercise 5 with a numerical example. Let be given the Hankel matrix of order $n = 4$:

$$H_3 = \begin{Vmatrix} i & 1 & -i & -1 \\ 1 & -i & -1 & i \\ -i & -1 & i & 1 \\ -1 & i & 1 & 2 \end{Vmatrix} .$$

Here $\rho = 2$, $r = 1$ (verify this!), i.e., $d = n-\rho = 2$. For arbitrary extensions "of one step" by means of elements α,β and "of two steps" by means of elements $\alpha,\beta,\gamma,\delta$ we have, respectively

$$\begin{vmatrix} i & 1 & -i & -1 & i \\ 1 & -i & -1 & i & 1 \\ -i & -1 & i & 1 & 2 \\ -1 & i & 1 & 2 & \alpha \\ i & 1 & 2 & \alpha & \beta \end{vmatrix} = 0, \quad \begin{vmatrix} i & 1 & -i & -1 & i & 1 \\ 1 & -i & -1 & i & 1 & 2 \\ -i & -1 & i & 1 & 2 & \alpha \\ -1 & i & 1 & 2 & \alpha & \beta \\ i & 1 & 2 & \alpha & \beta & \gamma \\ 1 & 2 & \alpha & \beta & \gamma & \delta \end{vmatrix} \neq 0.$$

Verify this!

7. For the case $r \geqq 1$ prove Frobenius' Theorem 11.2, not relying on Lemma 11.1 and Theorem 11.1 ([3], Ch.X, § 1o, Theorem 23).

HINT. With help of the Sylvester identity (S) of § 2 prove the relation

$$\tilde{D}_{\rho-1} = D_{r-1} T \begin{pmatrix} n-\rho+1 \cdots n-r \\ n-\rho+1 \cdots n-r \end{pmatrix},$$

where $T = T_{n-r-1}$ is the matrix, introduced in exercise 7 of § 1o, and therenpon apply the result of this exercise.

8. Kroneckers Theorem 11.6 deals with infinite Hankel matrics H_∞ of finite rank ρ. Prove that the infinite Hankel matrix

$$H_\infty = \|s_{i+j}\|_{i,j=0}^{\infty}$$

has finite rank ρ if and only if there exists ρ numbers $\alpha_0, \alpha_1, \cdots, \alpha_{\rho-1}$ such that

$$s_\nu = \sum_{j=0}^{\rho-1} \alpha_j s_{\nu-j-1} \quad (\nu=\rho, \rho+1, \cdots) \tag{11.4}$$

and ρ is minimal under those integers which has this property (cf.[3], Ch. XVI, § 1o, Theorem 7).

HINT. Apply Theorem 11.6 and also Theorem 9.2 (formula (9.4)) and its Corollary.

9. Conversely, show that Kroneckers Theorem 11.6 is also obtained as corollary of the result of exercise 8 (this is the way it is derived in [3], where the result of exercise 8 is established independently of Kroneckers Theorem).

1o. Derive Kroneckers Theorem 11.6 from his Theorems, which we presented in exercise 11 of § 1o and exercise 4 in § 2 [19].

HINT. Use here, that if ρ (> 0) is the finite rank of the matrix

$$H_\infty = \|s_{i+j}\|_{i,j=0}^{\infty},$$

then not all its elements are equal to zero, and thus (because of the Hankel structure) not all $D_{\nu-1}$ $(\nu = 1, 2, \cdots)$ are equal to zero; consider the maximal ν (≦ ρ) for which $D_{\nu-1} \neq 0$, and, having applied the mentioned theorems of Kronecker, prove that $\nu = \rho$.

11. If in the infinite Hankel matrix H_∞ the minor $D_{\rho-1} \neq 0$, but $D_\rho = D_{\rho+1} = \cdots = 0$, then the rank of the matrix H_∞ is finite and equal to ρ [19]. Prove this.

12. Let be given a proper rational quotient

$$R(z) = g(z)/h(z)$$

where

$$h(z) = a_o z^m + a_1 z^{m-1} + \cdots + a_m \qquad (a_o \neq 0),$$

$$g(z) = b_1 z^{m-1} + b_2 z^{m-2} + \cdots + b_m.$$

We expand R(z) in a series of negative powers of z

$$R(z) = \frac{g(z)}{h(z)} = \frac{s_o}{z} + \frac{s_1}{z^2} + \frac{s_2}{z^3} + \cdots \tag{11.5}$$

(it converges, evidently, outside an arbitrary disc $|z| \leq R$ which contains all poles of the function R(z) [1o], i.e., those values of z for which R(z) becomes infinite).

Prove, that the infinite Hankel matrix

$$H_\infty = \|s_{i+j}\|_{i,j=0}^{\infty}$$

has finite rank $\rho \leqq m$.

Conversely, if the rank of H_∞ is finite and equal to ρ, then the function R(z) defined by (11.5), is rational, and the number ρ coincides with the number of poles of R(z), counting each of these as many times as its order is (for the meaning of this concept see, for example, [1o]).

HINT. Write down the identity involving the coefficients a_μ $(\mu = 0,1,\cdots,m)$, b_ν $(\nu = 1,\cdots,m)$ and s_k $(k = 0,1,2,\cdots)$, starting with (11.5), and note, that they generate relations of the type (11.4) for $\rho = m$ (more explicitly, see [3], Ch. XVI, § 1o, Theorem 8).

NOTES

1) i.e. of its rows and columns. We note, that the inequality $r + k - n \leqq n - 2$ follows from (1o.6). Moreover, $r + k - n < k$, as $r < n$ (for $r = n$ one would have $k = 0$).

2) In paper [19] Frobenius introduces, besides the constant r, for each Hankel matrix also a constant k, defined (somewhat formally, as is clear now) by the identity $k = \rho - r$.

3) See also Exercise 7 at the end of this Section.

4) That was only noted for the first time in [4o]. As T. Ya. Azizov correctly observed to the author, it is, for values of r which are small compared with n, economical not to look for the rank $\rho = r+k$ of the matrix H_{n-1} by this rule but to calculate directly the component k in the (r,k)-characteristic by the methods of § 1o.

5) See, for example, [3], p. 469; there is also presented a proof which

is different from ours (see exercise 9 in the present section).

6) Here we formulate somewhat loosely, but the reader will understand that the construction of extensions $H_{n-m+1},\cdots,H_{n-2},H_{n-1}$ of the matrix H_{n-m} is meant.

7) Here we use again the loose formulation already mentioned above.

§ 12. HANKEL FORMS

12.1. A *Hankel form* of order n (> 0) is the name used for the quadratic form

$$H_{n-1}(x,x) = \sum_{i,j=0}^{n-1} s_{i+j}\xi_i\xi_j \tag{12.1}$$

with the corresponding Hankel matrix [1]

$$H_{n-1} = \| s_{i+j} \|_{i,j=0}^{n-1} \tag{12.2}$$

The regularities in the structure of Hankel matrices, which was revealed in §§ 9-11, allows to establish rather quickly a rule of Frobenius for the determination of the signature of the form (12.1), emerging (for this special class of forms) as a generalization of Jacobi's rule. This generalization is complete in the usual sense, as it extends to arbitrary forms (12.1) without any restriction of the kind usually imposed on their succesive principal minors $D_0,D_1,\cdots,D_{\rho-1}$ (ρ is the rank of the matrix H_{n-1}). As we have seen in § 8, difficulties arise here, when in the sequence of numbers

$$D_0,D_1,\cdots,D_{\rho-1} \tag{12.3}$$

there occur groups which contain more than two successive zeros, and also in the case $D_{\rho-1} = 0$. We shall show how these difficulties are surmounted for Hankel forms (12.1). For facilitation of our investigation, we shall divide it into several steps.

12.2 As usually, we shall consider along with the form (12.1) the truncated quadratic forms

$$H_\nu(x,x) = \sum_{i,j=0}^{\nu} s_{i+j}\xi_i\xi_j \qquad (\nu = 0,1,\cdots,n-1)$$

with the matrices $H_0,H_1,\cdots,H_{n-1}$ and the discriminants $D_0,D_1,\cdots,D_{n-1}$, respectively.

1° *Let the sequence* (12.3) *contain an isolated group of* p *zeros*

$$(D_{h-1} \neq 0), D_h = D_{h+1} = \cdots = D_{h+p-1} = 0, \ (D_{h+p} \neq 0), \tag{12.4}$$

where p *is an odd number:* $p = 2q-1$ $(q \geq 1)$. *Then the truncated forms* $H_{h-1}(x,x)$ *and* $H_{h+p}(x,x)$ *have the same signature.* Explicitly: *the form* $H_{h+p}(x,x)$ *contains in a canonical representation* q *positive and* q *negative squares more than the form* $H_{h-1}(x,x)$.

Indeed, for $q = 1$ $(p = 1)$ proposition 1° was established (moreover, not only for Hankel forms, but for arbitrary quadratic and Hermitian forms) already in § 8 (Theorem 8.2). Hence let $q > 1$. As the rank of the matrix H_{h+p} with determinant $D_{h+p} \neq 0$ is equal to its order $h+p+1$, the rank $\tilde{\rho}$ of the matrix H_{h+p-1} is not less than $h+p-1$ (Corollary to Lemma 6.1). But $D_{h+p-1} = 0$, so $\tilde{\rho} = h+p-1$ and the (r,k)-characteristic of this matrix has the shape $(h,p-1)$ (Theorem 11.1). From the condition $q > 1$ it follows that $p-1 > 0$, and hence the matrix H_{h+p-1} satisfies all conditions of Theorem 11.7. Because of the theorem the ranks of the extensions $H_h,\dots,H_{h+p-2},H_{h+p-1}$ of the matrix H_{h-1} will, in the process of construction, be increased only at the final $m = q-1$ $(k=p-1=2q-2)$ steps of extensions, and the each time by two units. As explained above, the last transition from H_{h+p-1} to H_{h+p} is also accompanied by a jump in the rank of two units. Hence (see Theorem 6.1) follows the validity of proposition 1°.

COROLLARY. *Under the conditions of proposition* 1° *one has for even* $q=2s\,(s>0)$

$$\operatorname{sign} D_{h+p} = \operatorname{sign} D_{h-1} \tag{12.5}$$

and for odd $q=2s-1\,(s>0)$

$$\operatorname{sign} D_{h+p} = -\operatorname{sign} D_{h-1}. \tag{12.6}$$

This follows from (4.3) and proposition 2° of § 5.

2° *Let in relations* (12.4) *the number* p *be even:* $p = 2q > 0$. *If* $\operatorname{sign} D_{h-1} = \operatorname{sign} D_{h+p}$, *then for even* q *the form* H_{h+p} *has* q+1 *positive squares and* q *negative squares and for odd* q *it has* q *positive and* q+1 *negative squares more than the form* $H_{h-1}(x,x)$. *In the case, where* $\operatorname{sign} D_{h-1} = -\operatorname{sign} D_{h-p}$ *one must interchange the roles of the positive and negative squares in the previous formulation.*

Again, for the proof it is only necessary to consider the transition from $H_{h-1}(x,x)$ to $H_{h+p-1}(x,x)$, since the final step - from $H_{h+p-1}(x,x)$ to $H_{h+p}(x,x)$ - is just as above accompanied by a jump in the rank of two units. Taking into account that the (r,k)-characteristic of the matrix H_{h+p-1} has the shape $(h,p-1)$, where $p-1 = 2q-1 > 0$, we apply Theorem 11.7 again, because of which at the transition $H_{h+q-1} \longrightarrow H_{h+q}$ the rank (for the first time) increases, and by one unit, and at each for the subsequent q-1 steps of transition form H_{h+1} to $(H_{h+2q-1}) = H_{h+p-1}$ it increases by two units.

Hence the increase of the rank on the whole way from $H_{h-1}(x,x)$ to $H_{h+p}(x,x)$ will be accompanied by the appearence of q "new" squares of one sign and q+1 squares of the opposite sign. It is clear (see (4.3) and proposition 2° of § 5) that for sign D_{h-1} = sign D_{h+p} and even q there appear q negative and q+1 positive squares, and for odd q, on the contrary, q positive and q+1 negative squares. It is also evident that for sign D_{h-1} = -sign D_{h+p} the situation is exactly the opposite.

12.3 Now really just a little is needed to establish the following rule.

LEMMA 12.1. *Let for the real Hankel form* $H_{n-1}(x,x)$ *of order* n *the sequence of successive principal minors contain an isolated group of* $p(\geqq 1)$ *zeros:*

$$(D_{h-1} \neq 0), \quad D_h = D_{n+1} = \cdots = D_{h+p-1} = 0, \quad (D_{h+p} \neq 0). \tag{12.7}$$

We prescribe for each of the zeros the sign plus or minus according to the rule

$$\operatorname{sign} D_{h-1+\nu} = (-1)^{\nu(\nu-1)/2} \operatorname{sign} D_{h-1} \quad (\nu = 1,2,\cdots,p). \tag{12.8}$$

Then the number of sign permanences $P(D_{h-1},D_h,\cdots,D_{h+p})$ *and the number of sign changes* $V(D_{h-1},D_h,\cdots,D_{h+p})$ *in the set* $D_{h-1},D_h,\cdots,D_{h+p}$ *will respectively be equal to the amount of positive and negative squares which additionally emerge at the transition form the form* $H_{h-1}(x,x)$ *to its "extension"* $H_{h+p}(x,x)$.

PROOF. For more clearness we shall consider the following tables:

I. $p = 2q - 1,\ q > 0$

I. a) $q = 2s,\ s > 0 \quad (p = 4s - 1)$

$\nu =$	1	2	3	4	5	$\cdots$	4s-2	4s-1	
$(D_{h-1})(-1)^{\frac{\nu(\nu-1)}{2}} =$	1	-1	-1	1	1	$\cdots$	-1	-1	(D_{h+p})

I. b) $q = 2s - 1,\ s > 0 \quad (p = 4s - 3)$

$\nu =$	1	2	3	4	5	$\cdots$	4s-4	4s-3	
$(D_{h-1})(-1)^{\frac{\nu(\nu-1)}{2}} =$	1	-1	-1	1	1	$\cdots$	1	1	(D_{h+p})

If one considers the relations (12.5) and (12.6), respectively, it is easy to see from the tables Ia) and Ib) that, having prescribed for the zero determinants $D_h,\cdots,D_{h+p-1}$ signs according to rule (12.8), we obtain

$$P(D_{h-1},D_h,\cdots,D_{h+p}) = q, \quad V(D_{h-1},D_h,\cdots,D_{h+p}) = q.$$

Comparing this with proposition 1°, we obtain the validity of the Lemma (for the case of odd $p = 2q-1$).

In the case of even p, we consider again tables

II. $p = 2q,\ q > 0$

II a) $q = 2s,\ s > 0 \quad (p = 4s)$

$\nu =$	1	2	3	4	5	$\cdots$	4s-1	4s	
$(D_{h-1})(-1)^{\frac{\nu(\nu-1)}{2}} =$	1	-1	-1	1	1	$\cdots$	-1	1	(D_{h+p})

II b) $q = 2s-1,\ s > 0 \quad (p = 4s-2)$

$\nu =$	1	2	3	4	5	$\cdots$	4s-3	4s-2	
$(D_{h-1})(-1)^{\frac{\nu(\nu-1)}{2}} =$	1	-1	-1	1	1	$\cdots$	1	-1	(D_{h+p})

From the tables IIa) and IIb) it is clear, that rule (12.8) leads to the following results

if sign D_{h+p} = sign D_{h-1} then

$$\left.\begin{aligned} P(D_{h-1},D_h,\cdots,D_{h+p}) &= q+1,\\ V(D_{h-1},D_h,\cdots,D_{h+p}) &= q \end{aligned}\right\} \text{ for even } q(=2s),$$

$$\left.\begin{aligned} P(D_{h-1},D_h,\cdots,D_{h+p}) &= q,\\ V(D_{h-1},D_h,\cdots,D_{h+p}) &= q+1 \end{aligned}\right\} \text{ for odd } q(=2s-1);$$

if sign D_{h+p} = -sign D_{h-1}, then

$$\left.\begin{aligned} P(D_{h-1},D_h,\cdots,D_{h+p}) &= q,\\ V(D_{h-1},D_h,\cdots,D_{h+p}) &= q+1 \end{aligned}\right\} \text{ for even } q(=2s),$$

$$\left.\begin{aligned} P(D_{h-1},D_h,\cdots,D_{h+p}) &= q+1,\\ V(D_{h-1})D_h,\cdots,D_{h+p}) &= q \end{aligned}\right\} \text{ for odd } q(=2s-1).$$

Comparing these results with proposition 2° we come to the conculusion, that also in the case of even $p(=2q)$ the validity of the Lemma is proved.

REMARK. Following F.R. Gantmaher [3], one may write down the result of Lemma 12.1 briefly in the shape of te following table:

	p *odd*	p *even*
$P_{h,p} = P(D_{h-1}, D_h, \cdots, D_{h+p})$	$\frac{p+1}{2}$	$\frac{p+1+\varepsilon}{2}$
$V_{h,p} = V(D_{h-1}, D_h, \cdots, D_{h+p})$	$\frac{p+1}{2}$	$\frac{p+1-\varepsilon}{2}$
$P_{h,p} - V_{h,p}$	0	ε

$$\varepsilon = (-1)^{p/2} \operatorname{sign} \frac{D_{h+p}}{D_{h-1}} . \tag{12.9}$$

12.4 Somewhat complicated is the analysis of the case, where $D_{\rho-1} = 0$, i.e., where in the (r,k)-characteristic of the matrix H_{n-1} we have $k > 0$, $\rho = r + k > r$.

3° *Let us in the* (r,k)-*characteristic of the matrix* H_{n-1} *(see* (12.2)) *of the form* (12.1) *have* $k > 0$, i.e., $\rho = r + k > r$. *Then for even* $k = 2m$ *the forms* $H_{r-1}(x,x)$ *and* $H_{n-1}(x,x)$ *have the same signature.* Explicitly: *the form* $H_{n-1}(x,x)$ *has in a canonical representation* m *positive and* m *negative squares more than the form* $H_{r-1}(x,x)$.

Indeed, at the reestablishment of the matrix H_{n-1} from the matrix H_{r-1} by means of the step by step construction of extensions, the total increase in rank from r by $\rho - r$ (=k) units takes place (see Theorem 11.7) only in the final m steps - with two units each time. Hence it follows (Theorem 6.1), that this increase will be accompanied with the appearance in the form (12.1) of m new positive and m new negative squares

4° *Let all conditions of proposition* 3° *be satisfied, except that the number* k *is odd:* $k = 2m - 1$. *We denote through* $s'_{2r+1}, s'_{2r+2}, \cdots, s'_{2n-1-k}$ (see (1o.3)) *the elements of the singular extensions* H'_{r+1}, $H'_{r+2}, \cdots$ *of the matrix* H_r. *Then the form* $H_{n-1}(x,x)$ *will have for* $s_{2n-1-k} > s'_{2n-1-k}$ m *positive squares and* m-1 *negative squares more than the form* $H_{r-1}(x,x)$; *in the case where* $s_{2n-1-k} < s'_{2n-1-k}$ *we have* m-1 *new positive squares and* m *new negative squares.*

For the proof we note above all, that again because of Theorem 11.7 the total increase in rank of k $(= \rho - r)$ units at the transition from H_{r-1} to H_{n-1} is made up of a first jump of one unit (at the transition from H_{n-m-1} to H_{n-m}) and m-1 further jumps of each two units. Hence it follows, that $H_{n-1}(x,x)$ in comparison to $H_{r-1}(x,x)$ contains additionally m squares of one sign and m-1 squares of the opposite sign.

For the more precise question of which sign will be the m squares and of which sign the m-1 squares, we shall return to formula (10.8), which is valid for odd $k = 2m-1$. From this is clear, that the form $H_{n-m}(x,x)$ can be represented in the shape of the sum

$$H_{n-m}(x,x) = H'_{n-m}(x,x) + (s_{2n-1-k} - s'_{2n-1-k})\xi^2_{n-m} \tag{12.10}$$

where $H'_{n-m}(x,x)$ is the form of rank r with the matrix

$$H'_{n-m} = \left\| \begin{array}{c|c} H_{n-m-1} & \begin{array}{c} s_{n-m} \\ \vdots \\ s_{2n-2-k} \end{array} \\ \hline s_{n-m} \cdots s_{2n-2-k} & s'_{2n-1-k} \end{array} \right\| .$$

If $H'_{n-m}(x,x)$ is reduced to the canonical shape, i.e., to a sum of r independent squares, then (12.10) is reduced to a representation of the form $H_{n-m}(x,x)$ in the shape of a sum of r+1 squares, which are independent, since the rank of the form $H_{n-m}(x,x)$ is exactly equal to r+1 (§ 5, proposition 3°). Hence follows, that for $s_{2n-1-k} > s'_{2n-1-k}$ the form $H_{n-m}(x,x)$ gains in comparison to the form $H_{n-m-1}(x,x)$ which has the same rank and signature as the form $H'_{n-m}(x,x)$ (see Theorem 6.2), one new positive square, and for $s_{2n-1-k} < s'_{2n-1-k}$ one new negative square. Since all further extensions, in m-1 steps, of the form $H_{n-m}(x,x)$ to the complete reestablishment of $H_{n-1}(x,x)$ (here we use again a loose formulation) do not change the signature because of Theorem 6.1 (the jumps in the rank are each equal to two units), proposition 4° is proved.

12.5 From the combination of propositions 3° und 4° with formula (11.2) is now obtained

LEMMA 12.2. *Let for the Hankel quadratic form* $H_{n-1}(x,x)$ *of rank* ρ *in the* (r,k)*-characteristic of its matrix the number* $k > 0$ *(i.e.* $\rho=r+k>r$*). We introduce in the considerations the minor* $\tilde{D}_{\rho-1}$, *defined in Lemma* 11.1 *and different from zero, and we prescribe for the zero determinants* $D_r = D_{r+1} = \cdots = D_{\rho-2} = 0$ *signs according to the rule*

$$\operatorname{sign} D_{r-1+\nu} = (-1)^{\nu(\nu-1)/2} \operatorname{sign} D_{r-1} \quad (\nu = 1,2,\cdots,k-1). \tag{12.11}$$

Then the number of positive squares of the form $H_{n-1}(x,x)$ *and the number of its negative squares surpass the corresponding numbers of the form* $H_{r-1}(x,x)$ *by the quantities* $P(D_{r-1},D_r,\cdots,D_{\rho-2},\tilde{D}_{\rho-1})$ and $V(D_{r-1},D_r,\cdots,D_{\rho-2},\tilde{D}_{\rho-1})$, *respectively.*[2)]

PROOF. Because of formula (11.2)

$$\tilde{D}_{\rho-1} = (-1)^m D_{r-1}(s_{2n-1-k} - s'_{2n-1-k})^{2m} \tag{12.12}$$

for even $k = 2m$ and

$$\tilde{D}_{\rho-1} = (-1)^{m-1} D_{r-1}(s_{2n-1-k} - s'_{2n-1-k})^{2m-1} \tag{12.13}$$

for odd $k = 2m-1$.

Thus, for even $k = 2m$ (see 12.12))

$$\begin{aligned} \operatorname{sign} \tilde{D}_{\rho-1} &= \operatorname{sign} D_{r-1} \quad (m \text{ even}), \\ \operatorname{sign} \tilde{D}_{\rho-1} &= -\operatorname{sign} D_{r-1} \quad (m \text{ odd}). \end{aligned} \tag{12.14}$$

For odd $k = 2m-1$ each of the relations

$$\operatorname{sign} \tilde{D}_{\rho-1} = \operatorname{sign} D_{r-1} \quad (m \text{ even}), \tag{12.15}$$

$$\operatorname{sign} \tilde{D}_{\rho-1} = -\operatorname{sign} D_{r-1} \quad (m \text{ odd}) \tag{12.16}$$

is valid if and only if (see (12.13))

$$s_{2n-1-k} < s'_{2n-1-k}.$$

To the opposite inequality

$$s_{2n-1-k} > s'_{2n-1-k}$$

is equivalent each of the relations

$$\operatorname{sign} \tilde{D}_{\rho-1} = -\operatorname{sign} D_{r-1} \quad (m \text{ even}) \tag{12.17}$$

$$\operatorname{sign} \tilde{D}_{\rho-1} = \operatorname{sign} D_{r-1} \quad (m \text{ odd}). \tag{12.18}$$

Now, as in the proof of Lemma 12.1 it remains to consider the tables

I. $k = 2m$, $m > 0$

I a) $m = 2t$, $t > 0$, $(k = 4t)$

$\nu =$	1	2	3	4	5	...	4t-2	4t-1	
$(D_{r-1})(-1)^{\frac{\nu(\nu-1)}{2}} =$	1	-1	-1	1	1	...	-1	-1	$(\tilde{D}_{\rho-1})$

I b) $m = 2t-1$, $t > 0$ $(k = 4t-2)$

$\nu =$	1	2	3	4	5	...	4t-4	4t-3	
$(D_{r-1})(-1)^{\frac{\nu(\nu-1)}{2}} =$	1	-1	-1	1	1	...	1	1	$(\tilde{D}_{\rho-1})$.

From the tables Ia) and Ib), if we have considered relations (12.14), we see, that, having preseribed for the zero determinants $D_r, D_{r+1}, \cdots, D_{\rho-2}$ signs according to the rule (12.11), we obtain

$$\begin{aligned} P(D_{r-1}, D_r, \cdots, D_{\rho-2}, \tilde{D}_{\rho-1}) &= m, \\ V(D_{r-1}, D_r, \cdots, D_{\rho-2}, \tilde{D}_{\rho-1}) &= m, \end{aligned}$$

and this coincides, because of 3°, with the statement of the Lemma.

II. $k = 2m - 1,\ m > 0$

II a) $m = 2t,\ t > 0\ (k = 4t - 1)$

$\nu =$	1	2	3	4	5	...	4t-3	4t-2	
$(D_{r-1})(-1)^{\frac{\nu(\nu-1)}{2}} =$	1	-1	-1	1	1	...	1	-1	$(\tilde{D}_{\rho-1})$

II b) $m = 2t - 1,\ t > 0\ (k = 4t - 3)$

$\nu =$	1	2	3	4	5	...	4t-5	4t-4	
$(D_{r-1})(-1)^{\frac{\nu(\nu-1)}{2}} =$	1	-1	-1	1	1	...	-1	1	$(\tilde{D}_{\rho-1})$

(we note, that table IIb) for $t = 1$, i.e. for $m = k = 1$, loses its meaning, but in this case it is not needed, since the statement of the Lemma then directly follows from proposition 4° and formula (12.13)).

Again having compared IIa) with (12.15) and (12.17) and IIb) with (12.16) and (12.18) we are convinced, that for $s_{2n-1-k} > s'_{2n-1-k}$

$$P(D_{r-1}, D_r, \dots, D_{\rho-2}, \tilde{D}_{\rho-1}) = m,$$
$$V(D_{r-1}, D_r, \dots, D_{\rho-2}, \tilde{D}_{\rho-1}) = m - 1,$$

and for $s_{2n-1-k} < s'_{2n-1-k}$

$$P(D_{r-1}, D_r, \dots, D_{\rho-2}, \tilde{D}_{\rho-1}) = m - 1,$$
$$V(D_{r-1}, D_r, \dots, D_{\rho-2}, \tilde{D}_{\rho-1}) = m,$$

i.e. (see 4°) of the correctness of the statement of Lemma 12.2.

REMARK 1. Comparing the reasoning, applied in the proofs of the Lemmas 12.1 and 12.2 (respectively of the proposition 1°, 2°, 3° and 4°) the attention is drawn to the seeming noncorrespondence in the rules for the cases of even and odd p on one hand and for even and odd k on the other hand. But this noncorrespondence disappears at once if one considers that in Lemma 12.2 (respectively in propositions 3° and 4°) it is not k that plays the role of the number of zeros p in Lemma 12.1 (respectively in propositions 1° and 2°) but $k - 1$: namely, the number of zeros in the set $(0 \neq)\ D_{r-1}, D_r = D_{r+1} = \dots = D_{r+k-2} = 0,\ \tilde{D}_{r+k-1}\ (\neq 0)$. In particular, as we have seen above, this is feasible also in the case $k = 1$, when the mentioned set is reduced to the pair $D_{r-1}, D_r (= \tilde{D}_{r+k-1})$, i.e., in general contains no zero.

REMARK 2. With regard to Remark 1 one can, having denoted $k - 1 \equiv p$, $r \equiv h$, express the result of Lemma 12.2 by the same table (12.9) (with the substitution of D_{h+p} by $\tilde{D}_{\rho-1}$) as the result of Lemma 12.1. We leave it to the reader to convince himself of this independently.

12.6. Now it remains only to make some observation in order to establish the fundemental Theorem of Frobenius.

Let (we preserve the same notation as in the previous subsections) in the set $D_{-1}(\equiv 1), D_o, \cdots, D_{\rho-1}$, moving from the left to the right, for the first time be met a group of zeros

$$(D_{h-1} \neq 0), D_h = D_{h+1} = \cdots = D_{h+p-1} = 0 \quad (D_{h+p} \neq 0). \tag{12.19}$$

This means that $D_{-1}D_o \cdots D_{h-1} \neq 0$, and hence holds for the form $H_{n-1}(x,x)$ Jacobi's rule (Theorem 8.1), because of which the amount of positive and negative squares of this form is equal to $P(D_{-1}, D_o, \cdots, D_{h-1})$ and $V(D_{-1}, D_o, \cdots, D_{h-1})$, respectively. Because of Lemma 12.1 this rule remains valid also for the form $H_{h+p}(x,x)$, if only to the zero determinants of the set (12.19) one prescribes signs according to the rule (12.8).

Advancing further, i.e., going over to the minors $D_{h+p+1}, D_{h+p+2}, \cdots$ (respectively to the forms $H_{h+p+1}(x,x), H_{h+p+2}(x,x), \cdots$) we can again apply Jacobi's rule as long as these minors are different from zero (see the footnote to the proof of Theorem 8.1 (Jacobi's signature rule)), and meeting the next (isolated) group of zeros, again apply Lemma 12.1 and so on.

Continuing this reasoning, we either reach the minor $D_{\rho-1} \neq 0$ and we will have calculated by means of our generalized rule of Jacobi the signature $\sigma = \pi - \nu$ of the form $H_{n-1}(x,x)$ (equal, because of proposition 1° of § 6 to the signature of the form $H_{\rho-1}(x,x)$, as these forms have one and the same rank ρ), or we find a final group of zeros

$$(D_{r-1} \neq 0), \; D_r = D_{r+1} = \cdots = D_{\rho-1} = 0,$$

on which one must apply Lemma 12.2.

Consequently is proved [3)]

THEOREM 12.1 (SIGNATURE RULE OF FROBENIUS). *Let the rank of the real Hankel form* $H_{n-1}(x,x)$ *be equal to* $\rho = r+k$, *where* (r,k) *is the characteristic of the matrix* H_{n-1} *of this form. We consider the set*

$$(1=) D_{-1}, D_o, D_1, \cdots, D_{\rho-2}, D^*_{\rho-1}$$

in which for $k = 0$ $(r = \rho)$ *is put* $D^*_{\rho-1} = D_{\rho-1}$, *and in the opposite case* $D^*_{\rho-1} = \tilde{D}_{\rho-1}$ (see Lemma 11.1). *To the zero determinants* $D_j (0 \leq j < \rho - 1)$, *if such are present, we prescribe signs according to the rules* (12.8) *and* (12.11). *Then the signature* $\sigma = \pi - \nu$ *of the form* $H_{n-1}(x,x)$ *is de-*

fined by the formulae

$$\pi = P(D_{-1}, D_0, \cdots, D_{\rho-2}, D^*_{\rho-1}),$$
$$\nu = V(D_{-1}, D_0, \cdots, D_{\rho-2}, D^*_{\rho-1}).$$

As useful exercise we leave it to the reader to verify,that the rule, formulated in Theorem 12.1, can be reformulated in an equivalent way as follows:

FROBENIUS' RULE [4] *We choose from the set* $D_{-1}, D_0, \cdots, D_{\rho-2}, D^*_{\rho-1}$ *only the minors*

$$D_{-1}, D_\alpha, D_\beta, D_\gamma, \cdots, D_\eta, D_\zeta, D^*_{\rho-1}$$

which are different from zero; we consider all differences of neighbouring indices

$$\alpha-(-1), \beta-\alpha, \gamma-\beta, \cdots, \zeta-\eta, (\rho-1)-\zeta$$

and retain of these only the odd ones; then (with $D^*_{\rho-1}$ *instead of* $D_{\rho-1}$ *for* $\mu = \rho-1$)

$$\sigma = \sum_{1 \leq (\mu-\lambda)\,\mathrm{odd}} (-1)^{\frac{1}{2}(\mu-\lambda-1)} \operatorname{sign}(D_\lambda D_\mu). \qquad (12.2o)$$

In particular, if in the righthand part of formula (12.2o) not a single summand is found, then $\sigma = 0$.

The PROOF is without great difficulty obtained by means of the Lemmas 12.1 and 12.2.

12.7 The abovementioned rule of Frobenius allows to reveal some new regularities in the distribution of zeros and plus and minus signs in the set of successive principal minors of a Hankel form.

THEOREM 12.2 *Let, as usual* π *and* ν *be the number of positive and negative squares, respectively, of the Hankel form* $H_{n-1}(x,x)$ *of rank* $\rho = \pi + \nu$, *and let* $\sigma = \pi - \nu$ *be its signature. We denote*

$$\kappa \equiv \min\{\pi, \nu\}.$$

Then in the set of successive principal minors

$$(1=)D_{-1}, D_0, D_1, \cdots, D_{\rho-1} \qquad (12.21)$$

of the form $H_{n-1}(x,x)$, *there are for* $D_{\rho-1} \neq 0$ *not more than* 2κ *zeros. If there are exactly* 2κ *zeros, then, after removal of these from the set (12.21), the signs in each of the remaining groups of numbers (not considering numbers which stand isolated) alternate strictly for* $\kappa = \pi$ *and coincide for* $\kappa = \nu$.

PROOF. 1) If there would be not less than $2\kappa+1$ zeros in (12.21), then, after these have been crossed out the set

$$D_{-1}, D_\alpha, D_\beta, D_\gamma, \cdots, D_\eta, D_\zeta, D_{\rho-1} \qquad (12.22)$$

would be obtained, which consists of not more than $\rho+1-(2\kappa+1)=\rho-2\kappa$ numbers, and hence in the sum

$$\sigma = \sum_{1\leq(\mu-\lambda)\,\mathrm{odd}} (-1)^{\frac{1}{2}(\mu-\lambda-1)} \operatorname{sign} D_\lambda D_\mu . \qquad (12.23)$$

there would, a forteriori, be left not more than $\rho-2\kappa-1$ nonzero summands. As their absolute value is always equal to one, one would have

$$|\pi-\nu| = |\sigma| \leq \rho-2\kappa-1 . \qquad (12.24)$$

On the other hand,

$$\begin{cases} \sigma = \rho-2\nu = 2\pi-\rho \\ -\sigma = \rho-2\pi . \end{cases}$$

Hence

$$\begin{cases} |\sigma| \geq \rho - 2\nu \\ |\sigma| \geq \rho - 2\pi \end{cases}$$

i.e.,

$$|\sigma| \geq \rho - 2\min\{\pi,\nu\} = \rho-2\kappa .$$

2) For $\kappa=\pi$ we have $\pi \leq \nu$, i.e.,

$$0 \geq \pi-\nu = \sigma = 2\pi-\rho = 2\kappa-\rho .$$

But if there are exactly 2κ zeros in the sequence (12.21), then the sequence (12.22) contains exactly $\rho-2\kappa+1$ numbers, and in the sum (12.23) there are not more than $\rho-2\kappa$ summands different from zero.

We note, that the identity $0=\sigma(=2\kappa-\rho)$ is impossible, since it would mean that in (12.22), which contains, as calculated above, exactly $\rho-2\kappa+1$ numbers, there wouldn't remain a single *minor*, but this contradicts the inequality $D_{\rho-1} \neq 0$.

So, $\sigma<0$. Then $\sigma=-(\rho-2\kappa)$ and, therefore, there are in the sum (12.23) exactly $\rho-2\kappa$ summands different from zero (and, moreover, all equal to (-1)). This means (see (12.23)) that the signs inside each of the nonzero groups in (12.21) are strictly alternating, as in these groups $\mu-\lambda-1=0$.

The case, where $\kappa=\nu$ $(\pi\geq\nu)$ is treated in an analogouous way.

EXAMPLES AND EXERCISES

1. We consider the Hankel form of order $n=6$

$$H_5(x,x) = \xi_0^2 + 2\xi_0\xi_3 + 2\xi_1\xi_2 + \xi_3^2 + 2\xi_1\xi_5 + 2\xi_2\xi_4 + 2\xi_4\xi_5 + \xi_5^2$$

with the matrix (cf. example 3 of § 11)

$$H_5 = \begin{Vmatrix} 1 & 0 & 0 & 1 & 0 & 0 \\ 0 & 0 & 1 & 0 & 0 & 1 \\ 0 & 1 & 0 & 0 & 1 & 0 \\ 1 & 0 & 0 & 1 & 0 & 0 \\ 0 & 0 & 1 & 0 & 0 & 1 \\ 0 & 1 & 0 & 0 & 1 & 1 \end{Vmatrix} .$$

Here $D_{-1} = 1$, $D_0 = 1$, $D_1 = 0$, $D_2 = -1$, $D_3 = D_4 = D_5 = 0$. Thus $r = 3$. In order to establish the rank ρ and the minor $D^*_{\rho-1}$ (see Theorem 12.1) we use the result of example 3 of § 11, according to which the matrix H_5 has rank ρ equal to 4 and $D^*_{\rho-1} = D^*_3 = -1$, so that

$$\{D_{-1}, D_0, D_1, D_2, D^*_3\} \equiv \{1,1,0,-1,-1\}. \qquad (12.25)$$

Hence, by rule (12.2o) (since $D_{\rho-1} = D_3 = 0$ one cannot apply the rules of § 8)

$$\sigma = \text{sign}(1\cdot 1) + \text{sign}(-1)(-1) = 2,$$

i.e. $\pi = \frac{1}{2}(\rho + \sigma) = 3$, $\nu = \frac{1}{2}(\rho + \sigma) = 1$.

2. We consider the Hankel form

$$H_3(x,x) = 2\xi_0\xi_3 + 2\xi_1\xi_2 - 2\xi_2^2 - 4\xi_1\xi_3 + 3\xi_3^2 .$$

Here

$$H_3 \quad \begin{Vmatrix} 0 & 0 & 0 & 1 \\ 0 & 0 & 1 & -2 \\ 0 & 1 & -2 & 0 \\ 1 & -2 & 0 & 3 \end{Vmatrix} ,$$

i.e., $D_{-1} = 1$, $D_0 = D_1 = D_2 = 0$, $D_3 = 1$. Thus $\rho = n = 4$ and by Frobenius' rule $\sigma = 0$ (the difference of the indices $3-(-1) = 4$ is even, and therefore there are in the sum (12.23) no summands different from zero). Hence $\pi = \nu = 2$.

We note, that none of the rules described in § 8 can be applied directly to the present example (three successive minors D_0, D_1, D_2 are equal to zero).

3. Find the rank ρ and the signature σ of the Hankel form

$$H_3(x,x) = 2\xi_1^2 - 2\xi_0\xi_1 + 12\xi_2^2 + 24\xi_1\xi_3 + $$
$$+ 4\xi_0\xi_2 - 1o\xi_1\xi_2 - 1o\xi_0\xi_3 - 58\xi_2\xi_3 - 3\xi_3^2 .$$

Solution. $\rho = 3$, $\sigma = -1$.

4. Prove, that Theorem 12.2 also remains valid for $D_{\rho-1} = 0$, if one substitutes in it $2\kappa + 1$ for the upper bound 2κ.

5. The upper bound 2κ in Theorem 12.2 (for $D_{\rho-1} \neq 0$) is sharp. This is already clear from the simple example

$$H_2(x,x) = 2\xi_0\xi_2 + \xi_1^2, \quad H_2 = \begin{Vmatrix} 0 & 0 & 1 \\ 0 & 1 & 0 \\ 1 & 0 & 0 \end{Vmatrix} .$$

Here $\rho = 3$, $D_{-1} = 1$, $D_0 = D_1 = 0$, $D_2 = 1$. By formula (12.2o) (or Theorem 8.3) $\sigma = 1$, so that $\pi = 2$, $\kappa = \nu = 1$. Hence $2\kappa = 2$, and the set D_{-1}, D_0, D_1, D_2 contains exactly two zeros, i.e., the bound is sharp.

6. Devise an example, which confirms that for $D_{\rho-1} = 0$ the bound $2\kappa+1$ (see exercise 4) is sharp as well.

7. Show, that Theorem 11.8 admits for real Hankel matrices the following refinement: besides the parameters $n(\geqq 2)$, $r \geqq 0$, $k > 0 (n > r + k)$ one can choose in advance the signature of the form $H_{n-1}(x,x) = \sum_{i,j=0}^{n-1} s_{i+j}\xi_i\xi_j$ with matrix H_{n-1}, taking for σ an arbitrary integer from the segement $[-r,r]$ with the same parity as $r + k$. To this end one should, in the case of even $k = 2m$, choose the nonsingular matrix H_{n-1}, of order r in step 1) of Theorem 11.8 at once in such a way, that the corresponding Hankel form $H_{r-1}(x,x)$ has signature σ. For odd $k = 2m + 1$ one should choose the signature of the form $H_{r-1}(x,x)$ to be equal to $\sigma - 1$ (or $\sigma + 1$) and take $s_{2(n-m)} > s'_{2(n-m)}$ (resp. $s_{2(n-m)} < s'_{2(n-m)}$) in step 5) of Theorem 11.8.

HINT. Use proposition 3° and 4°.

8. If in the (r,k)-characteristic of the Hankel form

$$H_{n-1}(x,x) = \sum_{i,j=0}^{n-1} s_{i+j}\xi_i\xi_j$$

with rank ρ the number $r < \rho$, then the change of variables

$$\eta_o = \xi_o, \eta_1 = \xi_1, \ldots, \eta_{r-1} = \xi_{r-1};$$
$$\eta_r = \xi_{n-\rho+r}, \eta_{r+1} = \xi_{n-\rho+r+1}, \ldots, \eta_{\rho-1} = \xi_{n-1};$$
$$\eta_\rho = \xi_r, \eta_{\rho+1} = \xi_{r+1}, \ldots, \eta_{n-1} = \xi_{n-\rho+r-1}$$

reduces the form $H_{n-1}(x,x)$ to the shape

$$H_{n-1}(x,x) \equiv A_{n-1}(x,x) = \sum_{i,j=0}^{n-1} a_{ij}\eta_i\eta_j .$$

Prove, that now in the set of successive principal minors of the obtained form the minor of order ρ, i.e.

$$\det \|a_{ij}\|_{i,j=0}^{\rho-1}$$

coincides with $\tilde{D}_{\rho-1}$, i.e. with the minor which occurs in Frobenius' rule (Theorem 12.1, see also Lemma 11.1) [3] [5].

9. Convince yourself that the Hankel form

$$H_5(x,x) = \xi_0^2 + \xi_3^2 + \xi_5^2 + 2(\xi_0\xi_3 + \xi_1\xi_2 + \xi_1\xi_5 + \xi_2\xi_4 + \xi_4\xi_5)$$

under the change of variables, suggested in exercise 8, turns into a form $A_5(y,y)$ which is not a Hankel form

HINT. Use the data of exercise 1.

1o. The infinite sequence of real numbers

$$s_0, s_1, s_2, \ldots$$

belongs, by definition, to the class $P_\kappa^{(H)}$, if for sufficiently large $m(>N)$ all Hankel forms

$$H_{m-1}(x,x) = \sum_{i,j=0}^{m-1} s_{i+j}\xi_i\xi_j$$

have in their canonical representation exactly κ positive squares.

If the Hankel form

$$H_{n-1}(x,x) = \sum_{i,j=0}^{n-1} s_{i+j}\xi_i\xi_j$$

of rank ρ has κ positive squares, then, in order that the set

$$s_0, s_1, \ldots, s_{2n-2}$$

of its coefficients admits an extension to an infinite sequence $\{s_\kappa\}$ in the class $P_\kappa^{(H)}$, it is necessary and sufficient that $D_{\rho-1} \neq 0$. If this condition is satisfied, then for $\rho < n$ the mentioned extension is defined uniquely but for $\rho = n$ there are infinitely many of such extensions (for references to the literature see below § 16, the note to exercise 8).

NOTES

1) Instead of the quadratic form (12.1) one can consider a Hermitian form $\sum_{i,j=0}^{n-1} s_{i+j}\xi_i\overline{\xi}_j$ (it is also called a Hankel form with the same matrix (12.2), cf. below § 19). It is easy to trace, that all theory, developed in § 12 for the quadratic forms (12.1), remains also valid for Hermitian Hankel forms.

2) For $k = 1$ this rule becomes empty, but the statement of the lemma remains valid.

3) In [3] a different proof of this theorem is adduced (Ch.X, § 1o, Theorem 24), of which the conclusive part (the case, where $r < \rho$) is, unfortunately, not correct (cf. exercises 8, 9 in the present section).

4) In this form, indeed, it was presented in the original memoir of Frobenius [19].

5) This result is used in [3] on p. 31o in the proof of Frobenius' rule for the reduction of the case $r < \rho$ to the more simple case $r = \rho$, which is analysed first. However, at this one overlooks the fact that the new form $A_{n-1}(y,y)$ is, in general, not a Hankel form (see exercise 9), i.e. the conclusions, drawn earlier for Hankel forms with $\rho = r$, cannot be applied to it (cf. Note 3).

REMARK. For the further development of the concept of the characteristic of Hankel matrices, and also for new applications of the theory of Hankel matrices and forms, see [2],[3],[7],[8],[14] of the additional list of references.

Chapter III

TOEPLITZ MATRICES AND FORMS

§ 13 TOEPLITZ MATRICES. SINGULAR EXTENSIONS.

13.1 A *Toeplitz matrix* of order $n(=1,2,\cdots)$ is the name for a matrix of the shape

$$T_{n-1} = \|c_{p-q}\|_{p,q=0}^{n-1} \tag{13.1}$$

where the c_p $(p=0, \pm 1, \cdots, \pm(n-1))$ are arbitrary complex numbers. More explicitly:

$$T_{n-1} = \left\| \begin{matrix} c_0 & c_{-1} & \cdots & c_{-n+2} & c_{-n+1} \\ c_1 & c_0 & \cdots & c_{-n+3} & c_{-n+2} \\ \cdot & \cdot & \cdots & \cdot & \cdot \\ c_{n-2} & c_{n-3} & \cdots & c_0 & c_{-1} \\ c_{n-1} & c_{n-2} & \cdots & c_1 & c_0 \end{matrix} \right\| .$$

As is clear from the structure of the matrix T_{n-1}, it is, in contrast to Hankel matrices, in general not a symmetric matrix (here all elements standing on the same diagonal parallel to the main diagonal - including the main diagonal itself - are identical, cf. Sec. 9.1). Only under the condition $c_p = \overline{c}_{-p}$ $(p=0,1,\cdots,n-1)$ the Toeplitz matrix T_{n-1} will be Hermitian, and in this case it generates a Hermitian Toeplitz form

$$T_{n-1}(x,x) = \sum_{p,q=0}^{n-1} c_{p-q}\xi_p\overline{\xi}_q .$$

To the investigation of such forms § 16 below is dedicated. But until there we are concerned with the investigation of Toeplitz matrices in general (without the condition of being Hermitian).

13.2 Just as with the Hankel matrices, the method of extension, already known to the reader, will be developed here.

An *extension* of the Toeplitz matrix T_{n-1} one calls each Toeplitz matrix

$$T_{n-1+\nu} = \|c_{p-q}\|_{p,q=0}^{n-1+\nu} \qquad (\nu = 1,2,\cdots),$$

for which the left upper corner ("block") coincides with the original matrix $T_{n-1} = \|c_{p-q}\|_{p,q=0}^{n-1}$.

A *singular extension* is such an extension of the matrix T_{n-1} that its rank coincides with the rank of the matrix T_{n-1}. We shall also consider *infinite extensions* $T_\infty = \|c_{p-q}\|_{p,q=0}^{\infty}$ of finite matrices T_{n-1}, among these also *singular infinite extensions* (if such exist), defining in this case the rank of an infinite matrix in the same way as in § 9.

Just as in § 9, we consider at first all extensions of the matrix T_{n-1} "by one step", i.e., the matrices

$$T_n = \left\| \begin{array}{cccc|c} c_0 & c_{-1} & \cdots & c_{-n+1} & c_{-n} \\ c_1 & c_0 & \cdots & c_{-n+2} & c_{-n+1} \\ \cdot & \cdot & \cdots & \cdot & \cdot \\ c_{n-1} & c_{n-2} & \cdots & c_0 & c_{-1} \\ \hline c_n & c_{n-1} & \cdots & c_1 & c_0 \end{array} \right\| .$$

As we see, each such extension is defined by a pair of numbers (c_n, c_{-n}). We recall, that also for Hankel matrices the analoguous extensions required that two numbers were given, but in the extended matrix these numbers were located completely differently - they filled out its lower righthand corner. This circumstance, but also the absence, in the general case of a Toeplitz matrix, of symmetry with respect to the main diagonal complicates to a certain extent the proofs of the extension theorems for Toeplitz Matrices, for which, in respect to the formulation and the general diagrams, the reasonings are completely analoguous to Theorems 9.1 and 9.2.

13.3 Again we bring into the considerations the successive principal minors of the matrix T_{n-1}:

$$D_k \equiv \det\|c_{p-q}\|_{p,q=0}^{k} \, (k=0,1,\cdots,n-1), \quad D_{-1} \equiv 1.$$

THEOREM 13.1 (FIRST EXTENSION THEOREM). *If the matrix* T_{n-1} *is non-singular* ($D_{n-1} \neq 0$), *then it has infinitely many singular extensions* T_n *of order* $n+1$.

PROOF. The determination of singular extensions T_n of the matrix T_{n-1} is reduced to solving the equation

$$(D_n(x,y)\equiv)\begin{vmatrix} c_0 & c_{-1} & \cdots & c_{-n+2} & c_{-n+1} & y \\ c_1 & c_0 & \cdots & c_{-n+3} & c_{-n+2} & c_{-n+1} \\ \cdot & \cdot & \cdots & \cdot & \cdot & \cdot \\ c_{n-2} & c_{n-3} & \cdots & c_0 & c_{-1} & c_{-2} \\ c_{n-1} & c_{n-2} & \cdots & c_1 & c_0 & c_{-1} \\ x & c_{n-1} & \cdots & c_2 & c_1 & c_0 \end{vmatrix} = 0. \qquad (13.2)$$

Applying to the determinant $D_n(x,y)$ the Sylvester Identity in the shape of (2.6) we have

$$D_n(x,y)D_{n-2} = D_{n-1}^2 - \begin{vmatrix} c_1 & c_0 & \cdots & c_{-n+2} \\ \cdot & \cdot & \cdots & \cdot \\ c_{n-1} & c_{n-2} & \cdots & c_0 \\ x & c_{n-1} & \cdots & c_1 \end{vmatrix} \begin{vmatrix} c_{-1} & \cdots & c_{-n+1} & y \\ c_0 & \cdots & c_{-n+2} & c_{-n+1} \\ \cdot & \cdots & \cdot & \cdot \\ c_{n-2} & \cdots & c_0 & c_{-1} \end{vmatrix}. \qquad (13.3)$$

We denote

$$a(x) \equiv \begin{vmatrix} c_1 & c_0 & \cdots & c_{-n+2} \\ \cdot & \cdot & \cdots & \cdot \\ c_{n-1} & c_{n-2} & \cdots & c_0 \\ x & c_{n-1} & \cdots & c_1 \end{vmatrix}, \quad b(y) \equiv \begin{vmatrix} c_{-1} & \cdots & c_{-n+1} & y \\ c_0 & \cdots & c_{-n+2} & c_{-n+1} \\ \cdot & \cdots & \cdot & \cdot \\ c_{n-2} & \cdots & c_0 & c_{-1} \end{vmatrix}. \qquad (13.4)$$

It is clear, that for $D_{n-2} \neq 0$ both the functions $a(x)$ and $b(y)$ are linear (with the coefficient D_{n-2} for x and y, respectively), and for $D_{n-2} = 0$ they do not depend on x and y, i.e., they are equal to the constants $a(0)$ $(\neq 0)$ and $b(0)$ $(\neq 0)$ (we recall that $D_{n-1} \neq 0$). Therefore, for $D_{n-2} \neq 0$ equation (13.2) which now because of (13.3) and (13.4) takes the shape

$$a(x)b(y) - D_{n-1}^2 = 0 \qquad (13.5)$$

has infinitely many solutions $\{x,y\}$. If however, $D_{n-2} = 0$, then equation (13.2) becomes linear:

$$(-1)^{n+2}b(0)x + (-1)^{n+2}a(0)y + D_n(0,0) = 0 \qquad (13.6)$$

As $a(0) \neq 0$ and $b(0) \neq 0$, this equation has again infinitely many solutions $\{x,y\}$.

Thus is proved that both for $D_{n-2} \neq 0$ as well as for $D_{n-2} = 0$ the matrix T_{n-1} has infinitely many singular extensions T_n.

REMARK 1. *If the Toeplitz matrix* T_{n-1} *is* Hermitian, *then under the conditions of Theorem* 13.1 *there exist among the singular extensions* T_n *infinitely many Hermitian extensions.*

Indeed, in this case all the minors $D_0, D_1, \cdots, D_{n-1}$ are real. If we substitute $x=\zeta$, $y=\overline{\zeta}$ in (13.3), we obtain instead of (13.3) the relation

$$D_n(\zeta,\overline{\zeta})D_{n-2} = D_{n-1}^2 - \left|\begin{vmatrix} c_1 & c_0 & \dots & c_{-n+2} \\ c_2 & c_1 & \dots & c_{-n+3} \\ \cdot & \cdot & \dots & \cdot \\ c_{n-1} & c_{n-2} & \dots & c_0 \\ \zeta & c_{n-1} & \dots & c_1 \end{vmatrix}\right|^2$$

For $D_{n-2} \neq 0$ the equation $D_n(\zeta,\overline{\zeta}) = 0$ therefore takes the shape

$$|\zeta D_{n-2} - z_0|^2 = D_{n-1}^2, \tag{13.7}$$

where $z_0 = -a(0)$ (cf.(13.4)).

Further, from formula (13.4) it is clear that in the considered case $b(\overline{\zeta}) = \overline{a(\overline{\zeta})}$ and, in particular,

$$(-1)^{n+2}b(0) = (-1)^{n+2}\overline{a(0)} \equiv \alpha.$$

Therefore, for $D_{n-2} = 0$ the equation $D_n(\zeta,\overline{\zeta}) = 0$, corresponding in this case to (13.6), turns into

$$\alpha\zeta + \overline{\alpha\zeta} + \gamma = 0 \quad (\overline{\gamma} = \gamma \equiv D_n(0,0)), \tag{13.8}$$

where $\alpha \neq 0$. Since the equations (13.7) and (13.8) define in the complex ζ-plane a circle with radius $|D_{n-1}/D_{n-2}|$ and a straight line, respectively, our claim is proved.

REMARK 2. *In the special case, where the matrix* T_{n-1} *is* real, *it has, under the conditions of Theorem* 13.1, *infinitely many real singular extensions.*

This is clear from the relations (13.4) - (13.6) in which all coefficients now are real.

REMARK 3. *If* T_{n-1} *is a real and Hermitian (symmetric) matrix, then it is also possible, under the conditions of Theorem* 13.1, *to construct singular extensions* T_n *of it, retaining these properties, but not in more than two different ways for* $D_{n-2} \neq 0$ *and in only one way if* $D_{n-2} = 0$.

Indeed, equation (13.7) now takes the shape

$$(\zeta D_{n-2} - z_0)^2 = D_{n-1}^2 \quad (z_0 = \overline{z}_0,\ D_{n-2} \neq 0),$$

whence $\zeta = D_{n-2}^{-1}(\pm|D_{n-1}| + z_0)$.

In the case $D_{n-2} = 0$ one must solve equation (13.8), which now has tha shape

$$2\alpha\zeta + \gamma = 0 \quad (\alpha = \overline{\alpha} \neq 0),$$

whence $\zeta = -\gamma/(2\alpha)$.

13.4 In the case where the matrix T_{n-1} is singular there holds

THEOREM 13.2 (SECOND EXTENSION THEOREM). *Let* T_{n-1} *be a singular Toeplitz matrix with rank* ρ $(<n)$. *If the principal minor* $D_{\rho-1} \neq 0$, *then there exists a unique pair of numbers* c_n, c_{-n} *which defines a singular extension* T_n *of the matrix* T_{n-1}.

PROOF. For $\rho = 0$ the verification of the theorem is trivial: $c_n = c_{-n} = 0$. Therefore, let $\rho > 0$. Since each $\rho + 1$ columns of the matrix T_{n-1} are linearly dependent, this implies, in particular, for the first $\rho + 1$ columns the existence of a nonzero system of constants $a_0, a_1, \cdots, a_\rho$, such that

$$a_0 c_p + a_1 c_{p-1} + \cdots + a_\rho c_{p-\rho} = 0 \quad (p = 0, 1, \cdots, n-1). \tag{13.9}$$

Restricting here to the values $p = 0, 1, \cdots, \rho$, we note that the numbers $a_0, a_1, \cdots, a_\rho$ are proportional to the cofactors of the elements of the first (and also of the last) row of the determinant D_ρ (cf. exercise 6 in § 9). Hence is follows, in the first place, that $a_0 \neq 0$ and $a_\rho \neq 0$ (since $D_{\rho-1} \neq 0$) and next, that the quotients a_ν / a_0 $(\nu = 1, 2, \cdots, \rho)$ are determined uniquely by the elements of the minor D_ρ, i.e., by the numbers $c_\nu (\nu = 0, \pm 1, \pm 2, \cdots, \pm \rho)$.

If T_n is the desired singular extension of the matrix T_{n-1}, i.e., it retains the rank ρ, then formulae (13.9) must remain valid also for $p = n$, i.e., the element

$$c_n = -\frac{a_1}{a_0} c_{n-1} - \frac{a_2}{a_0} c_{n-2} - \cdots - \frac{a_\rho}{a_0} c_{n-\rho} \tag{13.1o}$$

is defined in a unique way.

A completetely analoguous reasoning can be applied to the rows of the matrix T_{n-1} and the desired matrix T_n, which leads to

$$b_0 c_q + b_1 c_{q-1} + \cdots + b_\rho c_{\rho+p} = 0 \quad (q = 0, -1, \cdots, -n+1), \tag{13.11}$$

where the coefficients $b_0, b_1, \cdots, b_\rho$ are proportional to the cofactors of the elements of the first (and also of the last) column in the determinant D_ρ, so that again $b_0 \neq 0$, $b_\rho \neq 0$ and the element c_{-n} of the desired matrix T_n is defined uniquely:

$$c_{-n} = -\frac{b_1}{b_0} c_{-n+1} - \frac{b_2}{b_0} c_{-n+2} - \cdots - \frac{b_\rho}{b_0} c_{-n+\rho}. \tag{13.12}$$

Now it remains only to verify, that the numbers c_n and c_{-n}, obtained through formulae (13.1o) and (13.12), define a singular extension T_n. But the rank of the rectangular matrix

$$\tilde{T} = \left\| \begin{array}{cccc} c_0 & c_{-1} & \cdots & c_{-n+1} \\ c_1 & c_0 & \cdots & c_{-n+2} \\ \cdot & \cdot & \cdots & \cdot \\ c_{n-1} & c_{n-2} & \cdots & c_0 \\ \hline c_n & c_{n-1} & \cdots & c_1 \end{array} \right\| \qquad (13.13)$$

is equal to ρ, since it follows from (13.9) and (13.1o) that the last row of this matrix depends linearly on the preceding ρ rows (which belong to the matrix T_{n-1}), and that all rows of the matrix T_{n-1} depend on its independent first ρ rows.

Thus, as the rank of the matrix $\tilde{T}$ is equal to ρ, all columns of this matrix are also linearly dependent (since $D_{\rho-1} \neq 0$) on its first ρ columns. Having joined to $\tilde{T}$ just one column $\begin{pmatrix} c_{-n} \\ c_{-n+1} \\ \vdots \\ c_0 \end{pmatrix}$ we obtain the matrix T_n, and from formulae (13.11) and (13.12) we conclude, that this column also depends linearly on the previous ρ, and hence on the first ρ of its columns, i.e., the rank of the matrix T_n is equal to ρ.

The Theorem is proved.

COROLLARY. *Under the conditions of Theorem* 13.2 *there exist two uniquely defined sequences of numbers*

$$c_n, c_{n+1}, \cdots, \quad ; \quad c_{-n}, c_{-n-1}, \cdots \qquad (13.14)$$

such that all corresponding extensions

$$T_{n-1+\nu} = \| c_{p-q} \|_{p,q=0}^{n-1+\nu} \qquad (\nu = 1, 2, \cdots)$$

are singular (i.e., they retain the rank ρ*).*

The sequences (13.14) *are recursively defined through the relations*

$$a_0 c_p + a_1 c_{p-1} + \cdots + a_\rho c_{p-\rho} = 0 \qquad (p = n, n+1, \cdots)$$

$$b_0 c_q + b_1 c_{q+1} + \cdots + b_\rho c_{q+\rho} = 0 \qquad (q = -n, -n-1, \cdots)$$

where the coefficients $(0 \neq)\ a_0, a_1, \cdots, a_\rho\ (\neq 0)$ *(respectively* $(0 \neq)$ $b_0, b_1, \cdots, b_\rho\ (\neq 0)$*) are proportional to the cofactors of the elements in the first row (respectively the first column) of the determinant* D_ρ*; in particular, for* $\rho = 0$ *we have* $c_\nu = 0$ $(\nu = \pm n, \pm (n+1), \cdots)$.

In other words, the Corollary to Theorem 13.2 guarantees under the conditions of this Theorem the existence and unicity of an infinite singular extension T_∞ of the matrix T_{n-1}, which is defined through the formulae (13.1o) and (13.12).

REMARK 1. *If the matrix* T_{n-1} *is Hermitian* $(c_{-p} = \bar{c}_p,\ p = 0,1,\cdots,n-1)$, *then one can, under the conditions of Theorem* 13.2, *evidently assume that* $a_j = \bar{b}_j$ $(j = 0,1,\cdots,\rho)$. Hence follows for $p = n$, $q = -n$ from the relations (13.9) and (13.11) that $c_{-n} = \bar{c}_n$; putting then $p = n+1$, $q = -n-1$, we obtain $c_{-n-1} = \bar{c}_{n+1}$ and so on, i.e., *the singular extension* T_∞ *will also be a Hermitian matrix.*

It is clear, that an analoguous statement holds for a symmetric (complex) Toeplitz matrix.

REMARK 2. *For a real matrix* T_{n-1} *the extension* T_∞ *by Theorem* 13.2 *will,* evidently, *be real*; in particular, it follows from Remark 1, that a real symmetric matrix T_{n-1} determines under the conditions of Theorem 13.2 a singular extension T_∞ which turns out to be real and symmetric.

13.5 The formulae (13.9) and (13.11), which were established in the course of the proof of the second Extension Theorem (Theorem 13.2) allow to obtain a recurrence relation for certain minors of order ρ of the Toeplitz matrix T_{n-1} of rank ρ (>0). To this end we consider in the matrix

$$T_{n-1} = \left\| \begin{array}{c|ccc|c|ccc} c_0 & c_{-1} & \cdots & c_{-\rho+1} & c_{-\rho} & \cdots & c_{-n+1} \\ \hline c_1 & c_0 & \cdots & c_{-\rho+2} & c_{-\rho+1} & \cdots & c_{-n+2} \\ \cdot & \cdot & \cdots & \cdot & \cdot & \cdots & \cdot \\ c_{\rho-1} & c_{\rho-2} & \cdots & c_0 & c_{-1} & \cdots & c_{-n+\rho} \\ \hline c_\rho & c_{\rho-1} & \cdots & c_1 & c_0 & \cdots & c_{-n+\rho+1} \\ \hline \cdot & \cdot & \cdots & \cdot & \cdot & \cdots & \cdot \\ c_{n-1} & c_{n-1} & \cdots & c_{n-\rho} & c_{n-\rho-1} & \cdots & c_0 \end{array} \right\|$$

of rank ρ (>0) the minors of order ρ of the shape

$$\Delta_\rho^{(\omega)} = \begin{vmatrix} c_\omega & c_{\omega-1} & \cdots & c_{\omega-\rho+1} \\ c_{\omega+1} & c_\omega & \cdots & c_{\omega-\rho+2} \\ \cdot & \cdot & \cdots & \cdot \\ c_{\omega+\rho-1} & c_{\omega+\rho-2} & \cdots & c_\omega \end{vmatrix} \qquad (-n+\rho \leqq \omega \leqq n-\rho). \qquad (13.15)$$

We shall assume, that $\rho < n$, since for $\rho = n$ we have $\omega = 0$, $\Delta_n^{(0)} = D_{n-1}$, and further considerations are superfluous. We note, that always $\Delta_\rho^{(0)} = D_{\rho-1}$. Since the rank of the matrix T_{n-1} is equal to ρ, formulae (13.9) and (13.11) hold, and, further, if $D_{\rho-1} \neq 0$, then in these formulae $a_0 \neq 0$, $a_\rho \neq 0$, $b_0 \neq 0$, $b_\rho \neq 0$. Moreover, as was noted in Sec. 13.4, these coefficients are proportional to certain cofactors of the elements of the minor D_ρ, namely (in the just adopted notation):

$$\frac{a_o}{\Delta_\rho^{(o)}} = \frac{a_\rho}{(-1)^{\rho+2}\Delta_\rho^{(1)}}; \qquad \frac{a_o}{(-1)^{\rho+2}\Delta_\rho^{(-1)}} = \frac{a_\rho}{\Delta_\rho^{(o)}}; \tag{13.16}$$

$$\frac{b_o}{\Delta_\rho^{(o)}} = \frac{b_\rho}{(-1)^{\rho+2}\Delta_\rho^{(-1)}}; \qquad \frac{b_o}{(-1)^{\rho+2}\Delta_\rho^{(1)}} = \frac{b_\rho}{\Delta_\rho^{(o)}}; \tag{13.17}$$

From formulae (13.16) (and also from (13.17)) follows the relation

$$\Delta_\rho^{(1)}\Delta_\rho^{(-1)} = \left[\Delta_\rho^{(o)}\right]^2 \tag{13.18}$$

In particular, the condition $(D_{\rho-1} =)\ \Delta_\rho^{(o)} \neq 0$ implies the inequalities $\Delta_\rho^{(1)} \neq 0$ and $\Delta_\rho^{(-1)} \neq 0$ (cf. in this connection exercise 6 at the end of this section).

Now we consider an arbitrary integer ω $(-n+\rho \leqq \omega \leqq n-\rho-1)$. We assume for definiteness that $\omega < 0$, and we express (assuming $D_{\rho-1} \neq 0$, and hence $a_\rho \neq 0$) the last of the columns used in the determinant $\Delta_\rho^{(\omega)}$ (cf.(13.15)) by means of formulae (13.9) through the preceding ρ columns of this matrix [1])

$$c_{p-\rho} = -\frac{a_o}{a_\rho} c_p - \frac{a_1}{a_\rho} c_{p-1} - \cdots - \frac{a_{\rho-1}}{a_\rho} c_{p-\rho+1} \qquad (p = \omega+1, \omega+2, \cdots, \omega+\rho)$$

(here we have restricted ourselves to those values of the index p which are necessary to provide on the lefthand side of the formula all elements of the last column of the determinant $\Delta_\rho^{(\omega)}$). After inserting these expressions in $\Delta_\rho^{(\omega)}$ and applying the addition Theorem we obtain

$$\Delta_\rho^{(\omega)} = -\frac{a_o}{a_\rho}\begin{vmatrix} c_\omega & c_{\omega-1} & \cdots & c_{\omega-\rho+2} & \vdots & c_{\omega+1} \\ c_{\omega+1} & c_\omega & \cdots & c_{\omega-\rho+3} & \vdots & c_{\omega+2} \\ \cdot & \cdot & \cdots & \cdot & \vdots & \cdot \\ c_{\omega+\rho-1} & c_{\omega+\rho-2} & \cdots & c_{\omega+1} & \vdots & c_{\omega+\rho} \end{vmatrix} = (-1)^{\rho+2}\frac{a_o}{a_\rho}\Delta_\rho^{(\omega+1)}.$$

Taking into account relation (13.16) we obtain the recurrence formulae

$$\Delta_\rho^{(\omega)} = \frac{\Delta_\rho^{(o)}}{\Delta_\rho^{(1)}}\Delta_\rho^{(\omega+1)} \quad \text{or} \quad \Delta_\rho^{(\omega)} = \frac{\Delta_\rho^{(-1)}}{\Delta_\rho^{(o)}}\Delta_\rho^{(\omega+1)} \tag{13.19}$$

where $\omega = -1,-2,\cdots,-n+\rho$. These formulae hold also for $\omega = 0$, when the first of them is trivial, and the second one turns into (13.18).

Now if $(n-\rho-1 \geqq)\ \omega > 0$, then we arrive at same formulae (13.19), using the relations (13.11) and (13.17).

So is established:

1°. *If the rank of the Toeplitz matrix* T_{n-1} *is equal to* ρ $(0 < \rho < n)$

and $(\Delta_\rho^{(o)} =)\ D_{\rho-1} \neq 0$, *then* $\Delta_\rho^{(1)} \neq 0$, $\Delta_\rho^{(-1)} \neq 0$ *and*

$$\Delta_\rho^{(\omega)}\Delta_\rho^{(1)} = \Delta_\rho^{(\omega+1)}\Delta_\rho^{(o)};\quad \Delta_\rho^{(\omega)}\Delta_\rho^{(o)} = \Delta_\rho^{(\omega+1)}\Delta_\rho^{(-1)} \quad (-n+\rho\leqq\omega\leqq n-\rho-1) \qquad (13.2o)$$

For generalizations of this proposition cf. exercise 6 and 7 at the end of this section.

From proposition 1° we derive as an evident consequence:

2°. *Under the conditions of proposition* 1° *we have for arbitrary* ω $(-n+\rho \leqq \omega \leqq n-\rho)$

$$\Delta_\rho^{(\omega)} = \left[\frac{\Delta_\rho^{(1)}}{\Delta_\rho^{(o)}}\right]^\omega \Delta_\rho^{(o)} = \left[\frac{\Delta_\rho^{(o)}}{\Delta_\rho^{(-1)}}\right]^\omega \Delta_\rho^{(o)} \neq 0 \qquad (13.21)$$

(cf. exercise in § 9).

We need this proposition, which one could call the *Lemma on the translation* (of the nonzero "rank"-minor $\Delta_\rho^{(o)}$),in §§ 14 and 15.

EXAMPLES AND EXERCISES

1. The rank of the Toeplitz matrix of order 4

$$T_3 = \begin{Vmatrix} 0 & 0 & -1 & 0 \\ 2 & 0 & 0 & -1 \\ 0 & 2 & 0 & 0 \\ 0 & 0 & 2 & 0 \end{Vmatrix}$$

is equal to 3. According to Theorem 13.2 the matrix T_3 with the minor

$$D_2 = \begin{vmatrix} 0 & 0 & -1 \\ 2 & 0 & 0 \\ 0 & 2 & 0 \end{vmatrix} = -4 \neq 0$$

admits a unique singular extension T_4. Find it.

Solution. $$T_4 = \begin{Vmatrix} 0 & 0 & -1 & 0 & 0 \\ 2 & 0 & 0 & -1 & 0 \\ 0 & 2 & 0 & 0 & -1 \\ 0 & 0 & 2 & 0 & 0 \\ -4 & 0 & 0 & 2 & 0 \end{Vmatrix} .$$

HINT. Apply formulae (13.1o) and (13.12).

2. For the Hermitian Toeplitz matrix (of order $n = 2$)

$$T_1 = \begin{Vmatrix} -2 & 5+4i \\ 5-4i & -2 \end{Vmatrix}$$

the matrices

$$T_2 = \begin{Vmatrix} -2 & 5+4i & -3i \\ 5-4i & -2 & 5+4i \\ 3i & 5-4i & -2 \end{Vmatrix},$$

$$T_3 = \begin{Vmatrix} -2 & 5+4i & -3i & 7 \\ 5-4i & -2 & 5+4i & -3i \\ 3i & 5-4i & -2 & 5+4i \\ 7 & 3i & 5-4i & -2 \end{Vmatrix},$$

$$\tilde{T}_2 = \begin{Vmatrix} -2 & 5+4i & -4,5-1,5i \\ 5-4i & -2 & 5+4i \\ -4,5+1,5i & 5-4i & -2 \end{Vmatrix}$$

$$\tilde{T}_3 = \begin{Vmatrix} -2 & 5+4i & -4,5-1,5i & -i \\ 5-4i & -2 & 5+4i & -4,5-1,5i \\ -4,5+1,5i & 5-4i & -2 & 5+4i \\ i & -4,5+1,5i & 5-4i & -2 \end{Vmatrix}$$

are all Hermitian extensions. Moreover, T_3 is an extension of the matrix T_2 and $\tilde{T}_3$ an extension of the matrix $\tilde{T}_2$. For T_1 the matrix $\tilde{T}_2$ is a singular extension (verify this!).

HINT. The evaluation of the determinant $\tilde{D}_2 = \det \tilde{T}_2$ is most easily carried out (because of the Hermitian and Toeplitz structure of the matrix $\tilde{T}_2$) through the Sylvester Formula (2.6) - cf. Remark 1 to Theorem 13.1.

3. How many real singular extensions of order 4 has the Toeplitz matrix

$$T_3 = \begin{Vmatrix} 3 & 0 & 1 \\ 0 & 3 & 0 \\ 1 & 0 & 3 \end{Vmatrix} \quad ?$$

Find them.

4. Let the Toeplitz matrix T_{n-1} be (complex) symmetrical: $c_{-p} = c_p$ $(p=0,1,\cdots,n-1)$. Prove, that under the conditions of Theorem 13.1, the matrix T_{n-1} allows for $D_{n-2} \neq 0$ not more than two and for $D_{n-2} = 0$ only one symmetrical singular extension (cf. Remark 3 to Theorem 13.1).

HINT. Substitute in (13.2) (respectively in (13.13)) $x = y = \zeta$ and revise relations (13.4), (13.5) and (13.6) for this case.

5. Prove, that formula (13.18), developed above under the assumption $(D_{\rho-1} =)\ \Delta_\rho^{(o)} \neq 0$, also remains valid for $\Delta_\rho^{(o)} = 0$ (cf. the hint to exer-

cise 8 in § 9).

6. Adduce an example, where for a Toeplitz matrix with rank ρ and $\Delta_\rho^{(o)} = 0$ one of the determinants $\Delta_\rho^{(1)}$ or $\Delta_\rho^{(-1)}$ is nonzero. Show, that for Hermitian or (complex) symmetrical matrices (even not necessarily Toeplitz matrices) this is not possible - cf. the hint to exercise 8 to § 9).

7. Prove, that formulae (13.2o), eveloped under the assumption $\Delta_\rho^{(o)} \neq 0$, just as their analogon for Hankel matrices (see formulae (9.7)), can be generalized for the case, where $(D_{\rho-1} =)\ \Delta_\rho^{(o)} = 0$.

HINT. For $\Delta_\rho^{(1)} = \Delta_\rho^{(1)} = 0$ the statement is trivial; if one of these determinants is nonzero (see exercise 6), then apply formulae (13.9) and (13.11).

8. In proposition 2° it is, in particular, established, that from the inequality $D_{\rho-1} \neq 0$ (ρ is the rank of the Toeplitz matrix T_{n-1}) follow the inequalities $\Delta_\rho^{(\omega)} \neq 0$ $(-n+\rho \leqq \omega \leqq n-\rho)$ (for Hankel matrices the analoguous conclusion doesn't hold - cf. exercise 9 to § 9).

Prove, that a similar statement doesn't extend to arbitrary minors (of order ρ) of the matrix T_{n-1}, i.e., for $D_{\rho-1} \neq 0$ some minors of order ρ (not equal to the special minors $\Delta_\rho^{(\omega)}$) can be equal to zero.

NOTE

1) The negativity of the index ω here guarentees the availability in the matrix T_{n-1} of these ρ columns. However, to this end it suffices already that $\omega < n-\rho$. Now, if $\omega = n-\rho$, then the minor $\Delta_\rho^{(\omega)}$ occupies the lefthand lower corner of the matrix T_{n-1}, and our reasoning doesn't apply.

§ 14 THE (r,k,ℓ)-CHARACTERISTIC OF A TOEPLITZ MATRIX.

14.1 As in the case of Hankel matrices we shall develop the theory of Toeplitz matrices, starting with their entire-number-valued characteristic, which in the given case is also somewhat complicated because of the absence (in the general case) of symmetry.

For an arbitrary Toeplitz matrix T_{n-1} of order n and rank ρ $(0 \leqq \rho \leqq n)$ we consider the set of all its successive principal minors

$$(1=)\ D_{-1}, D_0, D_1, \cdots, D_{r-1}, D_r, \cdots, D_{n-1}.$$

Let in this set D_{r-1} be the last (reading from the left to the right) nonzero minor. In other words the *entire constant* r $(0 \leqq r \leqq \rho)$ *is defined*, just as in the case of Hankel matrices (cf. § 1o) *by the relations*

$$D_{r-1} \neq 0, \ D_{\nu-1} = 0 \qquad (\nu > r),$$

from which the second one for $r = n(=\rho)$, clearly, drops out.

We introduce two further entire constants, k, ℓ in the following way. For $r = \rho$, by definition $k = \ell = 0$. Thus, in particular, $k = \ell = 0$ for $\rho = 0$, since then $r = 0$, and also for $\rho = n$, when also $r = n$.

Now, let $r < \rho$ $(0 < \rho < n)$. We consider the "truncated" matrix

$$T_r = \| c_{p-q} \|_{p,q=0.}^{r}$$

The matrix T_r is singular $(D_r = 0)$, its rank is equal to r $(D_{r-1} \neq 0)$. Because of the second Extension Theorem there are two uniquely defined *infinite* sequences

$$c'_{r+1}, c'_{r+2}, \cdots; \quad c'_{-r-1}, c'_{-r-2}, \cdots , \tag{14.1}$$

giving the singular extension T'_∞ of the matrix T_r, and also its "intermediate" (finite) singular extensions $T'_{r+\nu}$ $(\nu = 1,2,\cdots)$ (it is clear that for $r = 0$ we have $T'_\infty = 0$).

Alongside with (14.1) we consider the *finite* sets of entries

$$c_{r+1}, c_{r+2}, \cdots, c_{n-1}; \quad c_{-r-1}, c_{-r-2}, \cdots, c_{-n+1} \tag{14.2}$$

from the matrix T_{n-1}. Each of these sets is nonempty, as $r < n-1$ (since $r < \rho < n$).

Comparing the sets (14.2) with the sequences (14.1) we define the constants k and ℓ through the relations

$$\begin{aligned} c_{r+1} &= c'_{r+1}, \cdots, c_{n-k-1} = c'_{n-k-1}; \quad c_{n-k} \neq c'_{n-k}; \\ c_{-r-1} &= c'_{-r-1}, \cdots, c_{-n+\ell+1} = c'_{-n+\ell+1}; \quad c_{-n+\ell} \neq c'_{-n+\ell}. \end{aligned} \tag{14.3}$$

Thus, for $r < \rho$ we have

$$0 \leq k, \ell \leq n - r - 1, \quad k + \ell > 0. \tag{14.4}$$

We stress that, though each of the numbers k and ℓ may individually turn out to be equal to zero, now $k + \ell > 0$, since for $k = \ell = 0$ the rank ρ of the matrix would coincide with r, contrary to the assumptions.

14.2 Just as in the case of Hankel matrices the meaning of the constants k and ℓ can be much better understood from the diagram

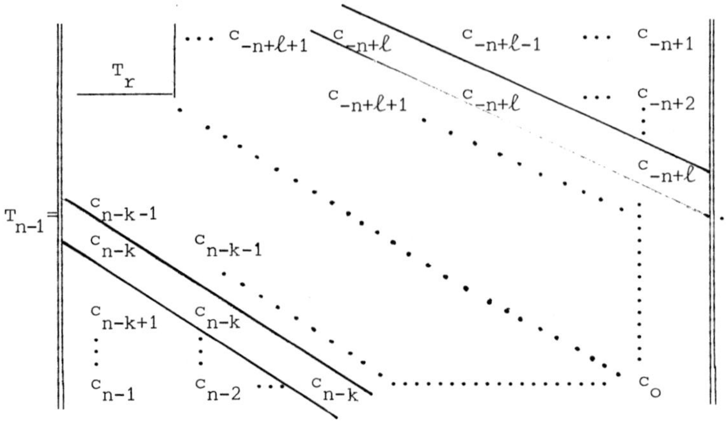

(14.5)

If one calls the entries c_ν ($\nu = 0, 1, \cdots, n-1$) of the matrix T_{n-1} which do not coincide with the corresponding entries c'_ν of the matrix T'_∞ "wrong" entries, then the numer k denotes at what "distance" (in the sense of the number of diagonals) from the lefthand lower corner of the matrix one finds the "wrong" diagonal which is nearest to the block T_r. In complete analogy one defines "wrong" elements in the set $c_{-\nu}$ ($\nu = 0, 1, \cdots, n-1$), and the constant ℓ denotes the "distance" from the first (after the block T_r) "wrong" diagonal to the righthand upper corner of the matrix T_{n-1}. In particular, for $k = 0$ (respectively $\ell = 0$) the lefthand lower (respectively the righthand upper) corner of the matrix T_{n-1} contains no "wrong" entries: $c_{r+1} = c'_{r+1}, \cdots, c_{n-1} = c'_{n-1}$ (respectively $c_{-r-1} = c'_{-r-1}, \cdots, c'_{-n+1} = c_{-n+1}$). If, moreover, also $r = 0$ then, since in this case $T'_\infty = 0$, we have

$$c'_0 = c'_{\pm 1} = c'_{\pm 2} = \cdots = 0 \qquad (14.6)$$

and for $k = 0$ ($\ell > 0$)

$$T_{n-1} = \begin{Vmatrix} 0 & 0 & \cdots & 0 & & c_{-n+\ell} & \cdots & c_{-n+1} \\ 0 & 0 & \cdots & 0 & 0 & & c_{-n+\ell} & \vdots \\ \cdots & \cdots & \cdots & \cdots & \cdots & & & \\ 0 & 0 & \cdots & 0 & 0 & 0 & & c_{-n+\ell} \\ 0 & 0 & \cdots & 0 & 0 & 0 & \cdot & 0 \\ \cdots & \cdots & \cdots & \cdots & \cdots & & & \\ 0 & 0 & \cdots & 0 & 0 & 0 & \cdot & 0 \end{Vmatrix}. \qquad (14.7)$$

An analoguous image is obtained for $r = \ell = 0$ ($k > 0$).

We call the ordered triple (r, k, ℓ) of the above-defined constants the (r, k, ℓ)-*characteristic* or simply the *characteristic* of the Toeplitz matrix T_{n-1} and the quantities r, k, ℓ the *components* of the (r, k, ℓ)-charac-

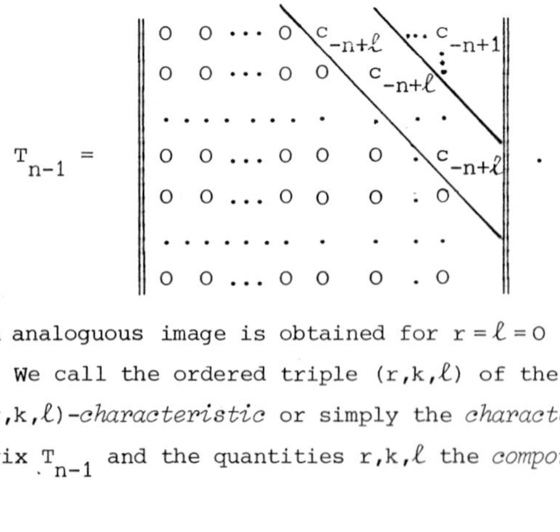

teristic. From Remark 1 to Theorem 13.2 it follows, that for Hermitian and symmetric Toeplitz matrices T_{n-1} always $k=\ell$, so in this case one can speak of the (r,k,k)-characteristic or, shorter, of the (r,k)-characteristic.

Finally, we note this obvious (see diagram (14.5)) fact:

1° *If in the* (r,k,ℓ)*-characteristic of the matrix* T_{n-1} *the component* $k>0$ *(respectively* $\ell>0$*) then, at the transition of the matrix* T_{n-1} *to an arbitrary extension* T_n *of it with* $\det T_n = 0$, *the first component* r *of the characteristic doesn't change but the second* k *(respectivly the third* ℓ*) increases by one.* (For $\det T_n \neq 0$ the characteristic of the matrix T_n will have the shape (n+1,0,0) - see Sec. 14.1).

14.3 In conclusion of the present section we use diagram (14.5) for the proof of another proposition which is needed in the sequel:

2° *In a matrix* T_{n-1} (see diagram (14.5)) *with* (r,k,ℓ)*- characteristic in which* $r>0$, *is an arbitrary mnior* $\Delta_r^{(\omega)}$ *of order* r *with the shape*

$$\Delta_r^{(\omega)} = \begin{vmatrix} c_\omega & c_{\omega-1} & \cdots & c_{\omega-r+1} \\ c_{\omega+1} & c_\omega & \cdots & c_{\omega-r+2} \\ \cdot & & \cdots & \cdot \\ c_{\omega+r-1} & c_{\omega+r-2} & \cdots & c_\omega \end{vmatrix},$$

nonzero, when the entire number ω *satisfies the inequality*

$$-n+\ell+r-1<\omega<n-k-r+1. \qquad (14.8)$$

Indeed, for $r=n$ the statement is trivial, since in this case $k=\ell=0$, so that $\omega=0$ and $\Delta_r^{(\omega)} = \Delta_n^{(0)} = D_{n-1} \neq 0$. Now let $r<n$. Conditions (14.8) give

$$\omega+r-1 < n-k, \quad \omega-r+1 > -n+\ell.$$

Comparing these inequalities with diagram (14.5), we convince ourselves, that the minor $\Delta_r^{(\omega)}$ doesn't "touch" the marked corners in diagram (14.5), i.e., it doesn't contain "wrong" entries of the matrix T_{n-1}. But then one may consider it as a minor of the matrix T'_∞ (or, if desired, of the "truncated" matrix T'_{n-1} which is obtained from it) of rank r. We substitute now $\rho=r$ and we note, that from (14.8) follows the condition

$$-n+r \leqq \omega \leqq n-r,$$

under which the lemma on the translation (§ 13, proposition 2°) is applicable. Because of this lemma $\Delta_r^{(\omega)} \neq 0$.

EXAMPLES AND EXERCISES

1. For the Toeplitz matrix

$$T_4 = \left\| \begin{matrix} 0 & 0 & 0 & 0 & 3-i \\ 0 & 0 & 0 & 0 & 0 \\ 0 & 0 & 0 & 0 & 0 \\ 5i & 0 & 0 & 0 & 0 \\ \sqrt{2} & 5i & 0 & 0 & 0 \end{matrix} \right\|$$

is in the (r,k,ℓ)-characteristic, obviously, $r=0$, since $D_0 = D_1 = D_2 = D_3 = D_4 = 0$. Hence the corresponding matrix T'_∞ (see Sec. 14.2) consist only of zeros. But then follows from the shape of the matrix T_4 (see (14.3) and diagram (14.5)), that for this matrix $k=2$, $\ell=1$. So the characteristic of the matrix T_4 looks like: (0,2,1).

2. Which is the (r,k,ℓ)-characteristic of the Toeplitz matrix

$$T_4 = \left\| \begin{matrix} 0 & 0 & -1 & 0 & 0 \\ 2 & 0 & 0 & -1 & 0 \\ 0 & 2 & 0 & 0 & -1 \\ 0 & 0 & 2 & 0 & 0 \\ 0 & 0 & 0 & 2 & 0 \end{matrix} \right\| ?$$

Solution. (3,1,0)

HINT. Cf. example 1 to § 13.

3. The Hermitian Toeplitz matrix

$$T_4 = \left\| \begin{matrix} 1 & i & -1 & -i & 1 \\ -i & 1 & i & -1 & -i \\ -1 & -i & 1 & i & -1 \\ i & -1 & -i & 1 & i \\ 1 & i & -1 & -i & 1 \end{matrix} \right\|$$

has rank $\rho = 1$ (why?). Hence, clearly, also $r=1$ and since $r=\rho$, $k=\ell=0$.

4. Find the (r,k)-characteristic of the symmetric Toeplitz matrices

$$T_2 = \left\| \begin{matrix} 1 & i & 0 \\ i & 1 & i \\ 0 & i & 1 \end{matrix} \right\|, \quad T_3 = \left\| \begin{matrix} 1 & 1 & 1 & 0 \\ 1 & 1 & 1 & 1 \\ 1 & 1 & 1 & 1 \\ 0 & 1 & 1 & 1 \end{matrix} \right\|, \quad T_4 = \left\| \begin{matrix} 0 & 1 & 0 & 1 & 1 \\ 1 & 0 & 1 & 0 & 1 \\ 0 & 1 & 0 & 1 & 0 \\ 1 & 0 & 1 & 0 & 1 \\ 1 & 1 & 0 & 1 & 0 \end{matrix} \right\| .$$

Solution. (3,0), (1,1), (2,1).

5. Calculate the (r,k,ℓ)-characteristic of the Toeplitz matrices

$$\tilde{T}_2 = \left\| \begin{matrix} -i & -2 & 4i \\ \frac{1}{2} & -i & -2 \\ -\frac{i}{4} & \frac{1}{2} & -i \end{matrix} \right\| , \quad \hat{T}_2 = \left\| \begin{matrix} -i & 2 & -4i \\ \frac{1}{2} & -i & 2 \\ -\frac{i}{4} & \frac{1}{2} & -i \end{matrix} \right\| ,$$

$$\hat{\tilde{T}}_2 = \left\| \begin{matrix} i & 2 & 4i \\ -\frac{1}{2} & i & 2 \\ -\frac{i}{4} & -\frac{1}{2} & i \end{matrix} \right\| , \quad \hat{\hat{T}}_2 = \left\| \begin{matrix} i & 2 & -4i \\ -\frac{1}{2} & i & 2 \\ -\frac{i}{4} & -\frac{1}{2} & i \end{matrix} \right\| .$$

Solution. (1,1,0); (2,0,0); (1,0,1); (1,0,0).

6. Let the matrix T_{n-1} ($\equiv H_{n-1}$) be a Toeplitz- and a Hankel matrix at the same time. Which (r,k)-characteristic can this matrix have:
a) as Toeplitz matrix; b) as Hankel matrix?

Solution. (0,0), (1,0), (2,0) (in both interpretations)

7. If T_{n-1} is a Toeplitz matrix with the characteristic (r,k,ℓ), then for the transposed matrix $(T_{n-1})^t$ the characteristic has the shape (r,ℓ,k). Prove this.

§ 15 THEOREMS ON THE RANK.

15.1 Just as in § 11, at the root of the present section there lies, in analogy to Lemma 11.1, a proposition on a nonzero minor, which now, generally speaking, turns out to be not a principal minor, i.e., it is not necessarily situated symmetrically with respect to the main diagonal of the initial matrix. Nevertheless, as will be clear from the Lemma 15.1 presented below, it can always be selected to be symmetrically situated (and so, because of the Toeplitz structure, to be symmetric) relative to the initial (and its own) auxiliary diagonal.

LEMMA 15.1. *Let* T_{n-1} *be a Toeplitz matrix of order* n *with a given* (r,k,ℓ)*-characteristic. Then*

$$r + k + \ell \leq n$$

and the matrix T_{n-1} *has some nonzero minor* $M_{k,\ell}^{(r)}$ *of order* $r+k+\ell$. *This minor can be chosen, generally speaking, in various ways, in particular the entries can be obtained from the following lines of the original matrix* T_{n-1}:

a) *from the first* $k+r$ *columns and the final* ℓ *columns, the first* ℓ *rows and the final* $r+k$ *rows;*

b) *from the first* k *columns and the final* $r+\ell$ *columns, the first* $r+\ell$ *rows and the final* k *rows.*

PROOF. Let the rank of the matrix T_{n-1} be equal to ρ. For $r=\rho$ the verification of the lemma is trivial, since in this case, by definition, $k=\ell=0$, and $M_{0,0}^{(\rho)} \equiv D_{\rho-1} = D_{r-1}$ $(\neq 0)$ is the desired minor of order $r(+0+0)$ (one can replace it by a minor, taken according to rule a) or b) from the formulation of the lemma, by means of proposition 2° from § 14 if $r>0$). We note, that for $\rho=r=0$ (then also $k=\ell=0$) the statement of the lemma doesn't make sence. In this case we agree to understand the "minor" $D_{-1} = 1$ to be the desired minor $M_{0,0}^{(0)}$.

So, let $0<\rho<n$ and $r<\rho$. We assume at first, that the condition $r+k+\ell \leqq n$ is realized. This means that the matrix T_{n-1} =

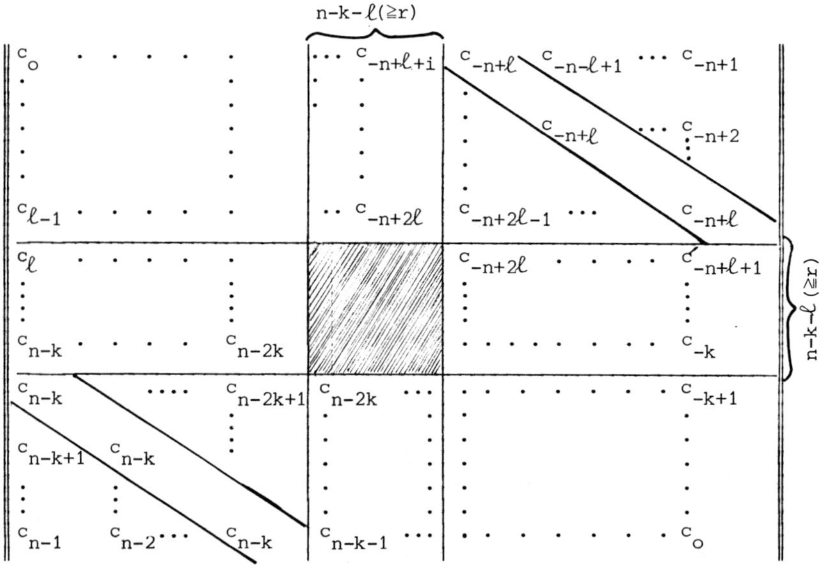

(15.1)

contains at least one minor $\Delta_r^{(\omega)}$ of order r of the shape

$$\Delta_r^{(\omega)} = \begin{Vmatrix} c_\omega & c_{\omega-1} & \cdots & c_{\omega-r+1} \\ c_{\omega+1} & c_\omega & \cdots & c_{\omega-r+2} \\ \cdot & \cdot & \cdots & \cdot \\ c_{\omega+r-1} & c_{\omega+r-2} & \cdots & c_\omega \end{Vmatrix} \tag{15.2}$$

where $\omega+r-1 < n-2k$, $\omega-r+1 > -n+2\ell$, i.e.

$$(-n+\ell+r-1 \leqq)\ -n+2\ell+r-1 < \omega < n-2k-r+1\ \ (\leqq n-k-r+1)$$

(in diagram (15.1) this minor is entirely situated in the hatched part). But then $\Delta_r^{(\omega)} \neq 0$ (§ 14, proposition 2°).

Now we consider the minor

$$M_{k,\ell}^{(r)} = \left\| \begin{array}{ccc|c|ccc} c_0 & \cdots & & \cdots & c_{-n+\ell} & \cdots & c_{-n+1} \\ \cdot & \cdots & & \cdots & \vdots & \ddots & \vdots \\ \cdot & \cdots & & \cdots & \vdots & \cdots & c_{-n+\ell} \\ \hline \cdot & \cdots & & & \cdot & \cdots & \cdot \\ \cdot & \cdots & & \Delta_r^{(\omega)} & \cdot & \cdots & \cdot \\ \cdot & \cdots & & & \cdot & \cdots & \cdot \\ \hline c_{n-k} & \cdots & & \cdots & \cdot & \cdots & \cdot \\ \vdots & \ddots & & \cdots & \cdot & \cdots & \cdot \\ c_{n-1} & \cdots & c_{n-k} & \cdots & \cdot & \cdots & c_0 \end{array} \right\|$$

(in the case $r = 0$ the square $\Delta_r^{(\omega)}$ is absent here). Applying proposition 2° of § 3 and formulae (14.3) we can consider $M_{k,\ell}^{(r)}$ not as a minor of the matrix T_{n-1}, but as being composed (see Theorem 13.2) from the matrix T'_{n-1}, which is a singular extension of the matrix T_r (of rank r), as a determinant $M_{k,\ell}^{(r)}(\xi,\eta)$ (see Lemma 3.1 in the shape of (3.7)), in which the numbers $\xi = c_{n-k}$ and $\eta = c_{-n+\ell}$ replace the elements c'_{n-k} and $c'_{-n+\ell}$ of the matrix T'_{n-1}, respectively. But then using the result of Lemma 3.1 in the shape of (3.8), we have

$$\begin{aligned} M_{k,\ell}^{(r)} &= M_{k,\ell}^{(r)}(c_{n-k}, c_{-n+\ell}) = \\ &= (-1)^{k\ell + r(k+\ell)} \Delta_r^{(\omega)} (c_{n-k} - c'_{n-k})^k (c_{-n+\ell} - c'_{-n+\ell})^\ell, \end{aligned} \tag{15.3}$$

whence because of (14.3) $M_{k,\ell}^{(r)} \neq 0$.

Now we note, that the minor $\Delta_r^{(\omega)}$ ($\neq 0$) of the shape (15.2), which enters into the structure of the determinant $M_{k,\ell}^{(r)}$, can be arbitrarily chosen in the part, which is hatched in diagram (15.1). In particular, having taken it in the lefthand lower (righthand upper) corner of this part, i.e., putting $\omega = n - 2k - r$ (respectively $\omega = -n + 2\ell + r$) we obtain rule a)(respectively rule b)) from the formulation of the lemma.

The obtained result indicates, among other things, the fact, that the identity

$$r + k + \ell = n \tag{15.4}$$

cannot hold in the considered case. Indeed, in the opposite case we would have

$$M_{k,\ell}^{(r)} = D_{n-1} \neq 0$$

i.e., $\rho = n = r$.

It remains to consider the case, where $r + k + \ell > n$, which, as we shall see, also turns out to be impossible (cf. the proof of Lemma 11.1). Indeed, assuming the opposite, we might throw out of the matrix T_{n-1} the final $(r + k + \ell - n)$ rows and columns (because of (14.4), whence $r + k + \ell - n \leq n - 2$). The matrix $T_{n-1-(r+k+\ell-n)} = T_{\tilde{n}-1}$ which will be obtained as result, would have the order

$$\begin{aligned} \tilde{n} &= 2n - r - k - \ell \geq 2n - r - 2\max\{k,\ell\} = \\ &= 2[n - \max\{k,\ell\}] - r \geq r \end{aligned} \tag{15.5}$$

(here we used $n - \max\{k,\ell\} \geq r$, cf. diagram (15.1)), so that in its characteristic $(\tilde{r},\tilde{k},\tilde{\ell})$

$$\begin{aligned} &\tilde{r} = r,\ \tilde{k} = k - (r+k+\ell-n) = n - r - \ell \\ &\tilde{\ell} = \ell - (r+k+\ell-n) = n - r - k \end{aligned} \tag{15.6}$$

where (see (14.4)) $\tilde{k} > 0$, $\tilde{\ell} > 0$, i.e., then rank $\tilde{\rho}$ of the matrix $T_{\tilde{n}-1}$ surpasses $\tilde{r}$ $(=r)$ (see Sec. 14.1). But from (15.6) follows

$$\begin{aligned} \tilde{r} + \tilde{k} + \tilde{\ell} &= r + (n-r-\ell) + (n-r-k) = \\ &= 2n - r - k - \ell = \tilde{n} \end{aligned}$$

which is, as we know (cf. (15.4)), not possible for $\tilde{r} < \tilde{\rho}$.

Lemma 15.1 is proved.

15.2 Some facts, which were discovered in the course of the proof of Lemma 15.1, allow to establish two propositions, which will show to be useful in the sequel.

We consider the quantities

$$E_{-1} \equiv 1,\ E_{p-1} = \begin{vmatrix} c_{-n+p} & c_{-n+p+1} & \cdots & c_{-n+1} \\ c_{-n+p+1} & c_{-n+p} & \cdots & c_{-n+2} \\ \cdot & \cdot & \cdots & \cdot \\ c_{-n+2p-1} & c_{-n+2p-2} & \cdots & c_{-n+p} \end{vmatrix} \qquad (p = 1,2,\cdots,n)$$

or, in the notation of § 1

$$E_{p-1} = T_{n-1}\begin{pmatrix} 1 & 2 & \cdots & p \\ n-p+1 & n-p+2 & \cdots & n \end{pmatrix} \qquad (p = 1,2,\cdots,n).$$

Thus, through (1≡) $E_{-1},E_0,\cdots,E_{n-1}$ are denoted the analoga to the successive principal minors (1≡) $D_{-1},D_0,D_1,\cdots,D_{n-1}$ moving (in contrast to the mentioned minors) not from the lefthand but from the righthand upper corner of the matrix T_{n-1}.

1° *For a Toeplitz matrix* T_{n-1} *with characteristic* (r,k,ℓ) *one has always*

$$E_{r+\ell-1} \neq 0, \quad E_{p-1} = 0 \qquad (p = r+\ell+1,\cdots,n).$$

Indeed, the minor $E_{r+\ell-1}$ is because of its definition obtained from the minor $M_{k,\ell}^{(r)}$, found in Lemma 15.1, if the latter is choosen according to rule b) of this lemma (i.e. for $\omega = -n+2\ell+r$) and then from it are discarded the first k columns and the final k rows:

$$M_{k,\ell}^{(r)} = \left|\begin{array}{ccc|ccc|ccc}
c_0 & \cdots & & & \cdots & & c_{-n+\ell} & \cdots & c_{-n+1} \\
\cdot & \cdots & & & \cdots & & \vdots & \ddots & \\
\cdot & \cdots & & & \cdots & & & \cdots & c_{-n+\ell} \\
\hline
\cdot & \cdots & & & & & \cdot & \cdots & \\
\cdot & \cdots & & & \Delta_r^{(\omega)} & & \cdot & \cdots & \\
\cdot & \cdots & & & & & \cdot & \cdots & \\
\hline
c_{n-k} & \cdots & & & \cdots & & \cdot & \cdots & \\
\vdots & \ddots & & & \cdots & & \cdot & \cdots & \\
& \cdots & c_{n-k} & & \cdots & & \cdot & \cdots & c_0
\end{array}\right| ,$$

$$E_{r+\ell-1} = \left|\begin{array}{ccc|ccc}
c_{\omega-\ell} & \cdots & c_{-n+\ell+1} & c_{-n+\ell} & \cdots & c_{-n+1} \\
\cdot & \cdots & \cdot & \cdot & \cdots & \cdot \\
c_{\omega-1} & \cdots & c_{\omega-r} & \cdot & \cdots & c_{-n+\ell} \\
\hline
c_{\omega} & \cdots & c_{\omega-r+1} & c_{\omega-r} & \cdots & c_{-n+\ell+1} \\
\cdot & \cdots & \cdot & \cdot & \cdots & \cdot \\
c_{\omega+r-1} & \cdots & c_{\omega} & c_{\omega-1} & \cdots & c_{-n+\ell+r}
\end{array}\right| ,$$

$$\omega = -n + 2\ell + r,$$

$$\Delta_r^{(\omega)} = \left|\begin{array}{ccc}
c_{-n+2\ell+r} & \cdots & c_{-n+2\ell+1} \\
\cdot & \cdots & \cdot \\
c_{-n+2\ell+2r-1} & \cdots & c_{-n+2\ell+r}
\end{array}\right| .$$

Thus, $E_{r+\ell-1}$ has exactly the same shape as $M_{k,\ell}^{(r)}$ for $k=0$. But this means, that one can calculate $E_{r+\ell-1}$ by the same rule[1] (15.3) as $M_{0,\ell}^{(r)}$:

$$E_{r+\ell-1} = (-1)^{r\ell}\Delta_r^{(-n+2\ell+r)}(c_{-n+\ell} - c'_{-n+\ell})^{\ell} \neq 0.$$

It remains to explain, why $E_{p-1} = 0$ for $p > r+\ell$. In this case the structure of the determinant E_{p-1} is as follows:

$$
E_{p-1} = \left|\begin{array}{ccc|c|cccc}
\multicolumn{3}{c|}{\overbrace{\hphantom{xxxxxx}}^{p-(r+\ell)}} & \overbrace{\hphantom{xx}}^{r} & \multicolumn{4}{c}{\overbrace{\hphantom{xxxxxx}}^{\ell}} \\
c_{-n+p} & & \cdots & \cdots & c_{-n+\ell} & \cdots & & c_{-n+1} \\
\cdot & & \cdots & \cdots & \cdot & \cdots & & \cdot \\
\cdot & & \cdots & \cdots & \cdot & \cdots & & c_{-n+\ell} \\
\hline
\cdot & \cdot & \cdot & & \cdot & \cdots & & \cdot \\
\cdot & \cdot & \cdot & \Delta_r^{(\omega)} & \cdot & \cdots & & \cdot \\
\cdot & \cdot & \cdot & & \cdot & \cdots & & \cdot \\
\hline
c_\sigma & & \cdot\ \cdot & \cdots & \cdot & \cdots & & \cdot \\
\cdot & & \cdot\ \cdot & \cdots & \cdot & \cdots & & \cdot \\
c_{-n+2p-1} & \cdot\ \cdot & c_\sigma & \cdots & \cdot & \cdots & & c_{-n+p}
\end{array}\right|
\begin{array}{l}
\left.\vphantom{\begin{array}{c}x\\x\\x\end{array}}\right\} \ell \\
\left.\vphantom{\begin{array}{c}x\\x\\x\end{array}}\right\} r \\
\left.\vphantom{\begin{array}{c}x\\x\\x\end{array}}\right\} p-(r+\ell)
\end{array}
$$

where $\omega = -n+2\ell+r$, and $\sigma = -n+p+r+\ell$. For $r = \rho$, where ρ is the rank of the matrix T_{n-1} (in this case $k = \ell = 0$) the order p of the minor E_{p-1} exceeds ρ $(p > r+\ell = \rho+\ell)$, i.e., $E_{p-1} = 0$. Now, if $r < \rho$, then as is evident from the proof of Lemma 15.1, $r+k+\ell < n$. Hence (we remind in addition, that $p \leqq n$)

$$\sigma = -n+p+r+\ell < p-k \leqq n-k,$$

i.e., the numbers c_σ, standing on the diagonal in the lefthand lower corner of the determinant, are "correct" elements of the matrix T_{n-1} : $c_\sigma = c'_\sigma$ (see (14.3)). But then, if one considers proposition 2° of § 3, the calculation of E_{p-1} in accordance to Lemma 3.1., i.e., by a formula which is analoguous to (15.3), yields $E_{p-1} = 0$.

The ascertainment of proposition 1° suggests an "asymmetry" between the components k and ℓ of the (r,k,ℓ)-characteristic of the matrix T_{n-1}. However, this apparent disparity between them disappears, if one introduces just another set of successive minors of the matrix T_{n-1}, starting the progression through it from the lefthand lower corner.

$$F_{-1} \equiv 1,\ F_{q-1} = T_{n-1}\begin{pmatrix} n-q+1 & n-q+2 & \cdots & n \\ 1 & 2 & \cdots & q \end{pmatrix} =$$

$$= \begin{vmatrix} c_{n-q} & c_{n-1-1} & \cdots & c_{n-2q+1} \\ c_{n-q+1} & c_{n-q} & \cdots & c_{n-2q+2} \\ \cdot & \cdot & \cdots & \cdot \\ c_{n-1} & c_{n-2} & \cdots & c_{n-q} \end{vmatrix} \qquad (q = 1,2,\cdots,n).$$

It is clear, that for these minors an analogon of proposition 1° holds, namely[2)]:

2° *For a Toeplitz matrix* T_{n-1} *with characteristic* (r,k,ℓ) *one always has*

$$F_{r+k-1} \neq 0,\ F_{q-1} = 0 \quad (q = r+k+1,\cdots,n).$$

15.3. To begin, we note that Lemma 15.1 is formulated (and proved) much more easily in that case, where the matrix T_{n-1} is Hermitian (or symmetrical). This relates also to the next propositions, to which we now come, postponing a more detailed consideration of Hermitian Toeplitz matrices to Sec. 15.5.

THEOREM 15.1 (FUNDAMENTAL THEOREM ON THE RANK). *If* T_{n-1} *is a Toeplitz matrix with a known* (r,k,ℓ)*-characteristic then its rank* ρ *can be found through the formula*

$$\rho = r + k + \ell\ . \tag{15.7}$$

PROOF. For $r = \rho$ formula (15.7) is trivial, since in this case $k = \ell = 0$.

Let $r < \rho$, i.e., $k + \ell > 0$. Clearly one can, without loss of generality, assume $k \geq \ell$. Because of (14.3) all matrices $T_{\nu-1} = \|c_{p-q}\|_{p,q=0}^{\nu-1}$ for $\nu = r, r+1,\cdots,n-k$ have rank r, but the matrix T_{n-k} has a rank $r_1 > r$. On the basis of the Corollary of Lemma 6.1, r_1 equals either $r+1$ or $r+2$.

If $k = \ell$ then by Lemma 15.1 the matrix T_{n-k}, which has the characteristic $(r,1,1)$, contains a minor of order $r+1+1 = r+2$, which is nonzero, i.e, $r_1 = r+2$. With the same reasoning (cf. § 14, proposition 1°) at the next step of extension of the matrix T_{n-k} by means of the elements c_{n-k+1} $(=c_{n-\ell-1})$ and $c_{-n+\ell-1}$ $(=c_{-n+k-1})$ (here we use again the loose formulation already mentioned in Ch. II) the rank again is raised by two units, and so on. As result after k $(=\ell)$ steps of extension we find that $\rho = r+2k = r+k+\ell$.

Now if $k > \ell$ (for $\ell > k$ the reasoning is analoguous), then (the rank of the matrix T_{n-k}) $r_1 = r+1$, since all rows of the matrix T_{n-k}, except the last one, can be expressed linearly through the first r of its rows (not containing "wrong" elements). So the characteristic of the matrix has the shape $(r,1,0)$. If $k = \ell+1$, then the matrix T_{n-k+1} will already have (see § 14, Proposition 1°) the characteristic $(r,2,1)$, and its rank surpasses r_1 $(=r+1)$ by two units (not more, because of the Corollary to Lemma 6.1, and not less, because of Lemma 15.1). In the case, where $k > \ell+1$, all rows of the matrix T_{n-k+1}, with exception of the last two, can be expressed linearly through the first r of its rows, i.e., the rank of the matrix T_{n-k+1} doesn't exceed $r+2 = r_1+1$. At the same time the characteristic of the matrix T_{n-k+1} in this case has the shape $(r,2,0)$, i.e., on account of Lemma 15.1 the rank of this matrix is exactly equal to $r+2 = r_1+1$.

Continuing this reasoning we find that at each of the first $k-\ell$ steps of extension of the matrix T_{n-k-1} by the pairs of elements

$$(c_{n-k}, c_{-n+k}), (c_{n-k+1}, c_{-n+k-1}), \dots, (c_{n-1}, c_{-n+1})$$

the rank of the corresponding matrices will be increased by one unit, and at each of the final (concluding) ℓ steps by two units. Hence follows, that the original matrix T_{n-1} has the rank

$$\rho = r + (k-\ell) + 2\ell = r + k + \ell.$$

The theorem is proved.

15.4. Now we consider some consequences of the fundamental Theorem on the rank.

THEOREM 15.2. [3] *If in the* (r,k,ℓ)*-characteristic of the Toeplitz matrix* T_{n-1} *of rank* ρ *the component* $r < \rho$*, then an arbitrary extension* T_n *of the matrix* T_{n-1} *by means of an arbitrary pair of elements* (c_n, c_{-n}) *has rank* $\rho_1 > \rho$*. More precisely, if* $k\ell > 0$*, then* $\rho_1 = \rho + 2$*; if* $k = 0$*,* $\ell > 0$ *(respectively* $k > 0$*,* $\ell = 0$*) then (in the notation of* § 14*) for* $c_n \neq c'_n$ *(respectively* $c_{-n} \neq c'_n$*) also* $\rho_1 = \rho + 2$*, but for* $c_n = c'_n$ *(respectively* $c_{-n} = c'_{-n}$*)* $\rho_1 = \rho + 1$.

PROOF. From the condition $r < \rho$ $(k + \ell > 0)$ follows, that the matrix T_{n-1} is singular, i.e., its rank ρ is not more than $n-1$: $\rho \leqq n-1$. Hence, if T_n is a nonsingular matrix, i.e., its rank $\rho_1 = n+1$, then, because of the Corollary of Lemma 6.1, $\rho = n-1$, and $\rho_1 = \rho + 2$. We note, that here either $k\ell > 0$ or $k\ell = 0$. In the latter case let, for example, $k = 0$, $\ell > 0$. Then necessarily $c_n \neq c'_n$.

Indeed, for $c_n = c'_n$ the matrix T_n would have $(n+1) - (\ell+1) = n - \ell$ $(> \rho - \ell = r)$ rows which do not contain "wrong" elements, i.e, rows, lying in the matrix T'_n of rank r, and therefore linearly dependent. But then $\det T_n = 0$ in contrast to our assumptions.

Now, let T_n be a singular matrix and let its characteristic have the shape $(\tilde{r}, \tilde{k}, \tilde{\ell})$. Then for $k\ell > 0$

$$\tilde{r} = r,\ \tilde{k} = k+1,\ \tilde{\ell} = \ell+1,$$

i.e., the matrix T_n has rank $\rho_1 = \tilde{r} + \tilde{k} + \tilde{\ell} = r + k + \ell + 2 = \rho + 2$.

If, however, $k = 0$, $\ell > 0$, then we have for $c_n \neq c'_n$

$$\tilde{r} = r,\ \tilde{k} = 0,\ \tilde{\ell} = \ell+1,$$

i.e., $\rho_1 = r + \ell + 1 = \rho + 1$.

In the case $k > 0$, $\ell = 0$ the reasoning is analoguous.

From Theorem 15.2 one obtains immediately a general criterium for the existence of singular extensions (see § 13) of arbitrary Toeplitz matrices, namely:

THEOREM 15.3. *In order that the Toeplitz matrix* T_{n-1} *of rank* ρ *admits a singular extension it is necessary and sufficient that the condition* $D_{\rho-1} \neq 0$ *is satisfied.*

PROOF. The sufficiency of the condition was established in the Theorems 13.1 and 13.2. The necessity follows from Theorem 15.2, since for $D_{\rho-1} = 0$ in the (r,k,ℓ)-characteristic of the matrix T_{n-1} the component r necessarily satisfies the inequality $r < \rho$, and hence T_{n-1} has no singular extension (T_n has already the rank $\rho+1$ or $\rho+2$).

From this result follows in turn the analogon of Kroneckers Theorem (see Theorem 11.6):

THEOREM 15.4. *If the infinite Toeplitz matrix* $T_\infty = \|c_{p-q}\|_{p,q=0}^{\infty}$ *has the finite rank* ρ, *then* $D_{\rho-1} \neq 0$.

PROOF. This is completely analoguous to the proof of Theorem 11.6.

15.5. We dedicate the final part of this section to the case, where the Toeplitz matrix is Hermitian, keeping in mind, in particular, applications to the theory of (Hermitian) Toeplitz forms.

At the end of Sec. 14.2 it was noted, that for a Hermitian Toeplitz matrix T_{n-1} the (r,k,ℓ)-characteristic always has the shape (r,k,k), i.e., one can speak about the (r,k)-characteristic. Hence the statement of the fundamental Theorem on the rank ρ for a Hermitian Toeplitz matrix now looks like

$$\rho = r + 2k. \tag{15.8}$$

Thus, the nonzero minor $M_{k,\ell}^{(r)}$ which emerges in Lemma 15.1 now (for $k=\ell$) can be chosen such, that it would be a principal minor, i.e., symmetrical with respect to the main diagonal of the matrix T_{n-1} (for this it suffices to take the minor $\Delta_r^{(\omega)}$ in formula (15.2) with $\omega = 0$ and situated in the hatched part of diagram (15.1), symmetrically with respect to the auxiliary diagonal). Then, clearly, $\Delta_r^{(0)} = D_{r-1}$ and since (see Remark 1 to Theorem 13.2)

$$c_{n-k} = \overline{c}_{-n+k}, \quad c'_{n-k} = \overline{c}'_{-n+k}$$

the minor $M_{k,k}^{(r)}$ turns out (see (3.9)) to be:

$$M_k^{(r)} \equiv M_{k,k}^{(r)} = (-1)^k D_{r-1} |c_{n-k} - c'_{n-k}|^{2k}. \tag{15.9}$$

The structure of the minor $M_k^{(r)}$ admits in the case of a Hermitian matrix T_{n-1} $(k = \ell)$ a new interpretation. Indeed, this minor now must consist of the principal minor D_{r-1}, bordered to the left and above with the initial k rows and columns and to the right and below with the final k rows and columns of the matrix T_{n-1}. But since the matrix T_{n-1} is a Toeplitz matrix, the minor D_{r-1} can be chosen also such, that (see diagram (15.1)) it stands adjacent to the initial k rows and columns i.e. (in the notations of § 1) such that

$$D_{r-1} = T_{n-1}\begin{pmatrix} k+1 & k+2 & \cdots & k+r \\ k+1 & k+2 & \cdots & k+r \end{pmatrix}$$

or in a diagram

$$M_k^{(r)} = \left|\begin{array}{ccc|ccc|ccc}
c_0 & \cdots & c_{-k+1} & c_{-k} & \cdots & c_{-k-\sigma} & c_\tau & \cdots & c_{-n+1} \\
\cdot & \cdots & \cdot & \cdot & \cdots & \cdot & \cdot & \cdots & \cdot \\
c_{k-1} & \cdots & c_0 & \cdot & \cdots & \cdot & \cdot & \cdots & c_\tau \\
\hline
c_k & \cdots & c_1 & c_0 & \cdots & c_{-\sigma} & c_\omega & \cdots & c_{\tau+1} \\
\cdot & \cdots & \cdot & \cdot & (D_\sigma) & \cdot & \cdot & \cdots & \cdot \\
c_{k+\sigma} & \cdots & c_r & c_\sigma & \cdots & c_0 & c_{\omega+\sigma} & \cdots & c_{\tau+r} \\
\hline
c_{-\tau} & \cdots & \cdot & \cdot & \cdots & \cdot & c_0 & \cdots & c_{-k+1} \\
\cdot & \cdots & \cdot & \cdot & \cdots & \cdot & \cdot & \cdots & \cdot \\
c_{n-1} & \cdots & c_{-\tau} & c_{-\tau-1} & \cdots & c_{-\tau-r} & c_{k-1} & \cdots & c_0
\end{array}\right|$$

where $\sigma = r-1$, $\tau = -n+k$, $\omega = -n+2k$. But this is, clearly, also the same as the minor, consisting of the first $r+k$ rows and columns of the matrix T_{n-1}, bordered with its final k rows and columns. This is in turn equivalent to a bordering of the minor D_{r-1}, taken in the lefthand upper corner of the matrix T_{n-1}, with the k rows and columns of the matrix T_{n-1} which are next to it, and then with its final k rows and columns:

$$M_k^{(r)} = \underbrace{\left|\begin{array}{ccc|ccc|ccc}
c_0 & \cdots & c_{-r+1} & c_{-r} & \cdots & c_{-r+k-1} & c_{-n+k} & \cdots & c_{-n+1} \\
\cdot & (D_{r-1}) & \cdot & \cdot & \cdots & \cdot & \cdot & & \cdot \\
c_{r-1} & \cdots & c_0 & \cdot & \cdots & \cdot & \cdot & \cdots & c_{-n+r} \\
\hline
c_r & \cdots & \cdot & c_0 & \cdots & \cdot & \cdot & \cdots & \cdot \\
\cdot & \cdots & \cdot & \cdot & & \cdot & \cdot & \cdots & \cdot \\
c_{r+k-1} & \cdots & \cdot & \cdot & \cdots & c_0 & \cdot & \cdots & \cdot \\
\hline
c_{n-k} & \cdots & \cdot & \cdot & \cdots & \cdot & c_0 & \cdots & \cdot \\
\cdot & & \cdot & \cdot & \cdots & \cdot & \cdot & & \cdot \\
c_{n-1} & \cdots & c_{n-r} & \cdot & \cdots & \cdot & \cdot & \cdots & c_0
\end{array}\right|}_{r \qquad\qquad k \qquad\qquad k}
\begin{array}{l} \}\, r \\ \}\, k \\ \}\, k \end{array}$$

This final variant of the interpretation of the minor $M_k^{(r)}$, namely, allows to formulate the following rule (algorithm) for discovering the rank ρ of the Hermitian Toeplitz matrix T_{n-1}. It is the analogon to Theorem 11.3.

THEOREM 15.5. *Let* T_{n-1} *be a Hermitian Toeplitz matrix, and let the number* r *be defined through the relations*

$$D_{r-1} \neq 0, \; D_s = 0 \qquad (s \geq r). \tag{15.10}$$

If $r = n$ *(i.e., the second of relations* (15.10) *is meaningless), then the rank of the matrix* T_{n-1} *is also equal to* $n : \rho = n$. *Now, if* $r < n$, *then, having bordered the minor* D_{r-1}, *standing in the lefthand upper corner of the matrix* T_{n-1}, *with the next row and column and the final row and column of this matrix, we form the minor* $\tilde{D}_{r+1}$ *of order* $r+2$; *then we form the minor* $\tilde{D}_{r+2}$ *of order* $r+4$, *having bordered* D_{r-1} *with the next two rows and columns and the final two rows and columns of the matrix* T_{n-1} *and so on, as long as this is possible, in view of the sizes of the minor* D_{r-1} *and the matrix* T_{n-1}. *We consider the maximal one (with respect to the order) under the minors* $D_{r-1+2\nu}$ $(\nu = 0,1,2,\cdots,$ $\tilde{D}_{r-1} \equiv D_{r-1})$ *which is nonzero. Then*

$$\max_{\tilde{D}_{r-1+2\nu} \neq 0} \nu = k$$

and $r + 2k = \rho$ *is the rank of the matrix* T_{n-1}.[4)]

Further, since for $r < \rho$ we have $(\ell =)$ $k > 0$ for a Hermitian matrix, in Theorem 15.2 always the case $k\ell$ $(=k^2) > 0$ occurs, and so

3° *An arbitrary extension* T_n *of a Hermitian Toeplitz matrix* T_{n-1} *with* $r < \rho$ *by an arbitrary pair* (c_n, c_{-n}) *of (not necessarily complex conjugated) numbers has rank* $\rho_1 = \rho + 2$.[5)]

In an analoguous way, one now would only have to consider the simple case $(k = \ell)$ in the proof of the fundamental Theorem on the rank. In view of the importance for the applications the reasoning used in this case (see the proof of Theorem 15.1) and in particular, its result, deserves to be presented in a separate proposition:

THEOREM 15.6. (ON THE JUMPS IN THE RANK OF A HERMITIAN TOEPLITZ MATRIX AT EXTENSION). *If in the* (r,k)*-characteristic of the Hermitian Toeplitz matrix* T_{n-1} *the component* $k > 0$, *then the truncated matrix* T_{n-k-1} *has rank* r, *and the matrices* $T_{n-k}, T_{n-k+1}, \cdots, T_{n-1}$ *have respectively the ranks* $r+2, r+4, \cdots, r+2k$.

In other words, *at the transition by means of extension from the matrix* T_{r-1} *of rank* r *to the matrix* T_{n-1} *of rank* ρ $(>r)$ *the rank increases only at the final* k *steps of extension, where each of these* k *steps is accompanied by an increase in the rank of two units.*

It is suggested to the reader to verify, as an exercise, that the Theorems 15.5 and 15.6, and also proposition 3°, also hold for (complex) symmetric Toeplitz matrices.

In conclusion (also as exercise) it is suggested to establish, in analogy to Theorem 11.8, the validity of the following proposition:

THEOREM 15.7 [32]. *The general shape of a Hermitian Toeplitz matrix* $T_{n-1} = \|c_{p-q}\|_{p,q=0}^{n-1}$ *of given order* n (≧3) *with a given* (r,k)*-characteristic* ($r \geq 0$, $k > 0$, $n > r + 2k$) *is determined in the following way:*

1) for $r \geq 1$ *is assigned an arbitrary nonsingular Hermitian Toeplitz matrix* T_{r-1} *of order* r *(for* $r = 0$ *this step drops out);*

2) *for* T_r *is chosen an arbitrary singular extension of the matrix* T_{r-1} *(for* $r = 0$ *we set* $T_0 = (0)$*);*

3) *the matrices* $T_{r+1}, T_{r+2}, \ldots, T_{n-k-1}$ *are determined uniquely as singular extensions of the matrix* T_r*;*

4) *the number* c'_{n-k}*, which gives the next singular extension* T'_{n-k} *(of order* $n-k+1$*) of the matrix* T_r*, is determined in a unique way;*

5) *this element* c'_{n-k} *in* T'_{n-k} *is replaced by an arbitrary* c_{n-k} $(\neq c'_{n-k})$ *and correspondingly,* $c'_{-n+k} = \overline{c'}_{n-k}$ *is replaced by* $\overline{c}_{n-k}$*; the matrix, which is obtained as result, is denoted by* T_{n-k}*;*

6) *for* $k = 1$ *the construction of the matrix* T_{n-1} *is completed; for* $k > 1$ *the remaining elements* $c_{n-k+1} (=\overline{c}_{-n+k-1})$, $c_{n-k+2} (=\overline{c}_{-n+k-2}, \ldots,$ $c_{n-1} = \overline{c}_{-n+1}$ *are chosen arbitrarily.*

It is clear, that an analoguous theorem can be formulated and proved also for (complex) symmetric Toeplitz matrices.

EXAMPLES AND EXERCISES

1. For the Toeplitz matrix (see exercise 2 of § 14)

$$T_4 = \begin{Vmatrix} 0 & 0 & -1 & 0 & 0 \\ 2 & 0 & 0 & -1 & 0 \\ 0 & 2 & 0 & 0 & -1 \\ 0 & 0 & 2 & 0 & 0 \\ 0 & 0 & 0 & 2 & 0 \end{Vmatrix}$$

of order 5 with (r,k,ℓ)-characteristic (3,1,0), a nonzero minor of order 4 ($r+k+\ell = 4$) can be chosen according to Lemma 15.1, for example, as

$$\begin{vmatrix} 2 & 0 & 0 & -1 \\ 0 & 2 & 0 & 0 \\ 0 & 0 & 2 & 0 \\ 0 & 0 & 0 & 2 \end{vmatrix} = 16 \quad \text{(variant a) of Lemma 15.1)}$$

or as

$$\begin{vmatrix} 0 & -1 & 0 & 0 \\ 2 & 0 & -1 & 0 \\ 0 & 0 & 0 & -1 \\ 0 & 0 & 2 & 0 \end{vmatrix} = 4 \quad \text{(variant b)).}$$

The rank of the matrix T_4 is equal to $\rho = r+k+\ell = 4$ (verify that $D_4 = |T_4| = 0$).

Further, in connection with propositions 1° and 2° we have

$$E_{r+\ell-1} = E_2 = \begin{vmatrix} -1 & 0 & 0 \\ 0 & -1 & 0 \\ 0 & 0 & -1 \end{vmatrix} = -1 \neq 0;$$

$$E_3 = \begin{vmatrix} 0 & -1 & 0 & 0 \\ 0 & 0 & -1 & 0 \\ 2 & 0 & 0 & -1 \\ 0 & 2 & 0 & 0 \end{vmatrix} = 0, \; E_4 = D_4 = 0;$$

$$F_{r+k-1} = F_3 = \begin{vmatrix} 2 & 0 & 0 & -1 \\ 0 & 2 & 0 & 0 \\ 0 & 0 & 2 & 0 \\ 0 & 0 & 0 & 2 \end{vmatrix} = 16 \neq 0, \; F_4 = D_4 = 0.$$

2. Convince yourself, that in the Toeplitz matrix (of order 6)

$$T_5 = \left\| \begin{matrix} 1 & 1 & 1 & \beta & \gamma & \delta \\ 1 & 1 & 1 & 1 & \beta & \gamma \\ 1 & 1 & 1 & 1 & 1 & \beta \\ 1 & 1 & 1 & 1 & 1 & 1 \\ 1 & 1 & 1 & 1 & 1 & 1 \\ \alpha & 1 & 1 & 1 & 1 & 1 \end{matrix} \right\| \quad (\alpha \neq 1, \; \beta \neq 1)$$

the minor of order 5

$$\begin{vmatrix} 1 & 1 & \beta & \gamma & \delta \\ 1 & 1 & 1 & \beta & \gamma \\ 1 & 1 & 1 & 1 & \beta \\ 1 & 1 & 1 & 1 & 1 \\ \alpha & 1 & 1 & 1 & 1 \end{vmatrix}$$

is nonzero. Compare this fact with Lemma 15.1, having calculated the (r,k,ℓ)-characteristic of the matrix T_5 beforehand. Study the conduct of the successive minors E_{q-1} and F_{q-1} of the matrix T_5 and explain it with the point of view of propositions 1° and 2°.

3. Calculate the rank ρ of the symmetric Toeplitz matrix of order 5

$$T_4 = \left\| \begin{array}{cc|ccc} 0 & 1 & 0 & 1 & 1 \\ 1 & 0 & 1 & 0 & 1 \\ \hline 0 & 1 & 0 & 1 & 0 \\ 1 & 0 & 1 & 0 & 1 \\ 1 & 1 & 0 & 1 & 0 \end{array} \right\|$$

with the rule of Theorem 15.5. Here $D_0 = 0$, $D_1 = -1$, $D_2 = D_3 = D_4 = 0$. Thus $r = 2$,

$$D_{r-1} = D_1 = \begin{vmatrix} 0 & 1 \\ 1 & 0 \end{vmatrix} .$$

Border this minor with the row and column next to it, and then with the final row and column of the matrix T_4. We obtain

$$\tilde{D}_3 = \left| \begin{array}{cc|cc} 1 & 1 & 0 & 1 \\ 1 & 0 & 1 & 1 \\ \hline 0 & 1 & 0 & 0 \\ 1 & 1 & 0 & 0 \end{array} \right| = 1\,(\neq 0).$$

As the size of the matrix doesn't allow further continuation of this algorithm, we have $\rho = 4$. Compare this result with the last of the answers in exercise 4 of § 14 and formula (15.8).

4. Find the rank ρ of the Hermitian Toeplitz matrix

$$T_4 = \begin{Vmatrix} 4 & 4i & 0 & 2i & -5 \\ -4i & 4 & 4i & 0 & 2i \\ 0 & -4i & 4 & 4i & 0 \\ -2i & 0 & -4i & 4 & 4i \\ -5 & -2i & 0 & -4i & 4 \end{Vmatrix} .$$

Solution. $\rho = 3$.

5. For the Toeplitz matrix T_4 of example 1 (order $n = 5$, (r,k,ℓ)-characteristic $(3,1,0)$, rank $\rho = 4$) we consider the extension

$$T_5 = \begin{Vmatrix} 0 & 0 & -1 & 0 & 0 & y \\ 2 & 0 & 0 & -1 & 0 & 0 \\ 0 & 2 & 0 & 0 & -1 & 0 \\ 0 & 0 & 2 & 0 & 0 & -1 \\ 0 & 0 & 0 & 2 & 0 & 0 \\ x & 0 & 0 & 0 & 2 & 0 \end{Vmatrix} .$$

Convince yourself, that for arbitrary x,y the matrix T_5 has rank $\tilde{\rho} > 4$. For what values of x,y will $\tilde{\rho} = 5$?

HINT. Use Theorem 15.2.

6. For the matrix

$$T_5 = \begin{Vmatrix} 1 & i & -1 & -i & -1 & i \\ -i & 1 & i & -1 & -i & -1 \\ -1 & -i & 1 & i & -1 & -i \\ i & -1 & -i & 1 & i & -1 \\ -1 & i & -1 & -i & 1 & i \\ -i & -1 & i & -1 & -i & 1 \end{Vmatrix}$$

(which is Hermitian and Toeplitz), is, as is not difficult to verify, in the (r,k)-characteristic the component r = 1, and $D_{r-1} = D_0 = 1$. Trace the jumps in the rank at the construction of the successive extensions of the matrix $T_r = T_1 = \begin{pmatrix} 1 & i \\ -i & 1 \end{pmatrix}$ (of rank r = 1):

$$T_2 = \begin{Vmatrix} 1 & i & -1 \\ -i & 1 & i \\ -1 & -i & 1 \end{Vmatrix}, \qquad T_3 = \begin{Vmatrix} 1 & i & -1 & -i \\ -i & 1 & i & -1 \\ -1 & -i & 1 & i \\ i & -1 & -i & 1 \end{Vmatrix},$$

and so on, until the complete reconstruction of the matrix T_5 and compare the result with Theorem 15.6. What is the value of the component k?

Solution. k = 2.

7. Prove the proposition (see [31], Theorem 4):

If for the Hermitian (or symmetric) Toeplitz matrix T_{n-1} *of rank* ρ *in the* (r,k)-*characteristic* $k > 0$ $(r < \rho < n)$ *and the defect of the matrix is equal to* d $(= n - \rho)$, *then after* d *steps of extension of this matrix*[6] *by means of arbitrary elements* $c_n = \overline{c}_{-n}$, $c_{n+1} = \overline{c}_{-n-1}, \ldots, c_{n+d-1} = \overline{c}_{-n-d+1}$ (respectively $c_n = c_{-n}$, $c_{n+1} = c_{-n-1}, \ldots, c_{n+d-1} = c_{-n-d+1}$) *we obtain (for the first time!) a nonsingular matrix* T_{n-1+d}.

HINT. Apply Theorem 15.2 (or its consequence - proposition 3°) and proposition 1° of § 14.

8. Do there exist singular extensions of the Toeplitz matrices:

$$T_1 = \left\| \begin{matrix} 0 & 3-2i \\ 0 & 0 \end{matrix} \right\| , \qquad T_2 = \left\| \begin{matrix} -i & -2 & 4i \\ 1/2 & -i & -2 \\ -i/4 & 1/2 & -i \end{matrix} \right\| ,$$

$$\hat{T}_2 = \left\| \begin{matrix} -i & 2 & -4i \\ 1/2 & -i & 2 \\ -i/4 & 1/2 & -i \end{matrix} \right\| , \qquad \hat{\hat{T}}_2 = \left\| \begin{matrix} i & 2 & 4i \\ -1/2 & i & 2 \\ -i/4 & -1/2 & i \end{matrix} \right\| ,$$

$$\hat{\hat{\hat{T}}}_2 = \left\| \begin{matrix} i & 2 & -4i \\ -1/2 & i & 2 \\ -i/4 & -1/2 & i \end{matrix} \right\| , \qquad T_3 = \left\| \begin{matrix} 1 & 1 & 1 & 0 \\ 1 & 1 & 1 & 1 \\ 1 & 1 & 1 & 1 \\ 0 & 1 & 1 & 1 \end{matrix} \right\|$$

(cf. Exercise 5 of § 14)?

HINT. Apply Theorem 15.3.

9. Formulate and prove a generalization of Theorem 15.7 for arbitrary Toeplitz matrices.

NOTES

1) For $r = \ell = 0$ we have, by definition, $E_{r+\ell-1} = E_{-1} = 1$. If $r = 0$, but $\ell > 0$, then, as is clear from definition (14.1) and relations (14.3), we have $c_{-n+p} = c'_{-n+p} = 0$ $(p < \ell)$, and $c_{-n+\ell} \neq c'_{-n+\ell} = 0$, so that in that case

$$E_{r+\ell-1} = E_{\ell-1} = (c_{-n+\ell})^{\ell} \neq 0$$

(see diagram (14.7)).

2) Propositions 1° and 2° will afterwards (see § 17) be used for the construction of a common theory of the characteristics of Toeplitz and Hankel matrices.

3) In the first publication on this Theorem ([37] Theorem 4, Corollary) the hypotheses were, unfortunately, misprinted; this was, however, corrected in the english translation.

4) Just as for Hankel matrices (see the note to Theorem 11.3) it is for small (in comparison to n) values of r easier to find the rank $\rho = r + 2k$ through a direct calculation of the component k in the (r,k)-characteristic by the method of § 14.

5) In fact, this proposition (see [31], Theorem 2) was originally fundamental for the whole theory of Hermitian Toeplitz matrices (and forms), developed in [31] and [32].

6) Cf. exercise 5 to § 11 and the note to it.

§ 16 HERMITIAN TOEPLITZ FORMS.

16.1. For an arbitrary Hermitian Toeplitz form

$$T_{n-1}(x,x) = \sum_{p,q=0}^{n-1} c_{p-q}\xi_p\bar{\xi}_q \quad (c_{-p} = \bar{c}_p,\ p = 0,1,\cdots,n-1) \tag{16.1}$$

with matrix $T_{n-1} = \|c_{p-q}\|_{p,q=0}^{n-1}$ of order $n(\geq 1)$ and rank ρ we occupy ourselves again with the study of the signature and we undertake to establish an analogon to the results of Frobenius in the theory of real Hankel forms (cf. § 12) (It is not difficult to trace, that all results of this § 16 remain valid for real quadratic Toeplitz forms as well). In our investigation there will emerge two, closely related, curious circumstances. In the first place, the distribution of possible zeros in the set of successive principal minors

$$(1=)\ D_{-1}, D_0, D_1, \cdots, D_{\rho-1} \tag{16.2}$$

now turns out to be subject to some "restriction", which we didn't encounter before (i.e., in the theory of Hankel forms). Secondly, precisely because of this restriction the theory turns out to be more well-proportioned and simple and finally its result - the signature rule - is absolutely elementary (and nevertheless it remained undiscovered until recently).

16.2. Following the same scheme as in § 12, we start with the case, where in (16.2) one encounters an isolated group of zeros arranged in succession

LEMMA 16.1. *If in the set* (16.2) *there is a group*

$$(D_{h-1} \neq 0),\ D_h = D_{h+1} = \cdots = D_{h+p-1} = 0 \quad (D_{h+p} \neq 0) \tag{16.3}$$

of p (>0) *zeros, standing isolated, then the number* p *is odd.*

PROOF. If one considers the "truncated" matrices $T_{h-1}, T_h, \cdots, T_{h+p-1}$, T_{h+p} of which the determinants appear in (16.3), then it is clear that the rank of the matrix T_{h+p} is equal to $h+p+1$, and the matrix T_{h+p-1} has a rank $\tilde{\rho}$ equal to $h+p-1$ (see the Corollary of Lemma 6.1).

In the $(\tilde{r},\tilde{k})$-characteristic of the Hermitian Toeplitz matrix T_{h+p-1} is, clearly, $\tilde{r} = h$, and $2\tilde{k} = \tilde{\rho} - \tilde{r} = (h+p-1) - h = p-1$. Hence follows, that $p = 2\tilde{k}+1$ is an odd number.

REMARK. The transition $T_{h+p-1} \longrightarrow T_{h+p}$ is accompanied by a jump in the rank of two units. For $p > 1$ (i.e. for $\tilde{k} > 0$ the transitions

$$T_{h-1} \longrightarrow T_h \longrightarrow \cdots \longrightarrow T_{h+p-1}$$

will, because of Theorem 15.6, only at each of the final $\tilde{k} = (p-1)/2$ steps be accompanied by an increase of the rank and then at each time by two units. Thus, in the arbitrary case ($p \geq 1$) if we put $p = 2q-1$ ($q > 0$), then the complete transition from T_{h-1} to T_{h+p} yields an increase of the rank by $2q$ units.

COROLLARY 1. *The signatures of the Hermitian forms* $T_{h-1}(x,x)$ *and* $T_{h+p}(x,x)$ *(with the matrices* T_{h-1} *and* T_{h+p}*, respectively) coincide. In this connection the form* $T_{h+p}(x,x)$ *has exactly* q *positive and* q *negative squares more, than the form* $T_{h-1}(x,x)$ $(p = 2q-1)$.

This statement follows directly from Theorem 6.1 and Lemma 16.1. Thence is in turn, with regard to formula (4.3), obtained

COROLLARY 2. *If* $p = 2q-1$ *then*

$$\operatorname{sign} D_{h+p} = (-1)^q \operatorname{sign} D_{h-1}. \tag{16.4}$$

In connection with this topic we note, that the restriction, imposed by Lemma 16.1, doesn't exist for Hankel matrices. Indeed, already the simple examples

$$H_2 = \begin{Vmatrix} 0 & 0 & 1 \\ 0 & 1 & 0 \\ 1 & 0 & 0 \end{Vmatrix}, \quad D_{-1} = 1;\ D_0 = D_1 = 0;\ D_2 = -1 \quad (p = 2),$$

$$H_3 = \begin{Vmatrix} 0 & 0 & 0 & 1 \\ 0 & 0 & 1 & 0 \\ 0 & 1 & 0 & 0 \\ 1 & 0 & 0 & 0 \end{Vmatrix}, \quad D_{-1} = 1;\ D_0 = D_1 = D_2 = 0;\ D_3 = 1 \quad (p = 3)$$

show, that for Hankel matrices the number p can be even as well as odd.

We note also, that the restriction, imposed by Lemma 16.1 neither remains valid for non-Hermitian Toeplitz matrices, which is clear even from the example

$$T_2 = \begin{Vmatrix} 0 & 1 & 0 \\ 0 & 0 & 1 \\ 1 & 0 & 0 \end{Vmatrix}, \quad D_{-1} = 1;\ D_0 = D_1 = 0;\ D_2 = 1 \quad (p = 2).$$

16.3. We proceed to consider the case where in the set (16.2) $D_{\rho-1} = 0$.

LEMMA 16.2. *Let for the Hermitian Toeplitz form* (16.1) *the* (r,k)-*characteristic of its matrix* T_{n-1} *be known. Then the truncated form*

$T_{r-1}(x,x)$ *and the "complete" form* $T_{n-1}(x,x)$ *have the same signature, where the form* T_{n-1} *has exactly* k *positive and* k *negative squares more than the form* $T_{r-1}(x,x)$.

PROOF. If the rank of the form $T_{n-1}(x,x)$ is equal to ρ and $r = \rho$ $(k=0)$, then the statement of the lemma is evident (see proposition 1° of § 6). Now, if $r < \rho$ $(k>0)$, then, by Theorem 15.6, the rank increases at the transition by extension from T_{r-1} to T_{n-1} only at the final k steps of extension, and by two units at each step. It suffices to apply Lemma 6.1.

16.4. From Lemma 16.1 and 16.2 one obtains without difficulty the simple signature rule for Hermitian Toeplitz forms:

THEOREM 16.1 (FUNDAMENTAL SIGNATURE THEOREM). *The signature* σ *of the form* (16.1) *can be calculated from the set of its successive principal minors*

$$(1=)\ D_{-1}, D_0, D_1, \cdots, D_{n-2}, D_{n-1} \tag{16.5}$$

of this form by means of the rule

$$\sigma = \sum_{\nu=0}^{n-1} \text{sign}\,(D_{\nu-1}D_\nu) \qquad (\text{sign}\ 0 = 0) \tag{16.6}$$

PROOF. We consider the (r,k)-characteristic of the matrix T_{n-1} of the form $T_{n-1}(x,x)$. If therein the component $k = 0$ (i.e. $r = \rho$, the rank of the form), and $D_\nu \neq 0$, $(\nu = 0,1,\cdots,\rho-1)$, then rule (16.6) simply coincides with Jacobi's rule (see Theorem 8.1),

$$\sum_{\nu=0}^{n-1} \text{sign}(D_{\nu-1}D_\nu) = \sum_{\nu=0}^{\rho-1} \text{sign}(D_{\nu-1}D_\nu) = \sum_{\nu=0}^{r} \text{sign}(D_{\nu-1}D_\nu).$$

Now let $k \geqq 0$, i.e., $r \leqq \rho$. We assume, that in the set (16.5) the first (reading from the left to the right) group of zeros has the shape

$$(D_{h-1} \neq 0),\ D_h = D_{h+1} = \cdots = D_{h+p-1} = 0 \quad (D_{h+p} \neq 0).$$

Then by Lemma 16.1 $p = 2q - 1$ $(q>0)$, and the form $T_{h+p}(x,x)$ has by Corollary 1 from Lemma 16.1 the same signature as $T_{h-1}(x,x)$. But for the latter, according to Jacobi's rule, the signature is equal to

$$\sum_{\nu=0}^{h-1} \text{sign}(D_{\nu-1}D_\nu) = \sum_{\nu=0}^{h+p} \text{sign}(D_{\nu-1}D_\nu).$$

If here $h+p = r-1$, then from Lemma 16.2 follows formula (16.6). Now, if $h+p < r-1$, then for $D_\nu \neq 0$ $(h+p \leqq \nu \leqq r-1)$ the signature of the form T_{r-1} is equal (cf. the note to the proof of Theorem 8.1) to

$$\sum_{\nu=0}^{r-1} \operatorname{sign}(D_{\nu-1}D_\nu) = \sum_{\nu=0}^{n-1} \operatorname{sign}(D_{\nu-1}D_\nu),$$

and, because of Lemma 16.2, that is also the signature σ. In the case where between $D_{h+p} \neq 0$ and $D_{r-1} \neq 0$ there are still further groups of zeros we arrive at the same result after renewed application of Lemma 16.1.

The theorem is proved.

REMARK. In rule (16.6) the following two peculiarities attract the attention. In the first place, in the way it is written down it is absolutely identical to Jacobi's rule, but in contrast to the latter, it can be applied without any restriction on the minors D_ν $(\nu = 0,1,\cdots,n)$. Further, unlike the rules of Jacobi (Theorem 8.1) and Frobenius (Theorem 12.1) it does not require the rank ρ of the form $T_{n-1}(x,x)$ to be known. The Toeplitz forms excluded, we do not know of any class of Hermitian of quadratic form for which it is possible to discover the signature σ from the set (16.5) without knowing the rank of the form.

16.5. It is clear, that knowledge of the signature σ of the form $T_{n-1}(x,x)$, in combination with the knowledge of its rank ρ (found, for example, by means of Theorem 15.5), allows to find out the numbers $\pi = \frac{1}{2}(\rho+\sigma)$ and $\nu = \frac{1}{2}(\rho-\sigma)$ of its positive and negative squares. However, just as in the case of Hankel forms the numbers π and ν can (knowing ρ) be found also directly by means of a Theorem, which is the analogon of 12.1.

Indeed, by using Corollaries 1 and 2 of Lemma 16.1, we obtain

1°. *Under the conditions of Lemma* 16.1 *we write for the zero minors from the set* (16.3) *the sign plus or minus according to the rule*

$$\operatorname{sign} D_{h-1+\nu} = (-1)^{\nu(\nu-1)/2} \operatorname{sign} D_{h-1} \tag{16.7}$$

Then the numbers of positive and negative squares of the form $T_{h+p}(x,x)$ *exceed the corresponding numbers for the form* $T_{h-1}(x,x)$ *by*

$$P(D_{h-1},D_h,\cdots,D_{h+p}) \text{ and } V(D_{h-1},D_h,\cdots,D_{h+p}),$$

respectively (see §§ 8 and 12 for the notations $P(\cdots)$ and $V(\cdots)$).

The proof is, because of the fact that $p = 2q-1$ $(q>0)$, quite identical to the proof of Lemma 12.1 in its elementary part I, if we consider, that the relations (12.5) and (12.6) used there coincide with (16.4).

2°. *Let in the* (r,k)*-characteristic of the matrix* T_{n-1} *of the form* (16.1) $r<\rho$ $(k>0)$. *We write for the zero determinants in the set*

$(D_{r-1} \neq 0)$, $D_r = D_{r+1} = \cdots = D_{\rho-2} = 0$ *signs according to the rule*

$$\operatorname{sign} D_{r-1+\nu} = (-1)^{\nu(\nu-1)/2} \operatorname{sign} D_{r-1} \quad (\nu = 1,2,\cdots,2k-1), \qquad (16.8)$$

and we replace the minor D_{r-1+2k} $(= D_{\rho-1} = 0)$ *by the nonzero minor* $\tilde{D}_{\rho-1} = M_k^{(r)}$ *(see* (15.9) *of order* ρ.

Then the numbers π *and* ν *of positive and negative squares of the form* $T_{n-1}(x,x)$ *exceed the corresponding numbers for the form* $T_{r-1}(x,x)$ *by*

$$P(D_{r-1}, D_r, \cdots, D_{\rho-2}, \tilde{D}_{\rho-1})$$

and

$$V(D_{r-1}, D_r, \cdots, D_{\rho-2}, \tilde{D}_{\rho-1})$$

respectively.

Here again the proof is, because of formula (15.9), in nothing different from the considerations in the simple case I a) of the proof of Lemma 12.2 (we leave the comparison of the different notations, occurring in our proposition 2° and those in the mentioned proof of Lemma 12.2 to the reader as an exercise).

Now, summing up propositions 1° and 2° we obtain a rule, which is the analogon of the statement of Theorem 12.1:

THEOREM 16.2. *Let the rank of the Hermitian Toeplitz form* $T_{n-1}(x,x)$ *be equal to* $\rho = r + 2k$, *where* (r,k) *is the characteristic of the matrix* T_{n-1}. *We consider the set*

$$(1 =)\ D_{-1}, D_0, D_1, \cdots, D_{\rho-2}, D^*_{\rho-1},$$

where for $k = 0$ $(r = \rho)$ *we substitute* $D^*_{\rho-1} = D_{\rho-1}$, *and in the opposite case* $D^*_{\rho-1} = \tilde{D}_{\rho-1} = (M_k^{(r)})$ (see proposition 2°). *For the zero determinants* D_{j-1} $(0 \leq j \leq \rho-1)$, *if such exist, we write signs according to the rules* (16.7) *and* (16.8). *Then the numbers* π *and* ν *of positive and negetive squares of the form* $T_{n-1}(x,x)$ *are respectively determined by the identities*

$$\left.\begin{aligned} \pi &= P(1, D_0, D_1, \cdots, D_{\rho-2}, D^*_{\rho-1}), \\ \nu &= V(1, D_0, D_1, \cdots, D_{\rho-2}, D^*_{\rho-1}). \end{aligned}\right\} \qquad (16.9)$$

REMARK. In the first publications [32,34] of the rule, formulated in Theorem 16.2, there appeared, instead of the factor $(-1)^{\nu(\nu-1)/2}$ (i.e., the same, as in Frobenius' rule for Hankel forms) the coefficient $(-1)^{\nu(\nu+1)/2}$. It is easy to understand, that precisely because of Lemma 16.1 and the relation $\rho = r + 2k$ these different factors yield in formulae (16.9) one and the same result.

EXAMPLES AND EXERCISES

1. We consider the Hermitian Toeplitz form (of order $n = 5$)

$$T_4(x,x) = 4(|\xi_0|^2 + |\xi_1|^2 + |\xi_2|^2 + |\xi_3|^2 + |\xi_4|^2) + 4i(\xi_0\overline{\xi}_1 + \xi_1\overline{\xi}_2 + \xi_2\overline{\xi}_3 + \xi_3\overline{\xi}_4) - 4i(\overline{\xi}_0\xi_1 + \overline{\xi}_1\xi_2 + \overline{\xi}_2\xi_3 + \overline{\xi}_3\xi_4) + 2i(\xi_0\overline{\xi}_3 + \xi_1\overline{\xi}_4) - 2i(\overline{\xi}_0\xi_3 + \overline{\xi}_1\xi_4) + - 5(\xi_0\overline{\xi}_4 + \overline{\xi}_0\xi_4)$$

with the matrix (cf. exercise 4 to § 15)

$$T_4 = \left\| \begin{matrix} 4 & 4i & 0 & 2i & -5 \\ -4i & 4 & 4i & 0 & 2i \\ 0 & -4i & 4 & 4i & 0 \\ -2i & 0 & -4i & 4 & 4i \\ -5 & -2i & 0 & -4i & 4 \end{matrix} \right\| .$$

Here

$$D_{-1} \equiv 1, \ D_0 = 4,$$

$$D_1 = \begin{vmatrix} 4 & 4i \\ -4i & 4 \end{vmatrix} = 0 \qquad D_2 = \begin{vmatrix} 4 & 4i & 0 \\ -4i & 4 & 4i \\ 0 & -4i & 4 \end{vmatrix} = -64,$$

and $D_3 = D_4 = 0$, as the matrix T_4 has rank 3 (cf. exercise 4 to § 15). Hence, in accordance to Theorem 16.1 (formula (16.16)) the signature is equal to

$$\sigma = \mathrm{sign}(D_{-1}D_0) = \mathrm{sign}(4) = +1.$$

As $\rho = 3$, one has $\pi = 2$, $\nu = 1$.

2. Calculate the signature of the real symmetric Toeplitz form (of order $n = 5$)

$$T_4(x,x) = \xi_0\xi_1 + \xi_0\xi_3 + \xi_0\xi_4 + \xi_1\xi_2 + \xi_1\xi_4 + \xi_2\xi_3 + \xi_3\xi_4.$$

Solution. $\sigma = 0$.

REMARK. With regard to the result of example 3 to § 15 ($\rho = 4$) we have $\pi = \nu = 2$.

3. Prove the analogon of Theorem 12.2 (cf. [37], Theorem 6):

Let π *and* ν *be the numbers of positive and negative squares of the Toeplitz form* $T_{n-1}(x,x)$, *respectively, and* $\kappa = \min\{\pi,\nu\}$. *Then in the set of successive principal minors*

$$(1 =)\ D_{-1}, D_0, D_1, \ldots, D_{\rho-1} \quad (\rho = \pi + \nu) \tag{16.1o}$$

of the form $T_{n-1}(x,x)$ *there are for* $D_{\rho-1} \neq 0$ *not more than* $2\kappa - 1$ *zeros. If there are exactly* $2\kappa - 1$ *zeros, then, after these have been removed from* (16.1o), *the signs inside each of the remaining groups of numbers (not regarding the numbers which stand isolated) are strictly alternating for* $\kappa = \pi$ *and for* $\kappa = \nu$ *they do not vary.*

HINT. Follow the way of the proof of Theorem 12.2, applying formula (16.6) instead of (12.2o).

4. Prove, that the Theorem of exercise 3 also remains valid for $D_{\rho-1} = 0$, if the upper bound $2\kappa - 1$ is replaced by 2κ (cf. exercise 4 to § 12).

5. Devise an example, which shows that the bounds $2\kappa - 1$ and 2κ in the exercises 3 and 4 respectively are sharp (cf. exercises 5 and 6 to § 12).

6. Show, that in Theorem 15.7 one can, besides the parameters n ($\geqq 3$), $r \geqq 0$, $k > 0$ ($n > r + 2k$), assign in advance the signature σ of the form

$$T_{n-1}(x,x) = \sum_{p,q=0}^{n-1} c_{p-q}\xi_p\overline{\xi}_q$$

with the matrix T_{n-1}, choosing for σ an arbitrary entire number of the same parity as r (and the same also as $\rho = r + 2k$) from the segment $[-r,r]$. For the construction of such a matrix T_{n-1} it is necessary and sufficient to select in step 1) of Theorem 15.7 the matrix T_{r-1} such that is satisfies the condition: the signature of the form $T_{r-1}(x,x)$ is equal to σ (cf. exercise 7 to § 12).

7. Verify the validity of the remark to Theorem 16.2.

8. The infinite sequence of complex numbers

$$(\overline{c}_0 =)\ c_0, c_1, c_2, \cdots$$

belongs, by definition, to the class $P_\kappa^{(T)}$, if for sufficiently large $m (> N)$ all Toeplitz forms

$$T_{m-1}(x,x) = \sum_{p,q=0}^{m-1} c_{p-q}\xi_p\overline{\xi}_q \qquad (c_{-p} = \overline{c}_p,\ p = 0,1,\cdots,m-1)$$

have in the canonical representation exactly κ positive squares.

If the Hermitian Toeplitz form

$$\sum_{p,q=0}^{n-1} c_{p-q}\xi_p\overline{\xi}_q$$

of rank ρ has κ positive squares, then in order that the finite set

$$c_0, c_1, \cdots, c_{n-1}$$

of its coefficients allows the exension to an infinite sequence $\{c_p\}_{p=0}^{\infty}$ of the class $P_\kappa^{(T)}$, it is necessary and sufficient, that the minor $D_{\rho-1} \neq 0$. If this condition is satisfied, then for $\rho < n$ the mentioned extension is defined uniquely, but for $\rho = n$, there exist infinitely many of such extensions [31] (cf. exercise 1o to § 12).

HINT. Use in the "necessary"-part proposition 3° and Theorem 6.1, and in the "sufficient"-part the Theorems 15.3, 6.2, 13.2 and 13.1.

REMARK. For more complete information on sequences of the class $P_\kappa^{(T)}$ (classification, integral representation, asymptotic behaviour, connections with the classical problem of moments etc.) and also for a detailed investigation of the extension problem, which was touched only in exercise 8, see in [31,33,36,42,43,51]. The analoguous information on the class $P_\kappa^{(H)}$ (see exercise 1o to § 12), is, though in considerably less extent, contained in [22,5o,41]. For the continual analoga to these problems see [45,54,21-23,59].

REMARK. For the further development of the concept of the characteristic of Toeplitz matrices, and also for new applications of the theory of Toeplitz matrices and forms see [5],[9] and [16] of the additional list of references.

Chapter IV

TRANSFORMATIONS OF TOEPLITZ AND HANKEL MATRICES AND FORMS

§ 17 MUTUAL TRANSFORMATIONS OF TOEPLITZ AND HANKEL MATRICES. RECALCULATION OF THE CHARACTERISTICS.

17.1. Over the full length of Ch. III we have traced the analogy between the properties of Toeplitz and Hankel matrices, and also the analogy in the behaviour of the corresponding forms, not forgetting, in truth, to underline also the sometimes essential differences which have turned out. The mentioned analogy becomes natural (at least in the aspects which touch the matrices) if one notes that each Toeplitz matrix

$$T_{n-1} = \left\| \begin{matrix} c_0 & c_{-1} & \cdots & c_{-n+2} & c_{-n+1} \\ c_1 & c_0 & \cdots & c_{-n+3} & c_{-n+2} \\ \cdot & \cdot & \cdots & \cdot & \cdot \\ c_{n-2} & c_{n-3} & \cdots & c_0 & c_{-1} \\ c_{n-1} & c_{n-2} & \cdots & c_1 & c_o \end{matrix} \right\|$$

by a simple rearrangement of the columns (rows) can be converted into a Hankel matrix and the other way round.

Indeed, it suffices, for example, in the matrix T_{n-1} to rearrange the colums in the reverse order, i.e., the last column on the first place, the penultimate one on the second place, and so on, and we obtain the Hankel matrix

$$H^I_{n-1} = \left\| \begin{matrix} c_{-n+1} & c_{-n+2} & \cdots & c_{-1} & c_0 \\ c_{-n+2} & c_{-n+3} & \cdots & c_0 & c_1 \\ \cdot & \cdot & \cdots & \cdot & \cdot \\ c_{-1} & c_0 & \cdots & c_{n-3} & c_{n-2} \\ c_0 & c_1 & \cdots & c_{n-2} & c_{n-1} \end{matrix} \right\| . \qquad (17.1)$$

The analoguous maniplulations on the rows of the matrix T_{n-1} also lead to a (in general, different) Hankel matrix

$$H^{\vdash\!\dashv}_{n-1} = \left\| \begin{matrix} c_{n-1} & c_{n-2} & \cdots & c_1 & c_0 \\ c_{n-2} & c_{n-3} & \cdots & c_0 & c_{-1} \\ \cdot & \cdot & \cdots & \cdot & \cdot \\ c_1 & c_0 & \cdots & c_{-n+3} & c_{-n+2} \\ c_0 & c_{-1} & \cdots & c_{-n+2} & c_{-n+1} \end{matrix} \right\|$$

If one introduces the notations

$$H^{\mathrm{I}}_{n-1} \equiv \| s^{\mathrm{I}}_{i+j} \|^{n-1}_{i,j=0}, \quad H^{\vdash\!\dashv}_{n-1} \equiv \| s^{\vdash\!\dashv}_{i+j} \|^{n-1}_{i,j=0}$$

then one easily notes that

$$s^{\mathrm{I}}_j = c_{j-(n-1)}, \quad s^{\vdash\!\dashv}_j = c_{-j+(n-1)} \quad (j = 0,1,2,\cdots,2n-2). \tag{17.2}$$

It is clear that one can, conversely, starting with a given Hankel matrix

$$H_{n-1} = \left\| \begin{matrix} s_0 & s_1 & \cdots & s_{n-2} & s_{n-1} \\ s_1 & s_2 & \cdots & s_{n-1} & s_n \\ \cdot & \cdot & \cdots & \cdot & \cdot \\ s_{n-2} & s_{n-1} & \cdots & s_{2n-4} & s_{2n-3} \\ s_{n-1} & s_n & \cdots & s_{2n-3} & s_{2n-2} \end{matrix} \right\|,$$

obtain, with aid of the formulae

$$c^{\mathrm{I}}_p = s_{p+n-1}; \quad c^{\vdash\!\dashv}_p = s_{-p+n-1} \quad (p = 0, \pm 1, \cdots, \pm (n-1)) \tag{17.3}$$

two Toeplitz matrices

$$T^{\mathrm{I}}_{n-1} = \| c^{\mathrm{I}}_{p-q} \|^{n-1}_{p,q=0} = \left\| \begin{matrix} s_{n-1} & s_{n-2} & \cdots & s_1 & s_0 \\ s_n & s_{n-1} & \cdots & s_2 & s_1 \\ \cdot & \cdot & \cdots & \cdot & \cdot \\ s_{2n-3} & s_{2n-4} & \cdots & s_{n-1} & s_{n-2} \\ s_{2n-2} & s_{2n-3} & \cdots & s_n & s_{n-1} \end{matrix} \right\|,$$

$$T^{\vdash\!\dashv}_{n-1} = \| c^{\vdash\!\dashv}_{p-q} \|^{n-1}_{p,q=0} = \left\| \begin{matrix} s_{n-1} & s_n & \cdots & s_{2n-3} & s_{2n-2} \\ s_{n-2} & s_{n-1} & \cdots & s_{2n-4} & s_{2n-3} \\ \cdot & \cdot & \cdots & \cdot & \cdot \\ s_1 & s_2 & \cdots & s_{n-1} & s_n \\ s_0 & s_1 & \cdots & s_{n-2} & s_{n-1} \end{matrix} \right\|,$$

which are mutually the transpose of each other

$$(T^{\vdash\!\dashv}_{n-1})^t = T^{\mathrm{I}}_{n-1}, \quad (T^{\mathrm{I}}_{n-1})^t = T^{\vdash\!\dashv}_{n-1}. \tag{17.4}$$

The transformations (17.2) and (17.3) can be described by means of the multiplication of the given matrices with the fixed matrix J_n of order n

$$J_n = \begin{Vmatrix} 0 & & & 1 \\ & & 1 & \\ & 1^{\cdot^{\cdot^{\cdot}}} & & \\ 1 & & & 0 \end{Vmatrix} . \tag{17.5}$$

Indeed, one verifies directly the relations

$$H^{I}_{n-1} = T_{n-1}J_n, \quad H^{\vdash\!\dashv}_{n-1} = J_n T_{n-1}; \tag{17.6}$$

$$T^{I}_{n-1} = H_{n-1}J_n, \quad T^{\vdash\!\dashv}_{n-1} = J_n H_{n-1}. \tag{17.7}$$

It is even more convenient to use these formulae, because the matrix J_n has remarkable properties: it is *Hermitian* (real and symmetric) and *unitary* at the same time:

$$J^*_n = J^t_n = J_n, \quad J^{-1}_n = J^*_n \; (=J_n)$$

and hence an *involution*

$$J^2_n = E. \tag{17.8}$$

In particular, the relation (17.4) follow from (17.7) and the symmetry of the matrices J_n and H_{n-1} $(J^t_n = J_n,\ H^t_{n-1} = H_{n-1})$.

17.2. In connection with the transformations introduced in Sec. 17.1 there arises the natural question: how are these transformations reflected in the characteristics of the respective matrices? Exactly formulated: is it possible to calculate from the given (r,k,ℓ)-characteristic of the Toeplitz matrix T_{n-1} the (r^I,k^I)-and $(r^{\vdash\!\dashv},k^{\vdash\!\dashv})$-characteristics of the Hankel matrices H^I_{n-1} and $H^{\vdash\!\dashv}_{n-1}$ and conversely? A positive answer to the first of these questions is obtained with help of the results of Sec. 15.2.

So, let the (r,k,ℓ)-characteristic of the Toeplitz matrix

$$T_{n-1} = \left\Vert \begin{array}{c|cccc} T_{r-1} & c_{-r} & \cdots & c_\tau & \cdots & c_{-n+1} \\ & \vdots & & & & \vdots \\ \hline c_r \;\cdots & c_0 & & & & \\ \vdots & & & & & c_\tau \\ c_\sigma & & & & & \vdots \\ \vdots & & & & & \\ c_{n-1} \;\cdots & c_\sigma & & & & c_0 \end{array} \right\Vert , \tag{17.9}$$

be known, where $\sigma = n-k$, $\tau = -n+\ell$. By the definition of the (r,k,ℓ)-characteristic (see Sec. 14.1) we have in the set of successive principal minors

$$D_{-1}(\equiv 1),\ D_{\nu-1} = \det \| c_{p-q} \|_{p,q=0}^{\nu-1} \qquad (\nu = 1,2,\cdots,n)$$

that $D_{r-1} \neq 0$, but $D_{\nu-1} = 0$ $(\nu > r)$.

Now we consider the sequence of minors (counting from the righthand upper corner of the matrix T_{n-1})

$$E_{-1}(\equiv 1), E_{p-1} = \begin{vmatrix} c_{-n+p} & c_{-n+p-1} & \cdots & c_{-n+1} \\ c_{-n+p+1} & c_{-n+p} & \cdots & c_{-n+2} \\ \cdot & \cdot & \cdots & \cdot \\ c_{-n+2p-1} & c_{-n+2p-2} & \cdots & c_{-n+p} \end{vmatrix} \quad (p = 1,2,\cdots,n), \quad (17.10)$$

introduced in Sec. 15.2. As was shown in proposition 1° of § 15

$$E_{r+\ell-1} \neq 0,\ E_{p-1} = 0 \quad (p > r + \ell). \qquad (17.11)$$

Now we note, that after rearrangement of the columns of the matrix T_{n-1} in converse order, i.e., after the transformation

$$H^I_{n-1} = T_{n-1}J_n,$$

the minors E_{p-1} of the matrix T_{n-1} $(p = 0,1,\cdots.n)$ coincide (up to the signs) with the corresponding successive principal minors of the Hankel matrix H^I_{n-1} (see (17.1)). But then it follows from formulae (17.11) (see the definition in Sec. 10.1) that the Hankel matrix H^I_{n-1} has in its (r^I,k^I)-characteristic a component r^I equal to

$$r^I = r + \ell. \qquad (17.12)$$

In order to determine the other component k^I it suffices to note, that the rank ρ of the matrix H^I_{n-1} is, clearly, the same as that of the matrix T_{n-1} and hence, according to the Theorems 11.1 and 15.1,

$$\rho = r^I + k^I = r + k + \ell,$$

i.e. (see (17.12))

$$k^I = k.$$

Thus we have established the proposition

1°. *If the Toeplitz matrix* T_{n-1} *with characteristic* (r,k,ℓ) *is connected with the Hankel matrix* H^I_{n-1} *through the transformation*

$$H^I_{n-1} = T_{n-1}J_n$$

(see (17.5)), *then the* (r^I,k^I)*-characteristic of the matrix* H^I_{n-1} *can be calculated by the rule*

$$r^I = r + \ell, \quad k^I = k \qquad (17.14)$$

In complete analogy (applying proposition 2° from § 15 instead of proposition 1° of the same section) one proves the proposition.

2°. *If the Toeplitz matrix* T_{n-1} *with characteristic* (r,k,ℓ) *is connected with the Hankel matrix* $H^{\vdash\!\dashv}_{n-1}$ *through the transformation*

$$H^{\vdash\!\dashv}_{n-1} = J_n T_{n-1} \tag{17.15}$$

(see (17.5)*), then the* $(r^{\vdash\!\dashv},k^{\vdash\!\dashv})$*-characteristic of the matrix* $H^{\vdash\!\dashv}_{n-1}$ *can be calculated by the rule*

$$r^{\vdash\!\dashv} = r + k, \quad k^{\vdash\!\dashv} = \ell \tag{17.16}$$

Since $k \geqq 0$, $\ell \geqq 0$ follows from proposition 1° and 2°

COROLLARY. *Under the transformations* (17.13) *and* (17.15) *one always has* $r^{I} \geqq r$ *and* $r^{\vdash\!\dashv} \geqq r$, *respectively.*

17.3. For the solution of the converse problem, i.e., the calculation of the characteristics of the Toeplitz matrices, obtained from a Hankel matrix H_{n-1} with a given (r,k)-characteristic trough the transformations of the shape (17.7), we must introduce additional tools. Indeed, already simple examples demonstrate, that the (r,k)-characteristic of the Hankel matrix H_{n-1} on its own doesn't, generally speaking, determine the (r^{I},k^{I},ℓ^{I})-characteristic (respectively the $(r^{\vdash\!\dashv},k^{\vdash\!\dashv},\ell^{\vdash\!\dashv})$-characteristic) of its transforms i.e., of the Toeplitz matrix $T^{I}_{n-1} = H_{n-1}J_n$ (respectively $T^{\vdash\!\dashv}_{n-1} = J_n H_{n-1}$).

EXAMPLE

$$(n=3)\left\{\begin{aligned} H_2 &= \begin{Vmatrix} 0 & 1 & 0 \\ 1 & 0 & 0 \\ 0 & 0 & 0 \end{Vmatrix}, \quad r = 2,\ k = 0\ (\rho = r+k = 2); \\ T^{I}_2 &= \begin{Vmatrix} 0 & 1 & 0 \\ 0 & 0 & 1 \\ 0 & 0 & 0 \end{Vmatrix}, \quad r^{I} = 0,\ k^{I} = 0,\ \ell^{I} = 2 \\ &\qquad (\rho = r^{I} + k^{I} + \ell^{I} = 2). \end{aligned}\right.$$

$$(n=3)\left\{\begin{aligned} \hat{H}_2 &= \begin{Vmatrix} 0 & 1 & 1 \\ 1 & 1 & 1 \\ 1 & 1 & 1 \end{Vmatrix}, \quad r = 2,\ k = 0\ (\rho = r+k = 2); \\ \hat{T}^{I}_2 &= \begin{Vmatrix} 1 & 1 & 0 \\ 1 & 1 & 1 \\ 1 & 1 & 1 \end{Vmatrix}, \quad r^{I} = 1,\ k^{I} = 0,\ \ell^{I} = 1 \\ &\qquad (\rho = r^{I} + k^{I} + \ell^{I} = 2). \end{aligned}\right.$$

So the Hankel matrices H_2 and $\hat{H}_2$ (both of order 3 and rank 2) having the same (r,k)-characteristic (2,0), turn into the Toeplitz matrices T^{I}_2 and $\hat{T}^{I}_2$ with the different characteristics (0,0,2) and (1,0,1), respectively.

Considering $T_2^H = (T_2^I)^t$ and $\hat{T}_2^H = (T_2^I)^t$, we obtain again Toeplitz matrices with the different characteristics (0,2,0) and (1,1,0), respectively.

Here has appeared a qualitative difference between the characteristics of Hankel- and Toeplitz matrices, defined in Chapter II and III, respectively. Indeed, the (r,k,ℓ)-characteristic of the Toeplitz matrix T_{n-1} reveals the "dynamics of the behaviour" of its elements at progress in two directions: along the main diagonal (the component r) and along the auxliary diagonal (the components k and ℓ). At the same time both components of the (r,k)-characteristics of the Hankel matrix H_{n-1} tell of the "dynamics of the behaviour" of its elements at movement in one direction only: along the main diagonal.

It is not difficult to eliminate the mentioned deficiency. Besides (and in analogy to) the component r [1)] in the (r,k)-characteristic of the Hankel matrix H_{n-1} we introduce another nonnegative numerical constant r_1, defined in the following way. We consider in the Hankel matrix

$$H_{n-1} = \left\| \begin{matrix} s_0 & s_1 & \cdots & s_{n-p} & \cdots & s_{n-2} & s_{n-1} \\ s_1 & s_2 & \cdots & s_{n-p+1} & \cdots & s_{n-1} & s_n \\ \cdot & \cdot & \cdots & \cdot & \cdots & \cdot & \cdot \\ s_{p-1} & s_p & \cdots & s_{n-1} & \cdots & s_{n+p-3} & s_{n+p-2} \\ \cdot & \cdot & \cdots & \cdot & \cdots & \cdot & \cdot \\ s_{n-1} & s_n & \cdots & s_{2n-p-1} & \cdots & s_{2n-3} & s_{2n-2} \end{matrix} \right\|$$

the successive minors of order p, already known to us (cf. the same considerations for Toeplitz matrices in Sec. 15.2, and also in (17.1o)):

$$E_{-1} \equiv 1, \quad E_{p-1} = \begin{vmatrix} s_{n-p} & \cdots & s_{n-2} & s_{n-1} \\ s_{n-p+1} & \cdots & s_{n-1} & s_n \\ \cdot & \cdots & \cdot & \cdot \\ s_{n-1} & \cdots & s_{n+p-3} & s_{n+p-2} \end{vmatrix} \qquad (17.17)$$

$$(p = 1,2,\cdots,n),$$

situated along the auxiliary diagonal, starting at the righthand upper corner of the matrix. The constant r_1 is now uniquely defined (cf. (1o.2)) through the relations

$$E_{r_1-1} \neq 0, \quad E_{p-1} = 0 \quad (p > r_1) \qquad (17.18)$$

We have

$$0 \leq r_1 \leq \rho$$

where ρ is the rank of the matrix H_{n-1}.

If one now transforms the Hankel matrix H_{n-1} into the Toeplitz matrix $T^{I}_{n-1} = H_{n-1}J_n$, i.e., rearranges the columns of H_{n-1} in reverse order, then one concludes, that the number r, is the component r^I in the (r^I, k^I, ℓ^I)-characteristic of the matrix T^I_{n-1}. Further (see (17.8)),

$$H_{n-1} = T^I_{n-1}J_n,$$

whence $k = k^I$ (proposition 1°). But since

$$r + k = (\rho =) r^I + k^I + \ell^I = r_1 + k + \ell^I$$

one has

$$\ell^I = r - r_1 \quad ^{2)}.$$

Completely analoguously one can consider in the Hankel matrix H_{n-1} the successive minors (introduced in Sec. 15.2 for Toeplitz matrices) "starting" in the lefthand lower corner of the matrix

$$F_{-1} \equiv 1, F_{p-1} = \begin{vmatrix} s_{n-p} & s_{n-p+1} & \cdots & s_{n-1} \\ s_{n-p+1} & s_{n-p+2} & \cdots & s_n \\ \cdot & \cdot & \cdots & \cdot \\ s_{n-1} & s_n & \cdots & s_{n+p-2} \end{vmatrix}$$
$$(p = 1, 2, \cdots, n),$$

which in the present case, because of the symmetry of the Hankel matrix relative to the main diagonal, coincide with the corresponding E_{p-1}:

$$F_{p-1} = E_{p-1} \quad (p = 0, 1, \cdots, n).$$

But after rearrangement of the rows of the matrix H_{n-1} in reverse order, i.e., at transition to the Toeplitz matrix

$$T^{\vdash\dashv}_{n-1} = J_n H_{n-1},$$

the minors F_{p-1} $(p = 0, 1, \cdots, n)$ turn (up to the signs) into the corresponding successive principal minors of the matrix $T^{\vdash\dashv}_{n-1}$, whence follows that in the $(r^{\vdash\dashv}, k^{\vdash\dashv}, \ell^{\vdash\dashv})$-characteristic of this matrix the number $r^{\vdash\dashv} = r_1$. Moreover, proposition 2° states (if one considers (17.8)) that $\ell^{\vdash\dashv} = k$, so that

$$r + k \ (=\rho) = r^{\vdash\dashv} + k^{\vdash\dashv} + \ell^{\vdash\dashv} = r_1 + k^{\vdash\dashv} + k$$

and

$$k^{\vdash\dashv} = r - r_1.$$

Thus is proved the proposition

3°. *If for the Hankel matrix* H_{n-1}, *besides its* (r,k)*-characteristic, also the quantity* r_1 *is introduced, given by the relations* (17.17) *and* (17.18), *then for the transformed (Toeplitz) matrices*

$$T^{I} = H_{n-1}J_n \quad \text{and} \quad T^{\vdash\!\dashv} = J_n H_{n-1}$$

the respective (r^I, k^I, ℓ^I)- *and* $(r^{\vdash\!\dashv}, k^{\vdash\!\dashv}, \ell^{\vdash\!\dashv})$-*characteristics can be calculted from the formulae*

$$r^I = r_1,\ k^I = k,\ \ell^I = r - r_1 \quad \text{and} \quad r^{\vdash\!\dashv} = r_1,\ k^{\vdash\!\dashv} = r - r_1,\ \ell^{\vdash\!\dashv} = k \qquad (17.19)$$

respectively.

REMARK. Formulae (17.19) indicate also that (in the notation used in proposition 3°)

$$r^I = r^{\vdash\!\dashv},\ k^I = \ell^{\vdash\!\dashv},\ \ell = k^{\vdash\!\dashv}. \qquad (17.2o)$$

But this was clear before, since (see 17.4)

$$T^{\vdash\!\dashv} = (T^I)^t$$

(cf. exercise 7 to § 14).

17.4 The arguments, applied in Sec. 17.2 and 17.3, suggest that the theory of the characteristics of Hankel- and Toeplitz matrices, developed in §§ 1o and 14, respectively, is not the only possible one.[3)]

Indeed, propositions 1° and 2° allow us to consider for the Toeplitz matrix T_{n-1} instead of the usual (r,k,ℓ)-characteristic two other characteristics of it, namely the pairs of numbers (r^I, k^I) and $(r^{\vdash\!\dashv}, k^{\vdash\!\dashv})$, respectively, defined through these propositions. Starting with any of these pairs one can develop a theory of extension of the matrices T_{n-1} (and their "corner blocks") "to the left and down" or "to the right and up" and a theory of the rank (ρ) (for example, the rule for the construction of a nonzero minor of order ρ - cf. Lemma 11.1) similar to that, which was established in §§ 9-11 for Hankel matrices.

In turn, one can for a given Hankel matrix H_{n-1}, instead of its (r,k,)-characteristic, consider any of the triplets

$$(r^I, k^I, \ell^I) \quad \text{and} \quad (r^{\vdash\!\dashv}, k^{\vdash\!\dashv}, \ell^{\vdash\!\dashv}) \qquad (17.21)$$

defined in proposition 3°, and on the basis of each of these construct a theory with §§ 13-15 as model. To this end, clearly, it will again be necessary to consider the extensions of the matrix H_{n-1} (and its corner blocks), but now not "to the right and down" but "to the left and down" and "to the right and up" respectively.

We note once again, that components of the characteristics (17.21) satisfy the relations (17.2o), i.e., in fact the characteristics differ only from each other in the mutual permutation of the second and the third components (see the remark to proposition 3°).

In order to obtain complete symmetry in the results it remains to note that so for we have only used three of the four corners of both

Hankel and Toeplitz matrices for the introduction of characteristics. In order to fill out this gap we consider for the Hankel matrix

$$H_{n-1} = \left\| \begin{array}{c|cccc|c|c} s_o & s_1 & \cdots & s_{n-3} & s_{n-2} & s_{n-1} \\ s_1 & s_2 & \cdots & s_{n-2} & s_{n-1} & s_n \\ \cdot & \cdot & \cdots & \cdot & \cdot & \cdot \\ s_{n-3} & s_{n-2} & \cdots & s_{2n-6} & s_{2n-5} & s_{2n-4} \\ s_{n-2} & s_{n-1} & \cdots & s_{2n-5} & s_{2n-4} & s_{2n-3} \\ s_{n-1} & s_n & \cdots & s_{2n-4} & s_{2n-3} & 2_{2n-2} \end{array} \right\|$$

its successive minors

$$G_{-1} \equiv 1, G_{p-1} = \left| \begin{array}{cccc} s_{2n-p} & \cdots & s_{2n-p-2} & s_{2n-p-1} \\ \cdot & \cdots & \cdot & \cdot \\ s_{2n-p-2} & \cdots & s_{2n-4} & s_{2n-3} \\ s_{2n-p-1} & \cdots & s_{2n-3} & s_{2n-2} \end{array} \right|$$

$$(p = 1,2,\cdots,n),$$

moving "to the left and up" from the lower righthand corner of the matrix and we define the nonnegative integer ${}^{t}r$ through the relations

$$G_{{}^{t}r-1} \neq 0, \; G_{p-1} = 0 \quad (p > {}^{t}r).$$

It is clear, that

$$0 \leqq {}^{t}r \leqq \rho,$$

where ρ is the rank of the matrix H_{n-1}, and that the pair

$$({}^{t}r, {}^{t}k) \quad ({}^{t}k = \rho - {}^{t}r)$$

itself represents the usual characteristic (in the sense of § 1o) of the Hankel matrix

$${}^{t}(H_{n-1}) = \left\| \begin{array}{ccccc} s_{2n-2} & s_{2n-3} & \cdots & s_n & s_{n-1} \\ s_{2n-3} & s_{2n-4} & \cdots & s_{n-1} & s_{n-2} \\ \cdot & \cdot & \cdots & \cdot & \cdot \\ s_n & s_{n-1} & \cdots & s_2 & s_1 \\ s_{n-1} & s_{n-2} & \cdots & s & s_o \end{array} \right\|$$

the "antitranspose" (i.e. the mirror image with respect to the auxiliary diagonal) related to the matrix H_{n-1}, which, incidentally, also explains our notation.

But one sees easily, that the transition from H_{n-1} to ${}^{t}(H_{n-1})$ can, in the present case, be realized with aid of the matrix J_n (see (17.5)) in the follwing way:

$${}^{t}(H_{n-1}) = J_n H_{n-1} J_n .$$

This allows us to compare, with the help of the proposition 2° and 3°, the usual (r,k)-characteristic of the matrix H_{n-1} with its $({}^t r, {}^t k)$-characteristic. To this end we consider at first the "intermediate" Toeplitz matrix

$$T^I_{n-1} = H_{n-1} J_n$$

for which the (r^I, k^I, ℓ^I)-characteristic is determined (proposition 3°) through the formulae

$$r^I = r, \; k^I = k, \; \ell^I = r - r_1,$$

and then we note, that

$${}^t(H_{n-1}) = J_n T^I_{n-1}.$$

Now, on the basis of 2°

$${}^t r = r^I + k^I, \; {}^t k = \ell^I,$$

i.e., finally

$${}^t r = r^I + k, \; {}^t k = r - r^I \; ({}^t r + {}^t k = r + k = \rho)$$

(we recall, that $r^I = r_1$ is determined through relations (17.18)).

We sum up all the facts discovered above in the shape of two theorems.

THEOREM 17.1. *To each Hankel matrix* H_{n-1} *of rank* ρ *there correspond four sets of numbers*

$$(r,k), ({}^t r, {}^t k), (r^I, k^I, \ell^I), (r^{\vdash\dashv}, k^{\vdash\dashv}, \ell^{\vdash\dashv}),$$

of which the first two characterize respectively H_{n-1} *and its antitransposed matrix* ${}^t(H_{n-1})$ *as Hankel matrices (in the sense of the definitions in* § 10*) and the other two characterize (in the sense of the definition of* § 14*) the Toeplitz matrices* $T^I = H_{n-1} J_n$ *and* $T^{\vdash\dashv} = J_n H_{n-1}$ (see (17.15)) *respectively. The components of the mentioned characteristics are connected through the relations.*

$${}^t r = r^I + k^I, \; {}^t k = r - r^I; \; k^I = k, \; \ell^I = r - r^I;$$

$$r^{\vdash\dashv} = r^I, \; k^{\vdash\dashv} = \ell^I, \; \ell^{\vdash\dashv} = k^I;$$

$$r + k = {}^t r + {}^t k = r^I + k^I + \ell^I = r^{\vdash\dashv} + k^{\vdash\dashv} + \ell^{\vdash\dashv} = \rho.$$

THEOREM 17.2. *To each Toeplitz matrix* T_{n-1} *of rank* ρ *there correspond four sets of numbers*

$$(r,k,\ell), \; (r^t, k^t, \ell^t); \; (r^I, k^I), \; (r^{\vdash\dashv}, k^{\vdash\dashv}),$$

of which the first two characterize respectively T_{n-1} *and its transposed matrix* $(T_{n-1})^t$ *as Toeplitz matrices (in the sense of the definitions in* § 14*), and the other two characterize (in the sense of the definitions of* § 10*) the Hankel matrices* $H^I_{n-1} = T_{n-1} J_n$ *and* $H^{\vdash\dashv}_{n-1} = J_n T_{n-1}$

(see (17.5)), *respectively. The components of the mentioned characteristics are connected through the relations*

$$r^t = r,\ k^t = \ell,\ \ell^t = k;$$

$$r^I = r+\ell,\ k^I = k;\ r^H = r+k,\ k^H = \ell;$$

$$r+k+\ell = r^t+k^t+\ell^t = r^I+k^I = r^H+k^H = \rho.$$

EXAMPLES AND EXERCISES

1. We consider the Toeplitz matrix

$$T_5 = \left\| \begin{array}{c|ccccc} 1 & 1 & 1 & \beta & \gamma & \delta \\ \hline 1 & 1 & 1 & 1 & \beta & \gamma \\ 1 & 1 & 1 & 1 & 1 & \beta \\ 1 & 1 & 1 & 1 & 1 & 1 \\ 1 & 1 & 1 & 1 & 1 & 1 \\ \alpha & 1 & 1 & 1 & 1 & 1 \end{array} \right\| \quad (\alpha \neq 1,\ \beta \neq 1)$$

from exercise 2 of § 15 with the (r,k,ℓ)-characteristic $(1,1,3)$. The rearrangement of its columns in reverse order gives the Hankel matrix

$$H_5^I = T_5 J_6 = \left\| \begin{array}{cccc|cc} \delta & \gamma & \beta & 1 & 1 & 1 \\ \gamma & \beta & 1 & 1 & 1 & 1 \\ \beta & 1 & 1 & 1 & 1 & 1 \\ 1 & 1 & 1 & 1 & 1 & 1 \\ \hline 1 & 1 & 1 & 1 & 1 & 1 \\ 1 & 1 & 1 & 1 & 1 & \alpha \end{array} \right\|$$

It is not difficult to find out that its successive principal minor of order four is nonzero:

$$\begin{vmatrix} \delta & \gamma & \beta & 1 \\ \gamma & \beta & 1 & 1 \\ \beta & 1 & 1 & 1 \\ 1 & 1 & 1 & 1 \end{vmatrix} = \begin{vmatrix} \delta & \gamma & \beta & 1 \\ \gamma & \beta & 1 & 1 \\ \beta & 1 & 1 & 1 \\ 1-\beta & 0 & 0 & 0 \end{vmatrix} = (\beta-1) \begin{vmatrix} \gamma & \beta & 1 \\ \beta & 1 & 1 \\ 1 & 1 & 1 \end{vmatrix} =$$

$$= -(\beta-1)^2 \begin{vmatrix} \beta & 1 \\ 1 & 1 \end{vmatrix} = -(\beta-1)^3 \neq 0,$$

but that all minors which follow it are equal to zero. Hence $r^I = 4$ $(= r+\ell)$, and $k^I = 1$ $(=k)$ (verify this!) - in accordance to proposition 1°.

2. If one rearranges in the matrix T_5 of example 1 the rows in reverse order, then one has for the obtained Hankel matrix

$$H_5^{\vdash\dashv} = J_6 T_5 = \left\| \begin{matrix} \alpha & 1 & 1 & 1 & 1 & 1 \\ 1 & 1 & 1 & 1 & 1 & 1 \\ 1 & 1 & 1 & 1 & 1 & 1 \\ 1 & 1 & 1 & 1 & 1 & \beta \\ 1 & 1 & 1 & 1 & \beta & \gamma \\ 1 & 1 & 1 & \beta & \gamma & \delta \end{matrix} \right\|$$

as is easily seen, that $r^{\vdash\dashv} = 2$ $(= r + k)$. Verify, that $k^{\vdash\dashv} = 3$ $(=\ell)$ - in accordance to proposition 2°.

3. We consider the Hankel matrix

$$H_4 = \left\| \begin{matrix} 0 & 1 & 0 & 1 & 0 \\ 1 & 0 & 1 & 0 & 1 \\ 0 & 1 & 0 & 1 & 0 \\ 1 & 0 & 1 & 0 & 0 \\ 0 & 1 & 0 & 0 & 0 \end{matrix} \right\|$$

with the (r,k)-characteristic (2,2) (see exercise 2 to § 10). According to the definitions (17.17) and (17.18) one has for this matrix

$$E_0 = 0,\ E_1 = \begin{vmatrix} 1 & 0 \\ 0 & 1 \end{vmatrix} = 1,\quad E_2 = E_3 = E_4 = 0,$$

i.e., $r_1 = 2$.

For the Toeplitz matrix

$$T_4^{I} = H_4 J_5 = \left\| \begin{matrix} 0 & 1 & 0 & 1 & 0 \\ 1 & 0 & 1 & 0 & 1 \\ 0 & 1 & 0 & 1 & 0 \\ 0 & 0 & 1 & 0 & 1 \\ 0 & 0 & 0 & 1 & 0 \end{matrix} \right\|$$

one has $r^{I} = 2$ $(=r_1)$, $k^{I} = 2$ $(=k)$, $\ell^{I} = 0$ $(=r - r_1)$ - in agreement with proposition 3°. According to the same proposition the Toeplitz matrix

$$T_4^{\vdash\dashv} = J_5 H_4 = \left\| \begin{matrix} 0 & 1 & 0 & 0 & 0 \\ 1 & 0 & 1 & 0 & 0 \\ 0 & 1 & 0 & 1 & 0 \\ 1 & 0 & 1 & 0 & 1 \\ 0 & 1 & 0 & 1 & 0 \end{matrix} \right\|$$

has $r^{\vdash\dashv} = 2$ $(=r_1)$, $k^{\vdash\dashv} = 0$ $(=r - r_1)$, $\ell^{\vdash\dashv} = 2$ $(=k)$.

4. Find for the Hankel matrix (cf. exercise 3 of § 10)

$$H_3 = \left\| \begin{array}{cccc} 0 & 4 & 0 & 1 \\ 4 & 0 & 1 & 0 \\ 0 & 1 & 0 & 1/4 \\ 1 & 0 & 1/4 & -6 \end{array} \right\|$$

all four characteristics which were mentioned in Theorem 17.1.

Solution. $(r,k) = (2,1), ({}^t r, {}^t k) = (3,0)$,
$(r^I, k^I, \ell^I) = (2,1,0)$, $(r^{\vdash\dashv}, k^{\vdash\dashv}, \ell^{\vdash\dashv}) = (2,0,1)$.

5. Find all four characteristics (see Theorem 17.2) of the Toeplitz matrix

$$T_2 = \left\| \begin{array}{ccc} i & 2 & 4i \\ -1/2 & i & 2 \\ -i/4 & -1/2 & i \end{array} \right\|$$

Solution. $(r,k,\ell) = (1,0,1)$, $(r^t, k^t, \ell^t) = (1,1,0)$;
$(r^I, k^I) = (2,0), (r^{\vdash\dashv}, k^{\vdash\dashv}) = (1,1)$.

NOTES.

1) We recall, that already Frobenius [19] payed attention to the role of the constant r for the first time (cf. note 2) to § 1o).

2) From this formula follows, in particular, that $r \geqq r_1$ (since $\ell^I \geqq 0$). However (if one considers that $r_1 = r^I$) this fact was already noted above in the Corollary to propositions 1° and 2° (we just recall, that in the present case the notations r and r^I have interchanged their roles).

3) The fact, namely, that it initially appeared in that shape (cf.[29, 3o,27]), psychologically stems from a tradition which goes back to Frobenius.

§ 18. INVERSION OF TOEPLITZ AND HANKEL MATRICES.

18.1. It is commonly known that the problem of the inversion of matrices, i.e., of determing for a given nonsingular matrix A ($|A| \neq 0$) its inverse matrix A^{-1}, is one of the central and difficult problems in matrix theory. The importance of its solution if only for individual classes of matrices both on the level of theory as well as in applied problems (the solution of systems of linear equations) is undoubted. Unfortunately, in spite of extensive literature, dealing with this question [1], the problem requires in many of its aspects a further and profound investigation.

In this section we explain at first recently obtained results of I.C. Gohberg and A.A. Semencul [25], and also of I.C. Gohberg and N.Ya. Krupnik [24] on the inversion of Toeplitz matrices T_n (for our convenience we work here, following the mentioned autors, with matrices of order $n+1$). Then we use these results for the solution by means of the methods of § 17 of the problem of the inversion of Hankel matrices H_n.

18.2. Central in the theory presented below is

THEOREM 18.1 (GOHBERG AND SEMENCUL). *If the Toeplitz matrix* $T_n = \|c_{p-q}\|_{p,q=0}^{n}$ *is such that each of the systems of equations*

$$\sum_{q=0}^{n} c_{p-q} x_q = \delta_{po} \qquad (p = 0,1,\cdots,n) \tag{18.1}$$

$$\sum_{q=0}^{n} c_{p-q} y_{q-n} = \delta_{pn} \qquad (p = 0,1,\cdots,n) \tag{18.2}$$

is solvable and satisfies the condition $x_o \neq 0$, *then the matrix* T_n *is nonsingular and its inverse matrix* T_n^{-1} *is constructed through the formula* 2)

$$T_n^{-1} = x_o^{-1} \left\{ \left\| \begin{matrix} x_o & 0 & \cdots & 0 \\ x_1 & x_o & \cdots & 0 \\ \cdot & \cdot & \cdots & \cdot \\ x_n & x_{n-1} & \cdots & x_o \end{matrix} \right\| \left\| \begin{matrix} y_o & y_{-1} & \cdots & y_{-n} \\ 0 & y_o & \cdots & y_{-n+1} \\ \cdot & \cdot & \cdots & \cdot \\ 0 & 0 & \cdots & y_o \end{matrix} \right\| - \right.$$

$$\left. - \left\| \begin{matrix} 0 & 0 & 0 & \cdots & 0 & 0 \\ y_{-n} & 0 & 0 & \cdots & 0 & 0 \\ y_{-n+1} & y_{-n} & 0 & \cdots & 0 & 0 \\ \cdot & \cdot & \cdot & \cdots & \cdot & \cdot \\ y_{-1} & y_{-2} & y_{-3} & \cdots & y_{-n} & 0 \end{matrix} \right\| \left\| \begin{matrix} 0 & x_n & x_{n-1} & \cdots & x_1 \\ 0 & 0 & x_n & \cdots & x_2 \\ \cdot & \cdot & \cdot & \cdots & \cdot \\ 0 & 0 & 0 & \cdots & x_n \\ 0 & 0 & 0 & \cdots & 0 \end{matrix} \right\| \right\} \tag{18.3}$$

PROOF. For $n = 0$ is the statement of theorem trivial (we just note, that in formula (18.3) the subtrahend between the braces is absent in this case). So, let $n > 0$. We prove the nonsingularity (invertibility) of the matrix T_n by contradiction.

Let $|T_n| = \det T_n = 0$. Then the rows $\Gamma_p = (c_p, c_{p-1}, \cdots, c_{p-n})$ of the matrix T_n $(p = 0,1,\cdots,n)$ are linearly dependent. Since the system (18.1) is solvable, and in its first equation (for $p = 0$) the righthand side is equal to one, the first row Γ_o of the matrix T_n cannot consist of zeros only: $\Gamma_0 \neq 0$. Thence and from the linear dependence of the rows follows, that at least one of the remaining rows $\Gamma_1, \Gamma_2, \cdots, \Gamma_n$ can be expressed linearly through the preceding rows. We denote with m (≥ 1) the largest of the numbers, for which

$$\Gamma_m = \sum_{p=0}^{m-1} \alpha_p \Gamma_p ,$$

where $\alpha_0, \alpha_1, \cdots, \alpha_{m-1}$ are some complex numbers. We shall show, that $m = n$. Let $m < n$. Since

$$c_{m-q} = \sum_{p=0}^{m-1} \alpha_p c_{p-q} \qquad (q = 0, 1, \cdots, n),$$

we obtain, after performing the index transformation $p = p'-1$, $q = q'-1$ and then returning to the previous notations, that

$$c_{m+1-q} = \sum_{p=1}^{m} \alpha_{p-1} c_{p-q} \qquad (q = 1, 2, \cdots, n). \tag{18.4}$$

Because of the solvability of the equation system (18.1) we have for each of its solutions $(x_0, x_1, \cdots, x_n)$, taking into account that $2 \leq m+1 \leq n$:

$$0 = \delta_{m+1,0} = \sum_{q=0}^{n} c_{m+1-q} x_q = \sum_{q=0}^{n} c_{m-1-q} x_q - \sum_{p=1}^{m} \alpha_{p-1} \delta_{po} =$$

$$= \sum_{q=0}^{n} c_{m+1-q} x_q - \sum_{p=1}^{m} \alpha_{p-1} \sum_{q=0}^{n} c_{p-q} x_q =$$

$$= \left(c_{m+1} - \sum_{p=1}^{m} \alpha_{p-1} c_p \right) x_0 + \sum_{q=1}^{n} \left(c_{m+1-q} - \sum_{p=1}^{m} \alpha_{p-1} c_{p-q} \right) x_q .$$

Because of identity (18.4) and the inequality $x_0 \neq 0$ follows thence, that

$$c_{m+1} = \sum_{p=1}^{m} \alpha_{p-1} c_p . \tag{18.5}$$

Now combining (18.4) and (18.5) we obtain

$$\Gamma_{m+1} = \sum_{p=1}^{m} \alpha_{p-1} \Gamma_p$$

but this contradicts the choice of the number m.

So, $m = n$, and, consequently, the final row Γ_n of the matrix T_n can be expressed linearly through the preceding rows. But this, in turn, contradicts the solvability of the system of equations (18.2), in which all righthand sides, except the last one, are equal to zero, and the last one equals one.

Thus, we have proved that the matrix T_n is invertible. Hence follows, that the solutions $\{x_0, x_1, \cdots, x_n\}$ and $\{y_{-n}, y_{-n+1}, \cdots, y_0\}$ of the systems (18.1) and (18.2), respectively, are uniquely defined. In particular,

$$x_0 = y_0 = \frac{|T_{n-1}|}{|T_n|} \qquad (18.6)$$

where $|T_{n-1}|$ is the determinant of the truncated matrix $T_{n-1} = \| c_{p-q} \|_{p,q=0}^{n-1}$.

We denote through $B = \| b_{jk} \|_{j,k=0}^{n}$ the matrix on the righthand side of identity (18.3). Using the multiplication rule for matrices, one easily verifies that the elements b_{jk} can be expressed through the formula

$$b_{jk} = x_0^{-1} \sum_{s=0}^{\min(j,k)} (x_{j-s} y_{s-k} - y_{-n-1+j-s} x_{n+1+s-k}) \qquad (18.7)$$
$$(j,k = 0,1,\cdots,n)$$

where for the uniformity of notation we have used $x_{n+1} = y_{-n-1} = 0$.

Now we transform this expression for b_{jk} in the following way. We isolate from the sum on the righthand side the summand, corresponding to the value $s = 0$ of the summation index, and in the remaining sum we carry out the index transformation $s = s' - 1$. Then we obtain

$$b_{jk} = x_0^{-1} (x_j y_{-k}) - y_{-n-1+j} x_{n+1-k}) + b_{j-1,k-1} \quad (j,k = 1,2,\cdots.n). \qquad (18.8)$$

Moreover, substituting in formula (18.7) for b_{jk} the index $j = 0$ $(k = 0)$ we have, with regard to identity (18.6)

$$b_{0k} = y_{-k}, \ b_{k0} = x_k \qquad (k = 0,1,2,\cdots,n). \qquad (18.9)$$

Then, substituting $j = n$, we find

$$b_{nk} = x_0^{-1} \sum_{s=0}^{k} (x_{n-s} y_{s-k} - y_{-1-s} x_{n+1+s-k}) =$$
$$= x_0^{-1} (x_n y_{-k} - y_{-1} x_{n+1-k} + x_{n-1} y_{1-k} - y_{-2} x_{n+2-k} + \cdots$$
$$\cdots + x_{n-k+2} y_{-2} - y_{-k+1} x_{n-1} + x_{n-k+1} y_{-1} - y_{-k} x_n +$$
$$+ x_{n-k} y_0 - y_{-1-k} x_{n+1}) = x_{n-k} \qquad (k = 0,1,2,\cdots,n),$$

where we have used identity (18.6) and $x_{n+1} = 0$. This result and its analogon (obtained for $k = n$) we combine in the shape of the formulae

$$b_{nk} = x_{n-k}, \ b_{kn} = y_{k-n} \quad (k = 0,1,\cdots,n). \qquad (18.1o)$$

We must verfiy that $T_n B = E$. At first we show, that the matrix

$$A = T_n B = \| a_{pq} \|_{p,q=0}^{n}$$

is a Toeplitz matrix. Indeed, by means of identity (18.8) we have for $p,q = 1,2,\cdots,n$

$$a_{pq} = \sum_{s=0}^{n} c_{p-s} b_{sq}$$

$$= c_p b_{oq} + \sum_{s=1}^{n} c_{p-s} [b_{s-1,q-1} + x_o^{-1}(x_s y_{-q} - y_{-n-1+s} x_{n+1-q})].$$

Converting this expression with regard to the formulae (18.6), (18.9) and (18.1o), we obtain

$$a_{pq} = x_o^{-1} c_p b_{oq} x_o + \sum_{s=0}^{n} c_{p-s-1} b_{s,q-1} - x_o^{-1} b_{n,q-1} c_{p-1-n} y_o +$$

$$+ x_o^{-1} b_{oq} \sum_{s=1}^{n} c_{p-s} x_s - x_o^{-1} b_{n,q-1} \sum_{s=0}^{n-1} c_{p-1-s} y_{s-n}.$$

Combining here the first summand with the fourth and the third with the fifth, we have

$$a_{pq} = \sum_{s=0}^{n} c_{p-1-s} b_{s,q-1} + x_o^{-1} b_{oq} \sum_{s=0}^{n} c_{p-s} x_s - x_o^{-1} \sum_{s=0}^{n} c_{(p-1)-s} y_{s-n},$$

i.e., taking into account (18.1) and (18.2),

$$a_{pq} = a_{p-1,q-1} \qquad (p,q = 1,2,\cdots,n).$$

But this means, that A is a Toeplitz matrix.

It remains to note, that with help of the identities (18.9) and (18.1o) in combination with (18.1) and (18.2) respectively

$$a_{po} = \sum_{s=0}^{n} c_{p-s} b_{so} = \sum_{s=0}^{n} c_{p-s} x_s = \delta_{po},$$

$$a_{pn} = \sum_{s=0}^{n} c_{p-s} b_{sn} = \sum_{s=0}^{n} c_{p-s} y_{s-n} = \delta_{pn},$$

i.e., the first and the last column of the matrix A coincide with the corresponding columns of the unit matrix $E = \|\delta_{pq}\|_{p,q=0}^{n}$. Hence follows, because of the Toeplitz structure of the matrix A, that these two matrices coincide completely: A = E.

The theorem is proved.

The condition $x_o \neq 0$ in Theorem 18.1 is essential - cf. for this the exercises 3-6 at the end of the present section.

18.3. It is easy to see (see (18.6)) that, if the conditions of Theorem 18.1 are satisfied, then also the "truncated" matrix

$$T_{n-1} = \|c_{p-q}\|_{p,q=0}^{n}$$

is invertible. Moreover, it turns out, that its inverse matrix T_{n-1}^{-1} can be constructed from the solutions of the equation systems (18.1) and (18.2) as well.

THEOREM 18.2. (GOHBERG AND SEMENCUL). *If there exist solutions* $\{x_o, x_1, x_2, \cdots, x_n\}$ *and* $\{y_{-n}, y_{-n+1}, \cdots, y_o\}$ *of the systems* (18.1) *and* (18.2) *and the condition* $x_o \neq 0$ *is satisfied, then the matrix* $T_{n-1} = \|c_{p-q}\|_{p,q=o}^{n-1}$ *is invertible and its inverse is constructed by the formula*

$$T_{n-1}^{-1} = x_o^{-1} \left\{ \begin{Vmatrix} x_o & 0 & \cdots & 0 \\ x_1 & x_o & \cdots & 0 \\ \cdot & \cdot & \cdots & \cdot \\ x_{n-1} & x_{n-2} & \cdots & x_o \end{Vmatrix} \begin{Vmatrix} y_o & y_{-1} & \cdots & y_{-n+1} \\ 0 & y_o & \cdots & y_{-n+2} \\ \cdot & \cdot & \cdots & \cdot \\ 0 & 0 & \cdots & y_o \end{Vmatrix} - \right.$$

$$\left. - \begin{Vmatrix} y_{-n} & 0 & \cdots & 0 \\ y_{-n+1} & y_{-n} & \cdots & 0 \\ \cdot & \cdot & \cdots & \cdot \\ y_{-1} & y_{-2} & \cdots & y_{-n} \end{Vmatrix} \begin{Vmatrix} x_n & x_{n-1} & \cdots & x_1 \\ 0 & x_n & \cdots & x_2 \\ \cdot & \cdot & \cdots & \cdot \\ 0 & 0 & \cdots & x_n \end{Vmatrix} \right\} . \qquad (18.11)$$

PROOF. Let us denote through $D = \|d_{jk}\|_{j,k=1}^{n-1}$ the matrix on the right-hand side of (18.11). A direct calculation gives (cf. (18.7))

$$d_{jk} = x_o^{-1} \sum_{s=o}^{\min(j,k)} (x_{j-s}y_{s-k} - y_{-n+j-s}x_{n+s-k}) \qquad (18.12)$$

for all $j,k = 0,1,\cdots,n-1$. Thence we find, using the identity $x_o = y_o$ (see (18.6))

$$d_{ko} = x_k - x_o^{-1} x_n y_{k-n}, \quad d_{ok} = y_{-k} - x_o^{-1} y_{-n} x_{n-k} \quad (k = 0,1,\cdots.n-1) \qquad (18.13)$$

$$d_{jk} = d_{j-1,k-1} + x_o^{-1}(x_j y_{-k} - y_{-n+j} x_{n-k}) \qquad (j,k = 1,2,\cdots,n-1) \qquad (18.14)$$

(cf. the analoguous calculations for b_{jk} in the proof of Theorem 18.1).

Further, we have from identity (18.12) for $j = n-1$ that for $k = 0,1,\cdots,n-1$

$$d_{n-1,k} = x_o^{-1} \sum_{s=o}^{k} (x_{n-1-s}y_{s-k} - y_{-1-s}x_{n+s-k}) =$$

$$= x_o^{-1}(x_{n-1}y_{-k} - y_{-1}x_{n-k} + x_{n-2}y_{1-k} - y_2 x_{n+1-k} + \cdots$$

$$\cdots + x_{n-k}y_{-1} - y_{-k}x_{n-1} + x_{n-1-k}y_o - y_{1-k}x_n) =$$

$$= x_{n-k-1} - x_o^{-1} x_n y_{-k-1}.$$

We write down this result together with its analogon, obtained from

(18.12) for k=n-1, seperately:

$$d_{n-1,k} = x_{n-k-1} - x_o^{-1} x_n y_{-k-1}; \quad d_{k,n-1} = y_{-n+k+1} - x_o^{-1} y_{-n} x_{k+1}$$
$$(k = 0,1,\cdots,n-1). \qquad (18.15)$$

Now we convince ourselves,that the first and the last column of the matrix $T_{n-1}D \equiv \|e_{jk}\|_{j,k=o}^{n-1}$ coincide with the corresponding columns of the unit matrix $E = \|\delta_{jk}\|_{j,k=o}^{n-1}$. Indeed, taking into account (18.13), (18.1), (18.2) and (18.6) we have

$$e_{jo} = \sum_{k=o}^{n-1} c_{j-k} d_{ko} = \sum_{k=o}^{n-1} c_{j-k} x_k - x_o^{-1} x_n \sum_{k=o}^{n-1} c_{j-k} y_{k-n} =$$
$$= \sum_{k=o}^{n} c_{j-k} x_k - c_{j-n} x_n - x_o^{-1} x_n \sum_{k=o}^{n} c_{j-k} y_{k-n} + x_o^{-1} x_n c_{j-n} y_o =$$
$$= \delta_{jo} - x_o^{-1} x_n \delta_{jn} = \delta_{jo} \qquad (j = 0,1,\cdots,n-1).$$

Analoguously (using formulae (18.15) as well) we obtain

$$e_{j,n-1} = \sum_{k=o}^{n-1} c_{j-k} d_{k,n-1} = \sum_{k=o}^{n-1} c_{j-k} (y_{-n+k+1} - x_o^{-1} y_{-n} x_{k+1}) =$$
$$= \sum_{k=1}^{n} c_{j+1-k} y_{k-n} - x_o^{-1} y_{-n} \sum_{k=1}^{n} c_{j+1-k} x_k =$$
$$= \sum_{k=o}^{n} c_{j+1-k} y_{k-n} - c_{j+1} y_{-n} - x_o^{-1} y_n \sum_{k=o}^{n} c_{j+1-k} x_k + x_o^{-1} y_{-n} c_{j+1} x_o =$$
$$= \delta_{j+1,n} - x_o^{-1} y_{-n} \delta_{j+1,o} = \delta_{j,n-1} \qquad (j = 0,1,\cdots,n-1).$$

For the completion of the proof it remains to prove, that the matrix $\|e_{jk}\|_{j,k=o}^{n-1}$ is a Toeplitz matrix. But from (18.14), (18.15) and (18.13) follows, that

$$e_{jk} = \sum_{s=o}^{n-1} c_{j-s} d_{sk} =$$
$$= c_j d_{ok} + \sum_{s=1}^{n-1} c_{j-s} [d_{s-1,k-1} + x_o^{-1} (x_s y_{-k} - y_{-n+s} x_{n-k})] =$$
$$= \sum_{s=o}^{n-2} c_{j-1-s} d_{s,k-1} + c_j d_{ok} + x_o^{-1} y_{-k} \sum_{s=1}^{n-1} c_{j-s} x_s -$$
$$- x_o^{-1} x_{n-k} \sum_{s=1}^{n-1} c_{j-s} y_{s-n} =$$
$$= \sum_{s=o}^{n-1} c_{j-1-s} d_{s,k-1} - c_{j-n} d_{n-1,k-1} + c_j d_{ok} +$$

$$+ x_o^{-1} y_{-k} \sum_{s=1}^{n-1} c_{j-s} x_s - x_o^{-1} x_{n-k} \sum_{s=1}^{n-1} c_{j-s} y_{s-n} =$$

$$= \sum_{s=0}^{n-1} c_{j-1-s} d_{s,k-1} - c_{j-n}(x_{n-k} - x_o^{-1} x_n y_{-k}) +$$

$$+ c_j (y_{-k} - x_o^{-1} y_{-n} x_{n-k}) + x_o^{-1} y_{-k} \sum_{s=1}^{n-1} c_{j-s} x_s -$$

$$- x_o^{-1} x_{n-k} \sum_{s=0}^{n-1} c_{j-s} y_{s-n} = e_{j-1,k-1} + x_o^{-1} y_{-k} \sum_{s=0}^{n} c_{j-s} x_s -$$

$$- x_o^{-1} x_{n-k} \sum_{s=0}^{n} c_{j-s} y_{s-n},$$

whence, because of (18.1) and (18.2) follows

$$e_{jk} = e_{j-1,k-1} \qquad (j,k = 1,2,\cdots,n-1).$$

The theorem is proved.

18.4. Turning to the problem of the inversion of arbitrary Toeplitz matrices T_n, we note that if its invertibility ($|T_n| \neq 0$) is known *a priori*, then the equation systems (18.1) and (18.2) are automatically uniquely solvable, and the sole restriction, imposed by Theorem 18.1 for the construction of the inverse T_n^{-1} from these solutions consists of the conditon $x_o \neq 0$, or, which the same (see (18.6)),

$$|T_{n-1}| \neq 0.$$

However, Theorem 18.2 gives the possibility to avoid the last restriction, i.e., to construct the matrix T_n^{-1} even in the case where $|T_{n-1}| = 0$.

To this end we consider the matrix

$$T_{n+1}(\zeta) = \left\| \begin{matrix} c_o & c_{-1} & \cdots & c_{-n} & \zeta \\ c_1 & c_o & \cdots & c_{-n+1} & c_{-n} \\ \cdot & \cdot & \cdots & \cdot & \cdot \\ c_n & c_{n-1} & \cdots & c_o & c_{-1} \\ c_{n+1} & c_n & & c_1 & c_o \end{matrix} \right\|,$$

where c_{n+1} is some fixed number, and ζ a complex parameter.

1°. *If the matrix* T_n *is invertible, then the matrix* $T_{n+1}(\zeta)$ *is also invertible for all values of* ζ *with, possibly, one exception.*

In fact, this proposition was already established in Chapter III in the proof of Theorem 13.1. Indeed, if one increases by one the order of all matrices and determinants considered there, then in the thus revised notations (13.4) the equation $|T_{n+1}(\zeta)| = 0$ can be rewritten for

$|T_{n-1} \neq 0|$ (cf. (13.5)) in the shape

$$a(c_{n+1})b(\zeta) - |T_n|^2 = 0,$$

where $a(\zeta)$ and $b(\zeta)$ are linear functions where the coefficients for ζ are in each equal to $|T_{n-1}|(\neq 0)$. Hence follows that the equation $|T_{n+1}(\zeta)| = 0$ has not more than one root $\zeta = \zeta_o$.

Now, if $|T_{n-1}| = 0$, then the functions $a(\zeta)$ and $b(\zeta)$ are nonzero constants $a(o)$ and $b(o)$, and the equation $|T_{n+1}(\zeta)| = 0$ takes the shape (cf. (13.6))

$$b(o)c_{n+1} + a(o)\zeta + C = 0$$

where C is some constant, i.e, this time (since $a(o) \neq 0$) it has exactly one root.

Thus, for all ζ, with exception, perhaps, of one, the matrix $T_{n+1}(\zeta)$ is invertible and it satisfies all conditions of Theorem 18.2 (since $|T_n| \neq 0$). But then the matrix T_n can be inverted by the rule, stated in this theorem.

18.5. With the help of Theorem 18.1 one can also find another method for the inversion of Toeplitz matrices.

THEOREM 18.3 (GOHBERG AND KRUPNIK). *If the Toeplitz matrix* $T_n = \|c_{p-q}\|_{p,q=o}^{n}$ *is such, that each of the systems*

$$\sum_{q=0}^{n} c_{p-q}x_q = \delta_{po} \qquad (p = 0,1,\cdots,n) \tag{18.16}$$

$$\sum_{q=0}^{n} c_{p-q}z_q = \delta_{p1} \qquad (p = 0,1,\cdots,n) \tag{18.17}$$

is solvable, and the condition $x_n \neq 0$ *is satisfied, then the matrix* T_n *is invertible and its inverse is constructed by the formula* [3)]

$$T_n^{-1} = x_n^{-1} \left\{ \begin{Vmatrix} x_ox_n & x_ox_{n-1} & \cdots & x_o^2 \\ x_1x_n & x_1x_{n-1} & \cdots & x_1x_o \\ \cdot & \cdot & \cdots & \cdot \\ x_n^2 & x_nx_{n-1} & \cdots & x_nx_o \end{Vmatrix} + \right.$$

$$+ \begin{Vmatrix} z_o & 0 & \cdots & 0 \\ z_1 & z_o & \cdots & 0 \\ \cdot & \cdot & \cdots & \cdot \\ z_n & z_{n-1} & \cdots & z_o \end{Vmatrix} \begin{Vmatrix} 0 & x_n & x_{n-1} & \cdots & x_2 & x_1 \\ 0 & 0 & x_n & \cdots & x_3 & x_2 \\ \cdot & \cdot & \cdot & \cdots & \cdot & \cdot \\ 0 & 0 & 0 & \cdots & 0 & x_n \\ 0 & 0 & 0 & \cdots & 0 & 0 \end{Vmatrix} -$$

$$- \begin{Vmatrix} x_o & 0 & \cdots & 0 \\ x_1 & x_o & \cdots & 0 \\ \cdot & \cdot & \cdots & \cdot \\ x_n & x_{n-1} & \cdots & x_o \end{Vmatrix} \begin{Vmatrix} 0 & z_n & z_{n-1} & \cdots & z_2 & z_1 \\ 0 & 0 & z_n & \cdots & z_3 & z_2 \\ \cdot & \cdot & \cdot & \cdots & \cdot & \cdot \\ 0 & 0 & 0 & \cdots & 0 & z_n \\ 0 & 0 & 0 & \cdots & 0 & 0 \end{Vmatrix} \Bigg\} \qquad (18.18)$$

PROOF. For $n = 0$ the theorem is trivial, therefore, let $n \geqq 1$. Again we prove the invertibility of the matrix T_n by contradiction. Let the rows $\Gamma_q = (c_q, c_{q-1}, \cdots, c_{q-n})$ $(q = 0,1,\cdots,n)$ of the matrix T_n be linearly dependent and let m be the least number such that the row Γ_m is a linear combination of the rows $\Gamma_{m+1}, \Gamma_{m+2}, \cdots, \Gamma_n$. From the existence of solutions of the systems (18.6) and (18.7) follows, that neither of the first two rows Γ_0 and Γ_1 is a linear combination of the remaining rows. So, $m \geqq 2$. For this m is, by assumption

$$\Gamma_m = \sum_{p=m+1}^{n} \alpha_p \Gamma_m$$

or

$$c_{m-q} = \sum_{p=m+1}^{n} \alpha_p c_{p-q} \quad (q = 0,1,\cdots,n). \qquad (18.19)$$

Now we extract from the system (18.16) the equations, corresponding to the values $p = m,m+1,\cdots,n-1$ of the index (for which, consequently, $\delta_{po} = 0$), having multiplicated each of them with α_{p+1}:

$$\alpha_{p+1} \sum_{q=0}^{n} c_{p-q} x_q = 0 \quad (p = m,m+1,\cdots,n-1).$$

Having added all these identities, we obtain

$$0 = \sum_{p=m}^{n-1} \alpha_{p+1} \sum_{q=0}^{n} c_{p-q} x_q = \sum_{q=0}^{n} \left(\sum_{p=m}^{n-1} \alpha_{p+1} c_{p-q} \right) x_q =$$

$$= \sum_{q=0}^{n} \left(\sum_{p=m+1}^{n} \alpha_p c_{p-q-1} \right) x_q = \sum_{q=0}^{n-1} c_{m-q-1} x_q + x_n \sum_{p=m+1}^{n} \alpha_p c_{p-n-1}.$$

We make it clear, that in the last of these identity transitions we used (for $q = 0,1,\cdots,n-1$) relations (18.19) (for $q = n$ one cannot use these because of the index shift of one unit, and therefore the respective summand is written sperately). Now we note, that with help of the relation (see (18.16))

$$\sum_{q=0}^{n} c_{m-q-1} x_q = \delta_{m-1,0} = 0$$

(we recall that $m \geqq 2$) the identity obtained above can be rewritten as

$$-c_{m-n-1}x_n + x_n \sum_{p=m+1}^{n} \alpha_p c_{p-n-1} = 0,$$

and since $x_n \neq 0$, we obtain at last

$$c_{m-n-1} = \sum_{p=m+1}^{n-1} \alpha_p c_{p-n-1}. \qquad (18.20)$$

From identities (18.19) and (18.20) we find

$$c_{m-q-1} = \sum_{p=m}^{n-1} \alpha_{p+1} c_{p-q} \quad (q = 0,1,\cdots,n)$$

i.e.

$$\Gamma_{m-1} = \sum_{p=m}^{n-1} \alpha_{p+1} \Gamma_p ,$$

which contradicts the assumtion regarding the number m(its minimality). So, the matrix T_n is invertible.

For the derivation of formula (18.18) we use Theorem 18.1, applying it to the perturbation

$$T_n(\varepsilon) = T_n - \varepsilon E$$

of the matrix T_n, where ε is a real parameter. We denote the entries of the Toeplitz matrix $T_n(\varepsilon)$ by $c_p(\varepsilon)$ $(p = 0, \pm 1, \cdots, \pm n)$. We choose $\delta > 0$ such, that for $0 < \varepsilon < \delta$ the conditions

$$|T_n(\varepsilon)| = \det T_n(\varepsilon) \neq 0, \quad |T_{n-1}(\varepsilon)| = \det T_{n-1}(\varepsilon) =$$

$$= \det \|c_{p-q}(\varepsilon)\|_{p,q=0}^{n-1} \neq 0 \qquad (18.21)$$

are satisfied. Such a choice is possible, since the polynomials $|T_n(\varepsilon)|$ and $|T_{n-1}(\varepsilon)|$ have only finitely many roots (cf. Sec. 1.2). Conditions (18.21) guarantee for arbitrary ε $(0 < \varepsilon < \delta)$ that the systems

$$\sum_{q=0}^{n} c_{p-q}(\varepsilon)x_q(\varepsilon) = \delta_{p0} \quad (p = 0,1,\cdots,n),$$

$$\sum_{q=0}^{n} c_{q-p}(\varepsilon)y_{-q}(\varepsilon) = \delta_{p0} \quad (p = 0,1,\cdots,n)$$

are solvable and that the condition

$$x_0(\varepsilon) = y_0(\varepsilon) \left(= \frac{|T_{n-1}(\varepsilon)|}{|T_n(\varepsilon)|} \right) \neq 0 \qquad (18.22)$$

is satisfied, i.e., that Theorem 18.1 can be applied (we point out, that the second of the linear systems written down differs from its "model" (18.2) only by the reversed order in which its successive equations are entered).

According to Theorem 18.2 (see its proof) the elements $b_{jk}(\varepsilon)$ of the matrix

$$(T_n(\varepsilon))^{-1} \equiv \| b_{jk}(\varepsilon) \|_{j,k=0}^{n} \quad (0 < \varepsilon < \delta)$$

are defined by the identities

$$b_{jk}(\varepsilon) = \frac{1}{x_0(\varepsilon)} \sum_{s=0}^{\min(j,k)} (x_{js}(\varepsilon) y_{s-k}(\varepsilon) - y_{-n+j-s-1}(\varepsilon) x_{n+s-k+1}(\varepsilon))$$
$$(j,k = 0,1,\cdots,n),$$

in which was assumed

$$x_{n+1}(\varepsilon) = y_{-n-1}(\varepsilon) \equiv 0.$$

But here it is convenient to us to rewrite these identities in the somewhat different shape

$$b_{jk}(\varepsilon) = \frac{1}{x_0(\varepsilon)} \left[\sum_{r=0}^{\min(k,j)} x_{j-r}(\varepsilon) y_{r-k}(\varepsilon) - \sum_{r=1}^{\min(k,j)} y_{j-n-r}(\varepsilon) x_{n-k+r}(\varepsilon) \right]$$
$$(j,k = 0,1,\cdots,n) \qquad (18.23)$$

where in the case $k = j = 0$ the second sum between the brackets must be read as zero. In particular, it follows from (18.23) that

$$b_{01}(\varepsilon) = y_{-1}(\varepsilon) \qquad (18.24)$$

and

$$b_{j1} = \frac{1}{x_0(\varepsilon)} \left| x_j(\varepsilon) y_{-1}(\varepsilon) + x_{j-1}(\varepsilon) y_0(\varepsilon) - y_{j-n-1}(\varepsilon) x_n(\varepsilon) \right|$$
$$(j = 1,2,\cdots,n). \qquad (18.25)$$

Inserting (18.24) into (18.25), we make an index transform, and, taking (18.22) into account, we obtain

$$b_{n+1-j,1}(\varepsilon) = \frac{1}{x_0(\varepsilon)} \left[x_{n+1-j}(\varepsilon) b_{01}(\varepsilon) + x_{n-j}(\varepsilon) x_0(\varepsilon) - x_n(\varepsilon) y_{-j}(\varepsilon) \right]$$
$$(j = 1,2,\cdots,n).$$

Consequently,

$$y_{-j}(\varepsilon) = \frac{1}{x_n(\varepsilon)} \left[x_{n+1-j}(\varepsilon) b_{01}(\varepsilon) + x_{n-j}(\varepsilon) x_0(\varepsilon) - x_0(\varepsilon) b_{n+1-j,1}(\varepsilon) \right]$$
$$(j = 1,2,\cdots,n). \qquad (18.26)$$

We point out, that by reducing (if necessary) $\delta > 0$, one can assume $x_n(\varepsilon) \neq 0$ $(0 < \varepsilon < \delta)$, as $x_n(0) = x_n \neq 0$ by the assumptions of the theorem, and the functions $x_j(\varepsilon), y_{-j}(\varepsilon)$ $(j = 0,1,\cdots,n)$ depend continuously on ε in a neighbourhood of zero. From (18.22) it is clear, that the functions $b_{jk}(\varepsilon)$ are also continuous in a neighbourhood of the point $\varepsilon = 0$, and the $b_{jk}(0)$ $(=b_{jk})$ are the entries of the matrix T_n^{-1} $(j,k = 0,1,\cdots,n)$.

We note, that because of (18.22), formulae (18.26) remain valid also for $j = 0$, if only we assume $b_{n+1,1}(\varepsilon)$ $(\equiv x_{n+1}(\varepsilon)) \equiv 0$, what we shall do.

We use formulae (18.26), inserting the expression for y_{-j} $(j = 0,1,\cdots,n)$ given by them into (18.23):

$$b_{jk} = \frac{1}{x_0(\varepsilon)} \left\{ \sum_{r=0}^{\min(k,j)} x_{j-r}(\varepsilon) \frac{1}{x_n(\varepsilon)} \left[x_{n+1+r-k}(\varepsilon) b_{01}(\varepsilon) + x_{n+r-k}(\varepsilon) x_0(\varepsilon) - x_0(\varepsilon) b_{n+1+r-k,1}(\varepsilon) \right] - \sum_{r=1}^{\min(k,j)} x_{n-k+r}(\varepsilon) \frac{1}{x_n(\varepsilon)} \left[x_{1+j-r}(\varepsilon) b_{01}(\varepsilon) + x_{j-r}(\varepsilon) x_0(\varepsilon) - x_0(\varepsilon) b_{j-r+1,1}(\varepsilon) \right] \right\} \quad (j,k = 0,1,\cdots,n),$$

where, we recall it, always is assumed that $\sum_{r=1}^{0} = 0$. It is convenient to us to extend this convention, by setting in the sequel generally $\sum_{\alpha}^{\beta} = 0$ for $\alpha > \beta$.

Transforming the obtained expression with regard to the accepted convention, we find

$$\begin{aligned} b_{jk}(\varepsilon) = \frac{1}{x_n(\varepsilon)} \Bigg[& x_j(\varepsilon) x_{n-k}(\varepsilon) + \sum_{r=1}^{\min(k,j)} x_{n-k+r}(\varepsilon) b_{j-r+1,1}(\varepsilon) - \\ & - \sum_{r=0}^{\min(k,j)} x_{j-r}(\varepsilon) b_{n+r-k+1,1}(\varepsilon) + \\ & + \frac{1}{x_0(\varepsilon)} \sum_{r=0}^{\min(k,j)} x_{j-r}(\varepsilon) x_{n+1+r-k}(\varepsilon) b_{01}(\varepsilon) - \\ & - \frac{1}{x_0(\varepsilon)} \sum_{r=1}^{\min(k,j)} x_{n-k+r}(\varepsilon) x_{1+j-r}(\varepsilon) b_{01}(\varepsilon) \Bigg] = \\ = \frac{1}{x_n(\varepsilon)} \Bigg\{ & x_j(\varepsilon) x_{n-k}(\varepsilon) + \sum_{r=0}^{\min(k,j)-1} \Big[x_{n-k+r+1}(\varepsilon) b_{j-r,1}(\varepsilon) - \\ & - x_{j-r}(\varepsilon) b_{n-k+r+1,1}(\varepsilon) \Big] - x_{j-\min(k,j)}(\varepsilon) b_{n+\min(k,j)-k+1,1}(\varepsilon) + \\ & + \frac{1}{x_0(\varepsilon)} x_{j-\min(k,j)}(\varepsilon) x_{n+\min(k,j)-k+1}(\varepsilon) b_{01}(\varepsilon) \Bigg\} \\ & (j,k = 0,1,\cdots,n). \end{aligned}$$

Now we note, that for $j \geqq k$ the last two summands between the braces vanish (since, because of our conventions, $b_{n+1,1}(\varepsilon) = 0$ and $x_{n+1}(\varepsilon) = 0$), and for $j < k$ these summands yield the sum

$$x_{n+j-k+1}(\varepsilon) b_{01}(\varepsilon) - x_0(\varepsilon) b_{n+j-k+1,1}(\varepsilon).$$

Taking into account, that

$$\min(k,j)-1 = \begin{cases} \min(j,k-1) & \text{for } k \leq j \\ \min(j,k-1)-1 & \text{for } k > j \end{cases}$$

we can finally rewrite the expression for $b_{jk}(\varepsilon)$ as

$$b_{jk}(\varepsilon) = \frac{1}{x_n(\varepsilon)} \left\{ x_j(\varepsilon)x_{n-k}(\varepsilon) + \sum_{r=0}^{\min(j,k-1)} \left[x_{n-k+r+1}(\varepsilon)b_{j-r,1}(\varepsilon) - \right.\right.$$

$$\left.\left. - x_{j-r}(\varepsilon)b_{n-k+r+1,1}(\varepsilon) \right] \right\} \quad (j,k = 0,1,\cdots,n). \qquad (18.27)$$

As was already recorded, $x_j(0) = x_j$ $(j = 0,1,\cdots,n)$. Now we note, that, due to the specific shape of the righthand part of the equation system (18.17), its solution $\{z_0,\cdots,z_n\}$ coincides with the second column (i.e., with the column with number $k = 1$) of the matrix $\|b_{jk}\|_{j,k=0}^{n}$, which is the inverse of the matrix T_n:

$$z_j = \frac{\begin{vmatrix} c_0 & \cdots & c_{-j+1} & 0 & c_{-j-1} & \cdots & c_{-n} \\ c_1 & \cdots & c_{-j+2} & 1 & c_{-j} & \cdots & c_{-n+1} \\ \cdot & \cdots & \cdot & \cdot & \cdot & \cdots & \cdot \\ c_n & \cdots & c_{-j+n+1} & 0 & c_{-j+n-1} & \cdots & c_0 \end{vmatrix}}{|T_n|} = b_{j1} = b_{j1}(0)$$

$$(j = 0,1,\cdots,n).$$

Setting $\varepsilon = 0$ in (18.27) we obtain (assuming $z_{n+1} = b_{n+1,1} = 0$)

$$b_{jk} = \left[\frac{1}{x_n} x_j x_{n-k} + \sum_{r=0}^{\min(j,k-1)} (x_{n-k+r+1}z_{j-r} - x_{j-r}z_{n-k+r+1}) \right]$$

$$(j,k = 0,1,\cdots,n).$$

But this identity is equivalent to formula (18.18).

The theorem is proved.

18.6. Just as Theorem 18.1 allowed to draw the conclusion, that under the conditions of this theorem not only the matrix T_n but also the truncated matrix T_{n-1} is invertible, Theorem 18.3 permits to make an analoguous inference in relation to the Toeplitz matrix (of order n)

$$T_{n-1}^{0,n} = \left\| \begin{matrix} c_1 & c_0 & c_{-1} & \cdots & c_{-n+2} \\ c_2 & c_1 & c_0 & \cdots & c_{-n+3} \\ \cdot & \cdot & \cdot & \cdots & \cdot \\ c_n & c_{n-1} & c_{n-2} & \cdots & c_1 \end{matrix} \right\|,$$

which is obtained from T_n by rejecting from the matrix T_n the first row (i.e. the row with index o) and the last column (with index n). Indeed, the solvability of the systems (18.16) and (18.17) together with the condition $x_n \neq 0$, is, clearly, equivalent to the relations

$$\det T_n \neq 0 \text{ and } \det T_{n-1}^{0,n} \neq 0.$$

In analogy to Theorem 18.2 one can here also state a more complete expression in relation to the matrix $(T_{n-1}^{0,n})^{-1}$:

THEOREM 18.4 (GOHBERG AND KRUPNIK). *If there exist solutions* $\{x_0, x_1, \ldots, x_n\}$ *and* $\{z_0, z_1, \ldots, z_n\}$ *of the systems* (18.16) *and* (18.17), *respectively, and the condition* $x_n \neq 0$ *is satisfied, then the Toeplitz matrix* $T_{n-1}^{0,n}$ *is invertible and its inverse matrix is constructed through the formula*

$$(T_{n-1}^{0,n})^{-1} = \frac{1}{x_n}\left\{\begin{Vmatrix} z_0 & 0 & \ldots & 0 \\ z_1 & z_0 & \ldots & 0 \\ \cdot & \cdot & \ldots & \cdot \\ z_{n-1} & z_{n-2} & \ldots & z_0 \end{Vmatrix}\begin{Vmatrix} x_n & x_{n-1} & \ldots & x_1 \\ 0 & x_n & \ldots & x_2 \\ \cdot & \cdot & \ldots & \cdot \\ 0 & 0 & \ldots & x_n \end{Vmatrix} - \right.$$
$$\left. - \begin{Vmatrix} x_0 & 0 & \ldots & 0 \\ x_1 & x_0 & \ldots & 0 \\ \cdot & \cdot & \ldots & \cdot \\ x_{n-1} & x_{n-2} & \ldots & x_0 \end{Vmatrix}\begin{Vmatrix} z_n & z_{n-1} & \ldots & z_1 \\ 0 & z_n & \ldots & z_2 \\ \cdot & \cdot & \ldots & \cdot \\ 0 & 0 & \ldots & z_n \end{Vmatrix}\right\}. \qquad (18.28)$$

PROOF. We form the square nonsingular matrix of order $n+1$:

$$M = \begin{Vmatrix} x_0 & 1 & 0 & \ldots & 0 \\ x_1 & 0 & 1 & \ldots & 0 \\ \cdot & \cdot & \cdot & \ldots & \cdot \\ x_{n-1} & 0 & 0 & \ldots & 1 \\ x_n & 0 & 0 & \ldots & 0 \end{Vmatrix} \qquad (\det M = x_n \neq 0).$$

It is easy to see, that because of the identities (18.16)

$$T_n M = \left\Vert\begin{array}{c|cccc} 1 & c_0 & c_{-1} & \ldots & c_{-n+1} \\ \hline 0 & & & & \\ \cdot & & T_{n-1}^{0,n} & & \\ \cdot & & & & \\ \cdot & & & & \\ 0 & & & & \end{array}\right\Vert .$$

Consequently, by the rule on the inversion of block-"quasi triangular" matrices (see [3], Ch. II, § 5)

$$M^{-1}T_n^{-1} = \left\| \begin{array}{c|ccc} 1 & * & * & \cdots & * \\ \hline 0 & & & & \\ \cdot & & & & \\ \cdot & & (T_{n-1}^{o,n})^{-1} & & \\ \cdot & & & & \\ o & & & & \end{array} \right\| \tag{18.29}$$

where, as usual, the elements which don't play an important role in the sequel, are denoted by asterisks. Identity (18.29) will be used for the derivation of formula (18.28).

According to Theorem 18.3 the matrix T_n^{-1} can be represented as

$$T_n^{-1} = \frac{1}{x_n}(K + R - S),$$

where (see (18.18))

$$K = \| x_j x_{n-k} \|_{j,k=o}^{n},$$

$$R = \left\| \begin{array}{cccc} z_o & 0 & \cdots & 0 \\ z_1 & z_o & \cdots & 0 \\ \cdot & \cdot & \cdots & \cdot \\ z_n & z_{n-1} & \cdots & z_o \end{array} \right\| \left\| \begin{array}{ccccc} 0 & x_n & x_{n-1} & \cdots & x_1 \\ 0 & 0 & x_n & \cdots & x_2 \\ \cdot & \cdot & \cdot & \cdots & \cdot \\ 0 & 0 & 0 & \cdots & 0 \end{array} \right\| ,$$

$$S = \left\| \begin{array}{cccc} x_o & 0 & \cdots & 0 \\ x_1 & x_o & \cdots & 0 \\ \cdot & \cdot & \cdots & \cdot \\ x_n & x_{n-1} & \cdots & x_o \end{array} \right\| \left\| \begin{array}{ccccc} 0 & z_n & z_{n-1} & \cdots & z_1 \\ 0 & 0 & z_n & \cdots & z_2 \\ \cdot & \cdot & \cdot & \cdots & \cdot \\ 0 & 0 & 0 & \cdots & 0 \end{array} \right\| .$$

The matrix M^{-1} looks like

$$M^{-1} = \frac{1}{x_n} \left\| \begin{array}{cccc|c} 0 & 0 & \cdots & 0 & 1 \\ \hline x_n & 0 & \cdots & 0 & -x_o \\ 0 & x_n & \cdots & 0 & -x_1 \\ \cdot & \cdot & \cdots & \cdot & \cdot \\ 0 & 0 & \cdots & x_n & -x_{n-1} \end{array} \right\| , \tag{18.3o}$$

as is easily verfied by forming the product $M^{-1}M$. One can also verify directely, that in the matrix $M^{-1}K$ the entries of all rows, except the first one, are equal to zero. Finally, one is easily convinced, that the difference R-S has the shape

$$R - S = x_n \cdot \left\| \begin{array}{c|c} 0 & \\ 0 & B \\ \cdot & \\ \cdot & \\ \cdot & \\ \hline 0 & 0 \ldots 0 \end{array} \right\| ,$$

where B is a square matrix of order n, which coincides with the right-hand side of formula (18.28).

Taking into account the identity (18.3o) and the shown properties of the entries of the matrices $M^{-1}K$ and R-S we obtain (again using the rule on the product of block matrices)

$$M^{-1}T_{n-1}^{-1} = M^{-1} \cdot \frac{1}{x_n} (R-S) + M^{-1} \cdot \frac{1}{x_n} K =$$

$$= \left\| \begin{array}{c|c} * & * \cdots * \\ \hline 0 & \\ 0 & \\ \cdot & B \\ \cdot & \\ \cdot & \\ 0 & \end{array} \right\| .$$

Comparing this result with (18.29), we arrive at the identity

$$(T_{n-1}^{o,n})^{-1} = B,$$

i.e., at formula (18.28).

The theorem is proved.

18.7. The analogy between the Theorems 18.2 and 18.4 extends even further. Namely, we can apply the latter Theorem (just as Theorem 18.2 was applied in Sec. 18.4) for the inversion of arbitrary nonsigular Toeplitz matrices $T_n = \|c_{p-q}\|_{p,q=o}^{n}$ ($|T_n| \neq 0$). To this end we consider the Toeplitz matrix (of order n+2)

$$\tilde{T}_\eta = \left\| \begin{array}{cccc|c} c_{-1} & c_{-2} & \cdots & c_{-n-1} & \eta \\ \hline c_o & c_{-1} & \cdots & c_{-n} & c_{-n-1} \\ \cdot & \cdot & \cdots & \cdot & \cdot \\ c_n & c_{n-1} & \cdots & c_o & c_{-1} \end{array} \right\| ,$$

where c_{-n-1} is some fixed complex number, and η a complex parameter. The function det $\tilde{T}_\eta$ is linear with respect to η with the coefficient $|T_n|$ ($\neq 0$) for η and hence it becomes zero only for one value η. For all other values η the matrix $\tilde{T}_\eta$ satisfies all conditions of Theorem 18.4, as it is invertible itself, and the role of the matrix $T_{n-1}^{o,n}$ of Theorem 18.4 here is taken by the also nonsingular matrix T_n. Consequently, the matrix T_n can now be inverted with help of formula (18.28).

18.8. Comparison of the Theorems 18.1 and 18.3 shows that in each of them for a nonsingular Toeplitz matrix $T_n = \|c_{p-q}\|_{p,q=0}^{n}$ the inverse matrix T_n^{-1} is defined by the solutions $\{x_0, x_1, \cdots, x_n\}$, $\{z_0, z_1, \cdots, z_n\}$ of systems of the shape

$$\sum_{q=0}^{n} c_{p-q} x_q = \delta_{p\mu}, \quad \sum_{q=0}^{n} c_{p-q} z_q = \delta_{p\nu}$$

$$(p = 0, 1, \cdots, n;\ \mu \neq \nu) \tag{18.31}$$

(in Theorem 18.1 for $\mu = 0$, $\nu = n$, in Theorem 18.3 for $\mu = 0$, $\nu = 1$). But in each of these theorems one had to subject these solutions to some additional condition ($x_0 \neq 0$ in Theorem 18.1, $x_n \neq 0$ in Theorem 18.3 - cf. exercises 3-6, 11 at the end of the section).

The natural question arises whether one cannot choose for each nonsingular matrix $T_n = \|c_{p-q}\|_{p,q=0}^{n}$ indices μ and ν $(0 \leqq \mu < \nu \leqq n)$ (for each matrix its own), such that the solutions of the systems (18.31) define uniquely the inverse matrix T_n^{-1} without any additional condition. A positive answer to this question would require the matrix T_n itself, i.e., all its elements c_p $(p = 0, \pm 1, \cdots, \pm n)$ to be uniquely defined by the solutions of (18.31). But already a simple example shows, that this is not always the case.

EXAMPLE. [24]. Let

$$T_3 = \begin{Vmatrix} 0 & 0 & 1 & 1 \\ 0 & 0 & 0 & 1 \\ 1 & 0 & 0 & 0 \\ 1 & 1 & 0 & 0 \end{Vmatrix} \qquad |T_3| \neq 0. \tag{18.32}$$

Assuming, let us say, $\mu = 0$, $\nu = 3$ (in accordance to Theorem 18.1) we are convinced, that the system (18.31) has in the present case the shape

$$\left.\begin{aligned} x_2 + x_3 &= 1, \\ x_3 &= 0, \\ x_0 \qquad &= 0 \\ x_0 + x_1 \qquad &= 0 \end{aligned}\right\} \qquad \left.\begin{aligned} z_2 + z_3 &= 0, \\ z_3 &= 0, \\ z_0 \qquad &= 0, \\ z_0 + z_1 \qquad &= 1, \end{aligned}\right\}$$

i.e., $x_0 = x_1 = 0$, $x_2 = 1$, $z_0 = z_2 = z_3 = 0$, $z_1 = 1$.

If one now uses the found solutions as coefficients, then the same equation systems (but related to the unknown c_p, $p = 0,\pm 1,\pm 2,\pm 3$) reappear as

$$\left.\begin{array}{l} c_{-2} = 1, \\ c_{-1} = 0, \\ c_0 = 0 \\ c_1 = 0; \end{array}\right\} , \quad \left.\begin{array}{l} c_{-1} = 0, \\ c_0 = 0, \\ c_1 = 0, \\ c_2 = 1, \end{array}\right\}$$

i.e., in this way not all the elements of the matrix T_3 are determined, namely c_{-3} and c_3 remain arbitrary. By this is, in particular, demonstrated that the here violated condition $x_o \neq 0$ in Theorem 18.1 is essential.

It is left to the reader to verify, that also for arbitrary other choices of μ and ν $(0 \leqq \mu < \nu \leqq 3)$ we discover a similar phenomenon: the matrix T_3 in (18.32) cannot be reconstructed from the corresponding systems (18.31) [4)].

18.9. In Theorem 18.1 is, by the first and final column of the matrix, which is the inverse of the given Toeplitz matrix T_n, the complete inverse matrix T_n^{-1} reestablished. One can raise the problem of reestablishing by these data the original matrix T_n. Here holds the following result [25].

THEOREM 18.5. *Let* $x_o, x_1, \cdots, x_n$ *and* $y_{-n}, y_{-n+1}, \cdots, y_o$ *be given systems of complex numbers and* $x_o \neq 0$. *For the existence of a Toeplitz matrix* $T_n = \|c_{p-q}\|^n_{p,1=o}$ *such, that*

$$\sum_{q=o}^{n} c_{p-q} x_q = \delta_{po}, \quad \sum_{q=o}^{n} c_{p-q} y_{q-n} = \delta_{pn} \quad (p = 0,1,\cdots,n), \tag{18.33}$$

it is necessary and sufficient, that the condition

$$x_o = y_o \tag{18.34}$$

is satisfied and that the matrix

$$P = \left\| \begin{array}{ccccc|ccccc} y_o & y_{-1} & y_{-2} & \cdots & y_{-n+1} & y_{-n} & 0 & \cdots & 0 & 0 \\ 0 & y_o & y_{-1} & \cdots & y_{-n+2} & y_{-n+1} & y_{-n} & \cdots & 0 & 0 \\ \cdot & \cdot & \cdot & \cdots & \cdot & \cdot & \cdot & \cdots & \cdot & \cdot \\ 0 & 0 & 0 & \cdots & y_o & y_{-1} & y_{-2} & \cdots & y_{n-+1} & y_{-n} \\ \hline x_n & x_{n-1} & x_{n-2} & \cdots & x_1 & x_o & 0 & \cdots & 0 & 0 \\ 0 & x_n & x_{n-1} & & & x_1 & x_o & \cdots & 0 & 0 \\ \cdot & \cdot & \cdot & \cdots & \cdot & \cdot & \cdot & \cdots & \cdot & \cdot \\ 0 & 0 & 0 & \cdots & x_n & x_{n-1} & x_{n-2} & \cdots & x_1 & x_o \end{array} \right\|$$

is nonsingular. If these conditions are fulfilled then the matrix T_n *is nonsingular and can be uniquely reestablished by formula* (18.3).

PROOF. The NECESSITY of condition (18.34) was established in Theorem 18.1. We show that the matrix P is nonsingular. To this end we divide it in four square blocks:

$$P = \begin{vmatrix} Y & U \\ V & X \end{vmatrix},$$

where

$$X = \begin{Vmatrix} x_o & 0 & \cdots & 0 \\ x_1 & x_0 & \cdots & 0 \\ \cdot & \cdot & \cdots & \cdot \\ x_{n-1} & x_{n-2} & \cdots & x_0 \end{Vmatrix}, \quad Y = \begin{Vmatrix} y_0 & y_{-1} & \cdots & y_{-n+1} \\ 0 & y_0 & \cdots & y_{-n+2} \\ \cdot & \cdot & \cdots & \cdot \\ 0 & 0 & \cdots & y_0 \end{Vmatrix},$$

$$U = \begin{Vmatrix} y_{-n} & 0 & \cdots & 0 \\ y_{-n+1} & y_{-n} & \cdots & 0 \\ \cdot & \cdot & \cdots & \cdot \\ y_{-1} & y_{-2} & \cdots & y_{-n} \end{Vmatrix}, \quad V = \begin{Vmatrix} x_n & x_{n-1} & \cdots & x_1 \\ 0 & x_n & \cdots & x_2 \\ \cdot & \cdot & \cdots & \cdot \\ 0 & 0 & \cdots & x_n \end{Vmatrix}.$$

By virtue of some proposition in the theory of determinants (see [3], Ch. II, § 5)

$$\det P = \det X \cdot \det(Y - UX^{-1}V)$$

(since $x_0 \neq 0$, the matrix X is invertible). It is easily verified, that the matrices X and U commute. Hence

$$\det P = \det(XY - UV).$$

Thence is clear, that P is nonsingular if and only if $XY - UV$ is nonsingular. But the nonsingularity of the latter was established in Theorem 18.2 (see formula (18.11).

SUFFICIENCY. One can rewrite the equation systems (18.33), after carrying out the index transformation $p = n + j$, $q = n + j - k$ in the second and $p = j$, $q = j - k$ in the first system, as a system of $2n + 2$ equations in the $2n + 1$ unknowns $c_{-n}, c_{-n+1}, \cdots, c_{-1}, c_0, c_1, \cdots, c_{n-1}, c_n$:

$$\left.\begin{aligned} &\sum_{k=j}^{j+n} y_{j-k} c_k = \delta_{j0} \qquad (j = -n, -n+1, \cdots, 0), \\ &\sum_{k=j-n}^{j} x_{j-k} c_k = \delta_{j0} \qquad (j = 0, 1, \cdots, n). \end{aligned}\right\} \tag{18.35}$$

The matrix Q of this system has the shape

$$Q = \left\| \begin{array}{ccccccccccc}
y_o & y_{-1} & y_{-2} & \cdots & y_{-n+1} & y_{-n} & 0 & \cdots & 0 & 0 & 0 \\
0 & y_o & y_{-1} & \cdots & y_{-n+2} & y_{-n+1} & y_{-n} & \cdots & 0 & 0 & 0 \\
\cdot & \cdot & \cdot & \cdots & \cdot & \cdot & \cdot & \cdots & \cdot & \cdot & \cdot \\
0 & 0 & 0 & \cdots & y_o & y_{-1} & y_{-2} & \cdots & y_{-n+1} & y_{-n} & 0 \\
\hline
0 & 0 & 0 & \cdots & 0 & y_o & y_{-1} & \cdots & y_{-n+2} & y_{-n+1} & y_{-n} \\
\hline
x_n & x_{n-1} & x_{n-2} & \cdots & x_1 & x_o & 0 & \cdots & 0 & 0 & 0 \\
0 & x_n & x_{n-1} & \cdots & x_2 & x_1 & x_o & \cdots & 0 & 0 & 0 \\
\cdot & \cdot & \cdot & \cdots & \cdot & \cdot & \cdot & \cdots & \cdot & \cdot & \cdot \\
0 & 0 & 0 & \cdots & x_n & x_{n-1} & x_{n-2} & \cdots & x_1 & x_o & 0 \\
0 & 0 & 0 & & 0 & x_n & x_{n-1} & \cdots & x_2 & x_1 & x_o
\end{array} \right\| .$$

If one discards in this matrix the $(n+1)$-st row, which is marked in the diagram, then a matrix is obtained with a determinant equal to $x_o \cdot \det P \neq 0$ (this can be calculated by developing it to the entries of the final column). Consequently, the rank of the matrix Q is maximal $(=2n+1)$.

Having joined with the matrix Q the column of the righthand sides of the systems (18.35), we obtain the augmented matrix $\tilde{Q}$. We calculate the determinant of this matrix $\tilde{Q}$ by developing it to the entries of the last column, which consists of zeros and two ones in the rows with index $n+1$ and $n+2$. We obtain

$$\det \tilde{Q} = (-1)^{n+1+2n+2} x_o \det P + (-1)^{n+2+2n+2} y_o \det P =$$
$$= (-1)^n (y_o - x_o) \det P.$$

Because of (18.34) we have $\det \tilde{Q} = 0$, i.e., the rank of the matrix $\tilde{Q}$ is the same as that of Q and by the Kronecker-Capelli-Theorem the system (18.35) is uniquely solvable. The remaining statements of Theorem 18.5 concerning the matrix $T_n = \|c_{p-q}\|_{p,q=0}^{n}$ follow from Theorem 18.1 [5)].

It is clear, that a problem, similar to the one solved by Theorem 18.5, can be formulated, not starting from Theorem 18.1 but from its analogon - Theorem 18.3. Now the question is to reestablish the matrix T_n from the first and the second column of its inverse matrix T_n^{-1}. Here holds

THEOREM 18.6 [24]. *Let* $x_o, x_1, \cdots, x_n$ *and* $z_o, z_1, \cdots, z_n$ *be given systems of complex numbers and let* $x_n \neq 0$. *For the existence of a Toeplitz matrix* $T_n = \|c_{p-q}\|_{p,q=0}^{n}$ *such, that*

$$\sum_{q=0}^{n} c_{p-q} x_q = \delta_{po}, \quad \sum_{q=0}^{n} c_{p-q} z_q = \delta_{p1} \quad (p = 0,1,\cdots,n) \tag{18.36}$$

it is necessary and sufficient, that the condition

$$z_n = x_{n-1} \tag{18.37}$$

is satisfied and that the matrix

$$N = \left\| \begin{array}{cccc|cccc} x_n & x_{n-1} & \dots & x_1 & x_o & 0 & 0 & \dots & 0 \\ 0 & x_n & \dots & x_2 & x_1 & x_o & 0 & \dots & 0 \\ \cdot & \cdot & \dots & \cdot & \cdot & \cdot & \cdot & \dots & \cdot \\ 0 & 0 & \dots & x_n & x_{n-1} & x_{n-2} & x_{n-3} & \dots & x_o \\ \hline z_n & z_{n-1} & \dots & z_1 & z_o & 0 & 0 & \dots & 0 \\ 0 & z_n & \dots & z_2 & z_1 & z_o & 0 & \dots & 0 \\ \cdot & \cdot & \dots & \cdot & \cdot & \cdot & \cdot & \dots & \cdot \\ 0 & 0 & \dots & z_n & z_{n-1} & z_{n-2} & z_{n-3} & \dots & z_o \end{array} \right\|$$

is nonsingular. If these conditions are fulfilled, then the matrix T_n *is invertible and it can be uniquely reestablished by formula* (18.18).

PROOF. The NECESSITIY of condition (18.37) follows from Theorem 18.3, by which the matrix T_n is nonsingular, and hence the systems (18.36) uniquely solvable, and it remains to evaluate z_n and x_{n-1} by Cramers rule.

We show the necessity of the condition det N $\neq$ 0. By Theorem 18.3 (formula (18.18))

$$T_n^{-1} = \frac{1}{x_n} (BC - DF)$$

where

$$B = \left\| \begin{array}{cccc} z_o & 0 & \dots & 0 \\ z_1 & z_o & \dots & 0 \\ \cdot & \cdot & \dots & \cdot \\ z_n & z_{n-1} & \dots & z_o \end{array} \right\|, \qquad C = \left\| \begin{array}{cccc} 0 & x_n & \dots & x_1 \\ \cdot & \cdot & \dots & \cdot \\ 0 & 0 & \dots & x_n \\ 0 & 0 & \dots & 0 \end{array} \right\|,$$

$$D = \left\| \begin{array}{cccc} x_o & 0 & \dots & 0 \\ x_1 & x_o & \dots & 0 \\ \cdot & \cdot & \dots & \cdot \\ x_n & x_{n-1} & \dots & x_o \end{array} \right\|,$$

$$F = \left\| \begin{array}{ccccc} 0 & z_n & z_{n-1} & \dots & z_1 \\ 0 & 0 & z_n & \dots & z_2 \\ \cdot & \cdot & \cdot & \dots & \cdot \\ 0 & 0 & 0 & \dots & z_n \\ 0 & 0 & 0 & \dots & 0 \end{array} \right\| - \left\| \begin{array}{cccc} x_n & x_{n-1} & \dots & x_o \\ 0 & 0 & \dots & 0 \\ \cdot & \cdot & \dots & \cdot \\ 0 & 0 & \dots & 0 \end{array} \right\| .$$

As BD = DB, we have, applying again some rule for the evaluation of determinants of block matrices ([3], Ch. II, § 5)

$$\det \begin{pmatrix} B & F \\ D & C \end{pmatrix} = \det\,(BC - DF).$$

But

$$\det \begin{pmatrix} B & F \\ D & C \end{pmatrix} =$$

$$= \begin{vmatrix} z_0 & 0 & \cdots & 0 & -x_n & z_n - x_{n-1} & z_{n-1} - x_{n-2} & \cdots & z_1 - x_0 \\ z_1 & z_0 & \cdots & 0 & 0 & 0 & z_n & \cdots & z_2 \\ \cdot & \cdot & \cdots & \cdot & \cdot & \cdot & \cdot & \cdots & \cdot \\ z_{n-1} & z_{n-2} & \cdots & 0 & 0 & 0 & 0 & \cdots & z_n \\ z_n & z_{n-1} & \cdots & z_0 & 0 & 0 & 0 & \cdots & 0 \\ x_0 & 0 & \cdots & 0 & 0 & x_n & x_{n-1} & \cdots & x_1 \\ x_1 & x_0 & \cdots & 0 & 0 & 0 & x_n & \cdots & x_2 \\ \cdot & \cdot & \cdots & \cdot & \cdot & \cdot & \cdot & \cdots & \cdot \\ x_n & x_{n-1} & \cdots & x_0 & 0 & 0 & 0 & \cdots & 0 \end{vmatrix}.$$

Developing this determinant to the entries in the (n+2)nd column, and thereupon moving the first column to the last place, we have

$$\left| \det \begin{pmatrix} B & F \\ D & C \end{pmatrix} \right| =$$

$$= |x_n| \cdot \left| \begin{vmatrix} z_0 & \cdots & 0 & 0 & z_n & \cdots & z_2 & z_1 \\ \cdot & \cdots & \cdot & \cdot & \cdot & \cdots & \cdot & \cdot \\ z_{n-1} & \cdots & z_0 & 0 & 0 & \cdots & 0 & z_n \\ \hline 0 & \cdots & 0 & x_n & x_{n-1} & \cdots & x_1 & x_0 \\ \hline x_0 & \cdots & 0 & 0 & x_n & \cdots & x_2 & x_1 \\ \cdot & \cdots & \cdot & \cdot & \cdot & \cdots & \cdot & \cdot \\ x_{n-1} & \cdots & x_0 & 0 & 0 & \cdots & 0 & x_n \end{vmatrix} \right| =$$

$$= |x_n|^2 \; |\det N|.$$

Thus

$$|\det N| = \frac{1}{|x_n|^2} \, |\det\,(BC - DF)| = \frac{1}{|x_n|^2} \, |\det T_n^{-1}| \neq 0.$$

SUFFICIENCY. Let $\det N \neq 0$ and $z_n = x_{n-1}$. Just as in the proof of Theorem 18.5, we consider the identities (18.36) as a system of $2n+2$ equations in $2n+1$ unknowns $c_{-n}, c_{-n+1}, \ldots, c_{-1}, c_0, c_1, \ldots, c_{n-1}, c_n$ with the matrix [6)]

$$G = \left\| \begin{array}{cccccccccc} x_n & x_{n-1} & \cdots & x_1 & x_o & 0 & 0 & \cdots & 0 \\ 0 & x_n & \cdots & x_2 & x_1 & x_o & 0 & \cdots & 0 \\ \cdot & \cdot & \cdots & \cdot & \cdot & \cdot & \cdot & \cdots & \cdot \\ 0 & 0 & \cdots & 0 & x_n & x_{n-1} & x_{n-2} & \cdots & x_0 \\ \hline z_n & z_{n-1} & \cdots & z_1 & z_0 & 0 & 0 & \cdots & 0 \\ \hline 0 & z_n & \cdots & z_2 & z_1 & z_0 & 0 & \cdots & 0 \\ \cdot & \cdot & \cdots & \cdot & \cdot & \cdot & \cdot & \cdots & \cdot \\ 0 & 0 & \cdots & 0 & z_n & z_{n-1} & z_{n-2} & \cdots & z_0 \end{array} \right\| .$$

If one discards in the matrix G the marked (n+2)-nd row, then the determinant of the remaining matrix will differ from det $N(\neq 0)$ only by a factor x_n. Hence the rank of the matrix G is maximal $(= 2n + 1)$.

The augmented matrix $\tilde{G}$ of the system (18.36), obtained by joining to G the column of the righthand side, has the shape

$$\tilde{G} = \left\| \begin{array}{ccccc|cccccc} x_n & x_{n-1} & \cdots & x_1 & x_o & 0 & 0 & \cdots & 0 & 1 \\ 0 & x_n & \cdots & x_2 & x_1 & x_o & 0 & \cdots & 0 & 0 \\ \cdot & \cdot & \cdots & \cdot & \cdot & \cdot & \cdot & \cdots & \cdot & \cdot \\ 0 & 0 & \cdots & 0 & x_n & x_{n-1} & x_{n-2} & \cdots & x_0 & 0 \\ \hline z_n & z_{n-1} & \cdots & z_1 & z_0 & 0 & 0 & \cdots & 0 & 0 \\ 0 & z_n & \cdots & z_2 & z_1 & z_0 & 0 & \cdots & 0 & 1 \\ \cdot & \cdot & \cdots & \cdot & \cdot & \cdot & \cdot & \cdots & \cdot & \cdot \\ 0 & 0 & \cdots & 0 & z_n & z_{n-1} & z_{n-2} & \cdots & z_0 & 0 \end{array} \right\| \equiv \left(\begin{array}{c|c} P & Q \\ \hline R & S \end{array} \right) .$$

As PR = RP, one has again ([3], Ch. II, § 5)

$$\det \tilde{G} = \det (PS - RQ).$$

But in the matrix $PS - RQ$ are, as one, applying (18.37), easily computes, all entries on the first row equal to zero, so that $\det \tilde{G} = 0$.

Thus, the system (18.36) is uniquely solvable in relation to the numbers c_p $(p = 0, \pm 1, \cdots, \pm n)$. Having compiled from these the Toeplitz matrix $T_n = \|c_{p-q}\|_{p,q=0}^{n}$, we see, that by Theorem 18.3 this matrix is invertible and it can be reestablished by formula (18.18).

18.1o. Turning to the problem of the inversion of Hankel matrices, we recall, that after multiplication of such a matrix on the right with the matrix

$$J_{n+1} = \left\| \begin{array}{ccccc} 0 & 0 & \cdots & 0 & 1 \\ 0 & 0 & \cdots & 1 & 0 \\ \cdot & \cdot & \cdots & \cdot & \cdot \\ 0 & 1 & \cdots & 0 & 0 \\ 1 & 0 & \cdots & 0 & 0 \end{array} \right\|$$

(see Sec. 17.1), also of order $n+1$, we obtain the Toeplitz matrix

$$H_n J_{n+1} = T_n^I = \| c_{p-q}^I \|_{p,q=0}^{n},$$

where the entries c_p^I are connected with the elements s_k appearing in the matrix H_n through the formulae (cf. (17.3))

$$c_p^I = s_{p+n} \qquad (p = 0, \pm 1, \cdots, \pm n). \tag{18.38}$$

We note further, that the invertibility of the matrix H_n is equivalent with the invertibility of T_n^I, and moreover (see (17.8))

$$(T_n^I)^{-1} = (H_n J_{n+1})^{-1} = J_{n+1} H_n^{-1},$$

or

$$H_n^{-1} = J_{n+1} (T_n^I)^{-1}. \tag{18.39}$$

Thus, one obtains for Hankel matrices as a direct corollary of Theorem 18.1.

THEOREM 18.7. *If the Hankel matrix* $H_n = \| s_{j+k} \|_{j,k=0}^{n}$ *is such, that each of the equation systems*

$$\sum_{q=0}^{n} s_{p-q+n} x_q = \delta_{po} \qquad (p = 0, 1, \cdots, n),$$

$$\sum_{q=0}^{n} s_{p-q+n} y_{q-n} = \delta_{pn} \qquad (p = 0, 1, \cdots, n)$$

is solvable and the condition $x_o \neq 0$ *is satisfied, then the matrix* H_n *is nonsingular, and its inverse is constructed by the formula*

$$H_n^{-1} = x_o^{-1} \left\{ \left\| \begin{matrix} o & o & \cdots & o & x_o \\ 0 & 0 & \cdots & x_o & x_1 \\ \cdot & \cdot & \cdots & \cdot & \cdot \\ x_o & x_1 & \cdots & x_{n-1} & x_n \end{matrix} \right\| \left\| \begin{matrix} y_o & y_{-1} & \cdots & y_{-n} \\ 0 & y_o & \cdots & y_{-n+1} \\ \cdot & \cdot & \cdots & \cdot \\ 0 & 0 & \cdots & y_o \end{matrix} \right\| - \right.$$

$$\left. - \left\| \begin{matrix} o & o & \cdots & o & o & o \\ o & o & \cdots & o & o & y_{-n} \\ 0 & 0 & \cdots & 0 & y_{-n} & y_{-n+1} \\ \cdot & \cdot & \cdots & \cdot & \cdot & \cdot \\ 0 & y_{-n} & \cdots & y_{-3} & y_{-2y-1} \end{matrix} \right\| \left\| \begin{matrix} o & x_n & x_{n-1} & \cdots & x_1 \\ 0 & 0 & x_n & \cdots & x_2 \\ \cdot & \cdot & \cdot & \cdots & \cdot \\ 0 & 0 & 0 & \cdots & x_n \\ o & o & o & \cdots & o \end{matrix} \right\| \right\}.$$

The PROOF reduce to the insertion of the expressions (18.38) for c_p^I (instead of c_p) in the systems (18.1) and (18.2) and the multiplication of both sides of formula (18.3) (in agreement with (18.39)) on the left with the matrix J_{n+1} (i.e.,to the rearrangement in reversed order of the columns of the lefthand factors of both products, standing between the braces in formula (18.4)).

It is clear, that one could (see Sec. 17.1) from the matrix H_n turn, not to T_n^I but to the Toeplitz matrix

$$T_n^{H} = J_{n+1}H_n = \| c^{H}_{p-q} \|_{p,q=0}^{n}$$

with the elements (cf. (17.3))

$$c_p^{H} = s_{-p+n} \qquad (p = 0, \pm 1, \cdots, \pm n). \tag{18.4o}$$

Then we would arrive at Theorem 18.7' (the "dual" of Theorem 18.7), which is obtained from Theorem 18.1, if the expressions (18.4o) for c_p^{H} are inserted in the formulae (18.1) and (18.2) instead of c_p, respectively, and the rows of the righthand factors in each of the two products, standing between the braces of formula (18.3) are written in reverse order[7].

We suggest to the reader to reformulate independently for Hankel matrices the Theorems 18.2 - 18.4, and each in two variants, using the formulae $H_n J_{n+1} = T^I$ and $J_{n+1}H_n = T_n^{H}$, resprectively[8].

EXAMPLES AND EXERCISES

1. For the Toeplitz matrix

$$T_3 = \begin{Vmatrix} 0 & 1 & -1 & i \\ i & 0 & 1 & -1 \\ 0 & i & 0 & 1 \\ 0 & 0 & i & 0 \end{Vmatrix}$$

we write down the systems (18.1) and (18.2):

$$\left.\begin{aligned} x_1 - x_2 + ix_3 &= 1, \\ ix_0 \quad + x_2 - x_3 &= 0, \\ ix_1 \quad + x_3 &= 0, \\ ix_2 &= 0; \end{aligned}\right\} \qquad \left.\begin{aligned} y_{-2} - y_{-1} + iy_0 &= 0, \\ iy_{-3} \quad + y_{-1} - y_0 &= 0, \\ iy_{-2} \quad + y_0 &= 0 \\ iy_{-1} &= 1. \end{aligned}\right\}$$

Its solutions are as follows:

$$x_0 = -\frac{1}{2}, \quad x_1 = \frac{1}{2}, \quad x_2 = 0, \quad x_3 = -\frac{i}{2};$$

$$y_{-3} = 1 + \frac{i}{2}, \; y_{-2} = -\frac{i}{2}, \; y_{-1} = -i, \; y_0 = -\frac{1}{2}.$$

The condition $x_0 \neq 0$ is satisfied. Hence (see (18.3))

$$T_3^{-1} = -2 \left\{ \begin{Vmatrix} -1/2 & 0 & 0 & 0 \\ 1/2 & -1/2 & 0 & 0 \\ 0 & 1/2 & -1/2 & 0 \\ -i/2 & 0 & 1/2 & -1/2 \end{Vmatrix} \begin{Vmatrix} -1/2 & -i & -i/2 & 1+i/2 \\ 0 & -1/2 & -i & -i/2 \\ 0 & 0 & -1/2 & -i \\ 0 & 0 & 0 & -1/2 \end{Vmatrix} - \right.$$

$$- \left.\begin{Vmatrix} 0 & 0 & 0 & 0 \\ 1+i/2 & 0 & 0 & 0 \\ -i/2 & 1+i/2 & 0 & 0 \\ -i & -i/2 & 1+i/2 & 0 \end{Vmatrix} \begin{Vmatrix} 0 & -i/2 & 0 & 1/2 \\ 0 & 0 & -i/2 & 0 \\ 0 & 0 & 0 & -i/2 \\ 0 & 0 & 0 & 0 \end{Vmatrix}\right\} =$$

$$= -2 \left\{\begin{Vmatrix} 1/4 & i/2 & i/4 & -1/2-i/4 \\ -1/4 & -i/2+1/4 & i/4 & 1/2+i/2 \\ 0 & -1/4 & -i/2+1/4 & i/4 \\ 1/4 & -1/2 & -1/2 & -i+1/2 \end{Vmatrix} \right. -$$

$$\left. - \begin{Vmatrix} 0 & 0 & 0 & 0 \\ 0 & -i/2+1/4 & 0 & 1/2+i/4 \\ 0 & -1/4 & -i/2+1/4 & -i/4 \\ 0 & -1/2 & -1/4 & -i+1/4 \end{Vmatrix} \right\} = \begin{Vmatrix} -1/2 & -i & -i/2 & 1+i/2 \\ 1/2 & 0 & -i/2 & -i/2 \\ 0 & 0 & 0 & -i \\ -i/2 & 0 & 1/2 & -1/2 \end{Vmatrix}$$

Verification:

$$T_3 T_3^{-1} = \begin{Vmatrix} 0 & 1 & -1 & i \\ i & 0 & 1 & -1 \\ 0 & i & 0 & 1 \\ 0 & 0 & i & 0 \end{Vmatrix} \begin{Vmatrix} -1/2 & -i & -i/2 & 1+i/2 \\ 1/2 & 0 & -i/2 & -i/2 \\ 0 & 0 & 0 & -i \\ -i/2 & 0 & 1/2 & -1/2 \end{Vmatrix} = \begin{Vmatrix} 1 & 0 & 0 & 0 \\ 0 & 1 & 0 & 0 \\ 0 & 0 & 1 & 0 \\ 0 & 0 & 0 & 1 \end{Vmatrix} .$$

2. For the matrix of example 1 we consider the truncated matrix

$$T_2 = \begin{Vmatrix} 0 & 1 & -1 \\ i & 0 & 1 \\ 0 & i & 0 \end{Vmatrix} .$$

Construct, using the data of example 1, the matrix T_2^{-1} by rule (18.11).

Solution. $$T_2^{-1} = \begin{Vmatrix} -i & -i & 1 \\ 0 & 0 & -i \\ -1 & 0 & -i \end{Vmatrix} .$$

3. Show, that in Theorem 18.1 the condition $x_0 \neq 0$ is essential - adduce an example, where the systems (18.1) and (18.2) are solvable, but $x_0 = 0$ and the matrix T_n turns out to be singular.

HINT. Such an example is already possible for $n = 2$, i.e., for Toeplitz matrices of order three.

4. Show, that for $n < 2$ it is not possible to construct an example, as required in exercise 3.

5. The condition $x_0 \neq 0$ in Theorem 18.1 is already essential in that respect, that, if it is violated, the matrix T_n can turn out to be non-

singular, but its inverse matrix T_n^{-1} may not be determined by the solutions of the systems (18.1) and (18.2). This was already shown at the end of Sec. 18.8 (see (18.32)) by the example of the matrix

$$T_3 = \begin{Vmatrix} 0 & 0 & 1 & 1 \\ 0 & 0 & 0 & 1 \\ 1 & 0 & 0 & 0 \\ 1 & 1 & 0 & 0 \end{Vmatrix}$$

Construct other examples of such kind.

6. Convince yourself, that for the nonsingular Toeplitz matrix

$$T_1 = \begin{Vmatrix} 0 & 1 \\ 1 & 0 \end{Vmatrix},$$

although $x_0 = 0$, this matrix (and thus, also its inverse) is nevertheless completely determined by the solutions of the systems (18.1) and (18.2) [9)]

7. Invert the matrix T_3 of example 5 by the method, explained in Sec. 18.4.

Solution. $$T_3^{-1} = \begin{Vmatrix} 0 & 0 & 1 & 0 \\ 0 & 0 & -1 & 1 \\ 1 & -1 & 0 & 0 \\ 0 & 1 & 0 & 0 \end{Vmatrix}.$$

8. Invert the matrix T_3 of example 1 by the method of Theorem 18.3 (by formula (18.18)).

9. Invert by the method of Theorem 18.4 the matrix

$$T_2^{0,3} = \begin{Vmatrix} i & 0 & 1 \\ 0 & i & 0 \\ 0 & 0 & i \end{Vmatrix}$$

(induced by the matrix T_3 of example 1)

Solution. $$(T_2^{0,3})^{-1} = \begin{Vmatrix} -i & 0 & 1 \\ 0 & -i & 0 \\ 0 & 0 & -i \end{Vmatrix}$$

1o. Invert the matrix T_3 of example 5 by the method, explained in Sec. 18.7.

Solution. See exercise 7.

11. Show. that the condition $x_n \neq 0$ in Theorem 18.3 is essential (cf. the exercises 3 and 5).

NOTES

1) The extend of the present monograph doesn't allow to reflect the results, relating to the numerous methods for the inversion of Toeplitz- and more general (the so-called *block-Toeplitz*-)matrices. These results have mainly an applied character (the construction of algorithms for the inversion of matrices on computers) and in most cases they are connected with the theory of filtration and extrapolation of scalar- and vector stationary random processes, where the correlated matrices are precisely also Toeplitz matrices (in the vector case block-Toeplitz matrices). For the orientation of the reader we refer here to the papers of N. Levinson [53], P. Wittle [61,62] , J. Durbin [17], V. Trench [6o], S. Zohar [63], L.M. Kutikov [49] and the quite recently appeared paper of S. Akaike [13]. We note, that in all these papers there are imposed on the Toeplitz matrices to be inverted rigid restrictions of the type of positive definiteness or the somewhat weaker requirement of strict nonsingularity, i.e., that all successive principal minors are nonzero, and for block Toeplitz matrices of the shape $\|T_{p-q}\|_{p,q=o}^{n-1}$, where the T_j $(j = 0,\pm 1,\cdots,\pm(n-1))$ are arbitrary square matrices, that all "truncated" matrices $\|T_{p-q}\|_{p,q=o}^{k}$ $(k = 0,1,\cdots,n-1)$ are invertible.

2) For a nonsingular matrix T_n the numbers $(x_o,x_1,\cdots,x_n)$ and $(y_o,y_{-1},\cdots,y_{-n})$, are, as is clear from the formulae (18.1) and (18.2), the entries of the first and the last columns of its inverse matrix T_n^{-1}. The fact, that the matrix T_n^{-1} is determined by these of its columns, was, apparently, ascertained for the first time (by transcendental methods) by G. Baxter and I. Hirschmann [14], who also obtained an explicit formula (different from (18.3)) for the reconstruction of the matrix T_n^{-1} from these columns. But in their paper the numbers $(x_o,x_1,\cdots,$ $x_n)$ and $(y_o,y_{-1},\cdots,y_{-n})$ were subjected to an additional restriction: the polynomials

$$x(\zeta) = \sum_{j=o}^{n} x_j\zeta^j \quad \text{and} \quad y(\zeta) \quad \sum_{j=o}^{n} y_{-j}\zeta^j$$

are not allowed to become zero in the unit circle $|\zeta| < 1$. Under the same restrictions formula (18.3) was discovered by A.A. Semencul [58], a result which was fundamental in the corresponding section of the book of I.C. Gohberg and I.A. Fel'dman [6] (Ch. III, § 6). In the redaction cited in the text, Theorem 18.1 and its proof where adopted from [25].

3) Thus, unlike in Theorem 18.1, here the question is to reconstruct the

matrix T_n^{-1} not from its first and last but from its first and second columns (cf. note [2] above).

[4] Recently, A.L. Sahnovič [55] published a method for the inversion of Toeplitz matrices by the solutions of equation systems of the type of (18.31), but with a different right side. Moreover, there was no additional restriction at all imposed on the solutions. Curiously, in the given case the direction of the development was inverse in comparison to that mentioned in the begin of the introduction of this book: A.L. Sahnovič obtained his own result for matrices in analogy to corresponding results of L.A. Sahnovič [56,57] for integral equations.

[5] As was remarked in [25], Theorem 18.5 was in the case of Hermitian Toeplitz matrices proved in an equivalent formulation but by a different method by M.G. Kreĭn [46]. The connection between this group of problems and the paper of M.G. Kreĭn [46] is traced in more detail in the book of I.C. Gohberg and I.A. Fel'dman [6] (Ch. III, § 6).

[6] Of course, the reader has noted, that in comparison to the system (18.35) here (if one only deals with the lefthand side) only the order of succession of the equations and the notations on the side of the coefficients have changed, and that this is also reflected in the difference between the matrices G and Q.

[7] It is interesting, that the rearrangement in reverse order of the rows (columns), i.e., the multiplication of the matrix on the left (right) with the matrix J_{n+1}, is applied in a broad fashion for the inversion by other methods of proper Toeplitz matrices and block-Toeplitz matrices (see, for example in the papers [63],[13], already mentioned above in note [1]).

[8] At the same time (already after the publication of the papers [24, 25]) another approach to the Theorems 18.1 and 18.3 and the corresponding results for Hankel matrices was discovered. This approach, based on the notion of the Bezoutiant of two polynomials, which is well-known in the algebra (see [8]), was developed by F.I. Lander [52].

[9] Thus, the condition $x_o \neq 0$ is, although it is essential, not necessary, and that not only for the invertibility of a Toeplitz matrix T_n (with was already clear from example 5), but also not for the fact, that the solutions of the systems (18.1) and 18.2) determine the inverse matrix.

§ 19. MUTUAL TRANSFORMATIONS OF TOEPLITZ AND HANKEL FORMS.

19.1. The results of § 17 concerning the mutual transformations of Toeplitz and Hankel matrices suggest their application to transformations of their respective Hermitian and quadratic forms. But on this way a disappointment awaits us. Before all we note, that for an arbitrary Hermitian Toeplitz form (of order n)

$$T_{n-1}(x,x) = \sum_{p,q=0}^{n-1} c_{p-q}\xi_p\overline{\xi}_q \quad (c_{-p} = \overline{c}_p,\ p = 0,1,\cdots,n-1) \tag{19.1}$$

the matrix $T_{n-1} = \|c_{p-q}\|_{p,q=0}^{n-1}$ after the transformations (see Sec. 17.1)

$$H_{n-1}^{I} = T_{n-1}J_n,\quad H_{n-1}^{\vdash\!\dashv} = J_nT_{n-1}$$

reduces to Hankel matrices H_{n-1}^{I}, $H_{n-1}^{\vdash\!\dashv}$ which are nonreal (and moreover, non-Hermitian - see Sec. 9.1), if the matrix T_{n-1} is nonreal. So we cannot, in the general case, form by means of H_{n-1}^{I} (or $H_{n-1}^{\vdash\!\dashv}$) a Hankel Hermitian or real quadratic form and compare it with the form (19.1).

Hence we restrict ourselves to real quadratic Toeplitz forms, i.e., forms

$$T_{n-1}(x,x) = \sum_{p,q=0}^{n-1} c_{p-q}\xi_p\xi_q \quad (c_{-p} = c_p,\ p = 0,1,\cdots,n-1) \tag{19.2}$$

with real symmetric Toeplitz matrices $T_{n-1} = \|c_{p-q}\|_{p,q=0}^{n-1}$. Now the Hankel matrix $H_{n-1} \equiv H_{n-1}^{I} = H_{n-1}^{\vdash\!\dashv} = \|s_{j+k}\|_{j,k=0}^{n-1}$ is real (and as always, symmetric) and hence it defines a quadratic Hankel form

$$\sum_{j,k=0}^{n-1} s_{j+k}\eta_j\eta_k \tag{19.3}$$

of which the coefficients s_k are connected to the coefficients c_p of the form (19.2) through the simple relations (see (17.2) and (17.3))

$$s_k = c_{k-(n-1)} = c_{(n-1)-k} \qquad (k = 0,1,\cdots,2n-2);$$
$$c_p = s_{p+(n-1)} = s_{-p+(n-1)} \qquad (p = 0,\pm 1,\cdots,\pm(n-1)).$$

Thus, the matrix $H_{n-1} = \|s_{j+k}\|_{j,k=0}^{n-1}$ is symmetric not only with respect to the main diganonal but also with respect to the auxiliary diagonal. Hence, we must restrict ourselves in the forms (19.3), namely, to those with such matrices, if we want to move also in the reserve direction - from Hankel to Toeplitz matrices.

It is necessary to raise the question: how are the basic invariants of the forms - rank and signature - transformed under the above indicated transformations, after what rules are they recalculated? But such a formulation turns out to be incorrect. Already simple examples show, that the rank and the signature of, let us say, the form (19.2) do, generally speaking, not determine the signature of the correspon- ding form (19.3) (the ranks of the forms (19.2) and (19.3), it is clear, do coincide, since the matrices T_{n-1} and H_{n-1} of these forms differ from each other only in the order of succession of their rows (columns)).

EXAMPLE. Let the Toeplitz form $T_2(x,x)$ (of order n=3) be defined by the matrix

$$T_2 = \begin{Vmatrix} 1 & 1 & 0 \\ 1 & 1 & 1 \\ 0 & 1 & 1 \end{Vmatrix}$$

with the successive principal minors $D_{-1} = 1$, $D_0 = 1$, $D_1 = 0$, $D_2 = -1$. Its rank and signature are respectively equal to: $\rho = 3$, $\sigma_{T_2} = 1$ (by ru- le (16.6) or Theorem 8.2). The corresponding Hankel matrix $H_2 = J_3T_2$ has the shape

$$H_2 = \begin{Vmatrix} 0 & 1 & 1 \\ 1 & 1 & 1 \\ 1 & 1 & 0 \end{Vmatrix},$$

its successive principal minors are equal to:

$$D_{-1} = 1,\ D_0 = 0,\ D_1 = -1,\ D_2 = 1,$$

so that (by Frobenius' rule (12.2o) or Theorem 8.2) $\sigma_{H_2} = -1$.

Now we consider the Toeplitz matrix

$$\tilde{T}_2 = \begin{Vmatrix} 1 & 0 & 2 \\ 0 & 1 & 0 \\ 2 & 0 & 1 \end{Vmatrix}$$

with again the order $n = 3$ and the rank $\rho = 3$. The signature of the corresponding form $\tilde{T}(x,x)$ (since $D_{-1} = 1$, $D_0 = 1$, $D_1 = 1$, $D_2 = -3$) is equal to: $\sigma_{\tilde{T}_2} = 1$, just as that of the form $T_2(x,x)$.

But the Hankel form $\tilde{H}_2(x,x)$ with the matrix

$$(J_3\tilde{T}_2 =)\tilde{H}_2 = \begin{Vmatrix} 2 & 0 & 1 \\ 0 & 1 & 0 \\ 1 & 0 & 2 \end{Vmatrix}$$

and the successive principal minors $D_{-1} = 1$, $D_0 = 2$, $D_1 = 2$, $D_2 = 3$ has signature equal to: $\sigma_{\tilde{H}_2} = 3$.

19.2. The negative result, found in Sec. 19.1, shouldn't discourage us. In reality, it turns out that there exist nonsingular linear transformations of independent variables, transforming arbitrary Hermitian Toeplitz forms into Hermitian Hankel forms and backwards (here we admit a somewhat loose formulation - for the exact formulation see below). We mean the transformations of E. Fischer and G. Frobenius, to which we turn right now.

Let a and b be arbitrary complex numbers, different from zero, and n a natural number. We assign a linear transformation of (complex) parameters $(\xi_p,\xi_1,\cdots,\xi_{n-1})$ to (complex) parameters $(\eta_0,\eta_1,\cdots,\eta_{n-1})$ in the following way. We consider the identity

$$\xi_0 + \xi_1\varepsilon + \xi_2\varepsilon^2 + \cdots + \xi_{n-1}\varepsilon^{n-1} \equiv$$

$$\equiv (a+\bar{a}\varepsilon)^{n-1}\eta_0 + (a+\bar{a}\varepsilon)^{n-2}(b+\bar{b}\varepsilon)\eta_1 + \cdots + (b+\bar{b}\varepsilon)^{n-1}\eta_{n-1} \qquad (F.-F.)$$

of two polynomials, in which ε is an independent parameter. If we open up in the polynomial, standing on the right, all parentheses, and thereafter compare the coefficients of the equal powers of ε on the left-hand and righthand side, then we obtain relations of the shape

$$\xi_p = \sum_{j=0}^{n-1} \alpha_{pj}\eta_j \qquad (p = 0,1,\cdots, n-1) \qquad (19.4)$$

giving the desired linear transformation, which we shall call the *transformation of Fisher-Frobenius* [1] (abbreviated the *transformation* (F.-F.)).

1° *If the condition*

$$\Delta \equiv \bar{a}b - a\bar{b} \neq 0 \qquad (19.5)$$

is satisfied, then the transformation (19.4) *is nonsingular.*

Indeed, under the condition (19.5) the mutually inverse linear-fractional transformations

$$\theta = \frac{b+\bar{b}\varepsilon}{a+\bar{a}\varepsilon}, \qquad \varepsilon = \frac{a\theta - b}{\bar{b}-\bar{a}\theta}$$

make sense, by means of which the relation (F.-F.) can be rewritten as

$$(a\bar{b}-\bar{a}b)^{n-1}\sum_{j=0}^{n-1}\theta^j\eta_j \equiv \sum_{p=0}^{n-1}(\bar{b}-\bar{a}\theta)^{n-1-p}(a\theta-b)^p\xi_p. \qquad (F.-F.\ bis)$$

Comparing here the coefficients of the equal powers of θ, we obtain the relations

$$\eta_j = \sum_{p=0}^{n-1}\beta_{jp}\xi_p \qquad (j = 0,1,\cdots,n-1) \qquad (19.6)$$

which yield the inverse of the linear transformation (19.4).

19.3 Now we find explicit expressions for the coefficents α_{pj} of the transformation (19.4).

2° *The coefficients of transformation* (19.4) *can be calculated by the formula*

$$\alpha_{pj} = a^{n-1-j} b^{j-p} \bar{b}^{p} \sum_{\mu=0}^{n-1-j} C^{\mu}_{n-1-j} C^{p-\mu}_{j} \left(\frac{\bar{a}b}{a\bar{b}}\right)^{\mu} \qquad (p,j = 0,1,\cdots,n-1), \qquad (19.7)$$

where

$$C^{t}_{s} = \begin{cases} \dfrac{s!}{t!(s-t)!} & \textit{for } s \geqq t \geqq 0, \\ 0 & \textit{for all other } s,t. \end{cases} \qquad (19.8)$$

Indeed, we insert expression (19.4) in the equations (F.-F.) which generate it:

$$\sum_{p=0}^{n-1} \left(\sum_{j=0}^{n-1} \alpha_{pj} \eta_j \right) \varepsilon^{p} = \sum_{j=0}^{n-1} (a + \bar{a}\varepsilon)^{n-1-j} (b + \bar{b}\varepsilon)^{j} \eta_j .$$

Changing the order of summation on the lefthand side and comparing the corresponding coefficients for arbitrary values η_j we obtain

$$\sum_{p=0}^{n-1} \alpha_{pj} \varepsilon^{p} = (a + \bar{a}\varepsilon)^{n-1-j} (b + \bar{b}\varepsilon)^{j} \quad (j = 0,1,\cdots,n-1) \qquad (19.9)$$

or

$$\sum_{p=0}^{n-1} \alpha_{pj} \varepsilon^{p} = a^{n-1-j} b^{j} \sum_{\mu=0}^{n-1-j} C^{\mu}_{n-1-j} \left(\frac{\bar{a}}{a}\right)^{\mu} \varepsilon^{\mu} \sum_{\nu=0}^{j} C^{\nu}_{j} \left(\frac{\bar{b}}{b}\right)^{\nu} \varepsilon^{\nu} .$$

Now we carry out the multiplication on the right hand side and compare the coefficients of ε^{p} on both sides:

$$\alpha_{pj} = a^{n-1-j} b^{j} \sum_{\substack{\mu=0 \\ (\mu+\nu=p)}}^{n-1-j} \sum_{\nu=0}^{j} C^{\mu}_{n-1-j} C^{\nu}_{j} \left(\frac{\bar{a}}{a}\right)^{\mu} \left(\frac{\bar{b}}{b}\right)^{\nu} \qquad (p,j = 0,1,\cdots,n-1).$$

It remains to substitute $p-\mu$ for ν, and then, taking into account the convention (19.8), we obtain (19.7).

In the case, where $\Delta = a\bar{b} - \bar{a}b \neq 0$, one can, in an analoguous way, calculate the coefficients β_{jp} of the (inverse) transformation (19.6).

3° *There hold (with the notations* (19.8)) *the formulae*

$$\beta_{jb} = \frac{1}{\Delta^{n-1}} \bar{b}^{n-1-p} (-b)^{p-j} a^{j} \sum_{\mu=0}^{n-1-p} C^{\mu}_{n-1-p} C^{j-\mu}_{p} \left(\frac{\bar{a}b}{a\bar{b}}\right)^{\mu} \qquad (j,p = 0,1,\cdots,n-1). \qquad (19.10)$$

For the proof we insert in (F.-F. bis) the expressions (19.6)

$$\Delta^{n-1} \sum_{j=0}^{n-1} \left(\sum_{p=0}^{n-1} \beta_{jp} \xi_p \right) \theta^{j} = \sum_{p=0}^{n-1} (\bar{b} - \bar{a}\theta)^{n-1-p} (a\theta - b)^{p} \xi_p .$$

Comparison of the coefficients of ξ_p yields

$$\sum_{j=0}^{n-1} \beta_{jp}\theta^j = \frac{1}{\Delta^{n-1}}(\bar{b}-\bar{a}\theta)^{n-1-p}(a\theta-b)^p \quad (p = 0,1,\cdots,n-1) \tag{19.11}$$

whence (19.1o) is obtained without difficulty [2).]

From formula (19.7) one easily discoveres a property of the coefficients α_{pj} and β_{jp} of the transformations (19.4) and (19.6), respectively, which is useful for further calculations.

4° *For* p,j = o,1,$\cdots$,n-1 *one has*

$$\alpha_{n-1-p,j} = \bar{\alpha}_{pj} \tag{19.12}$$

$$\beta_{j,n-1-p} = \bar{\beta}_{jp} \tag{19.13}$$

For the proof of relations (19.12) it suffices to substitute everywhere in formula (19.7) n-1-p instead of p, and then perform the transformation of the summation indices: $\mu' = n-1-j-\mu$. It is somewhat complicated to derive (19.13) from (19.1o) (cf. exercise 5 at the end of this section), but one can obtain it at once from 19.12, if one remembers, that the matrices $\|\alpha_{pj}\|_{p,j=o}^{n-1}$ and $\|\beta_{jp}\|_{j,p=o}^{n-1}$ are mutually inverse (since the transformations (19.4) and (19.6) are such). This simple verification is also left to the reader. Finally, one can obtain both formulae (19.12) and (19.13) without resorting, in general, to the explicit formulae (19.7) and (19.1o) for the coefficients α_{pj} and β_{jp} - see exercise 6 at the end of this section.

19.4 Now we shall explain, how an arbitrary Toeplitz form

$$T_{n-1}(x,x) = \sum_{p,q=0}^{n-1} c_{p-q}\xi_p\bar{\xi}_q \tag{19.14}$$

where

$$c_{-p} = \bar{c}_p \qquad (p = 0,1,\cdots,n-1) \tag{19.15}$$

transforms as result of the parameter change (19.4). To this end we insert the expressions (19.4) into (19.14):

$$T_{n-1}(x,x) = \sum_{p,q=0}^{n-1} c_{p-q} \sum_{j=0}^{n-1}\alpha_{pj}\eta_j \sum_{k=0}^{n-1}\bar{\alpha}_{qk}\bar{\eta}_k = \sum_{j,k=0}^{n-1} s_{jk}\eta_j\bar{\eta}_k, \tag{19.16}$$

where

$$s_{jk} = \sum_{p,q=0}^{n-1} c_{p-q}\alpha_{pj}\bar{\alpha}_{qk} \qquad (j,k = 0,1,\cdots,n-1).$$

Hence is clear at once, that the matrix $\|s_{jk}\|_{j,k=o}^{n-1}$ is Hermitian. Indeed in view of (19.5), we have

$$\bar{s}_{jk} = \sum_{p,q=0}^{n-1} \bar{c}_{p-q}\bar{\alpha}_{pj}\alpha_{qk} = \sum_{p,q=0}^{n-1} c_{q-p}\alpha_{qk}\bar{\alpha}_{pj} = s_{kj} \quad (j,k = 0,1,\cdots,n-1).$$

Now we show, that this matrix is a Hankel matrix. To this end we calculate an explicit shape for the coefficients s_{jk}, using formulae (19.7). We have

$$s_{jk} = a^{n-1-j}b^j\bar{a}^{n-1-k}\bar{b}^k \sum_{p,q=0}^{n-1} c_{p-q} b^{-(p-q)}\bar{b}^{p-q} \times$$

$$\times \sum_{\mu=0}^{n-1-j} C^{\mu}_{n-1-j}C^{p-\mu}_{j}\left(\frac{a\bar{b}}{\bar{a}b}\right)^{\mu} \sum_{\nu=0}^{n-1-k} C^{\nu}_{n-1-k}C^{q-\nu}_{k}\left(\frac{\bar{a}b}{a\bar{b}}\right)^{\nu}.$$

In the latter of the sums $\left(\sum_{\nu=0}^{n-1-k}\right)$ we perform the transformation of the summation index: $\nu = n-1-k-\nu'$ and we use the formula

$$C^t_s = C^{s-t}_s$$

(which holds for arbitrary numbers s and t because of (19.8)). Then, returning to the original designation of the summation index (and taking (19.8) into account all the time), we obtain

$$s_{jk} = a^{n-1-j}b^j\bar{a}^{n-1-k}\bar{b}^k \sum_{p,q=0}^{n-1} c_{p-q} b^{-(p-q)}\bar{b}^{p-q} \times$$

$$\times \sum_{\mu=0}^{n-1-j} C^{\mu}_{n-1-j}C^{p-\mu}_{j}\left(\frac{a\bar{b}}{\bar{a}b}\right)^{\mu} \sum_{\nu=0}^{n-1-k} C^{\nu}_{n-1-k}C^{n-1-q-\nu}_{k}\left(\frac{a\bar{b}}{\bar{a}b}\right)^{n-1-k-\nu} =$$

$$= a^{2(n-1)-(j+k)}b^{j+k-(n-1)}\bar{b}^{n-1} \sum_{\mu=0}^{n-1-j}\sum_{\nu=0}^{n-1-k} C^{\mu}_{n-1-j}C^{\nu}_{n-1-k}\left(\frac{\bar{a}b}{a\bar{b}}\right)^{\mu+\nu} \times$$

$$\times \sum_{p,q=0}^{n-1} c_{p-q}\left(\frac{\bar{b}}{b}\right)^{p-q} C^{p-\mu}_{j}C^{n-1-q-\nu}_{k} \quad (j,k = 0,1,\cdots,n-1).$$

Now we group together in the latter of the sums $\left(\sum_{p,q=0}^{n-1}\right)$ the terms, in which the difference

$$p-q \ (\equiv\sigma)$$

retains one and the same value. We obtain

$$s_{jk} = a^{2(n-1)}\left(\frac{b}{a}\right)^{j+k}\left(\frac{\bar{b}}{b}\right)^{n-1} \sum_{\mu=0}^{n-1-j}\sum_{\nu=0}^{n-1-k} C^{\mu}_{n-1-j}C^{\nu}_{n-1-k}\left(\frac{\bar{a}b}{a\bar{b}}\right)^{\mu+\nu} \times$$

$$\times \sum_{\sigma=-(n-1)}^{n-1}\left(\sum_{p-q=\sigma} C^{p-\mu}_{j}C^{m-1-q-\nu}_{k}\right) c_{\sigma}\left(\frac{\bar{b}}{b}\right)^{\sigma}$$

$$(j,k = 0,1,\cdots,n-1).$$

For the calculation of the sum $\left(\sum_{p-q=\sigma}\right)$, standing between the parentheses we use the formula

$$C^0_rC^m_s + C^1_rC^{m-1}_s + \cdots + C^m_rC^0_s = C^m_{r+s} \qquad (19.17)$$

well-known from combinatorics (see, for example, [12], p. 56). If we take into account (19.8) and then the fact, that $p,q = 0,1,\cdots,n-1$, then it follows from (19.17) that

$$\sum_{p-q=\sigma} C_j^{p-\mu} C_k^{n-1-q-\nu} = C_{j+k}^{\sigma+(n-1)-(\mu+\nu)}$$

$(j,k = 0,1,\cdots,n-1;\ \mu = 0,1,\cdots,n-1-j;\ \nu = 0,1,\cdots,n-1-k;$
$\sigma = 0,\pm 1,\cdots,\pm(n-1))$.

Thus

$$s_{jk} = a^{2(n-1)} \left(\frac{b}{a}\right)^{j+k} \left(\frac{\bar b}{b}\right)^{n-1} \sum_{\sigma=-(n-1)}^{n-1} c_\sigma \left(\frac{\bar b}{b}\right)^{\sigma} \times$$

$$\times \sum_{\mu=0}^{n-1-j} \sum_{\nu=0}^{n-1-k} C_{n-1-j}^{\mu} C_{n-1-k}^{\nu} \left(\frac{\bar a b}{a \bar b}\right)^{\mu+\nu} C_{j+k}^{\sigma+(n-1)-(\mu+\nu)}$$

$$(j,k = 0,1,\cdots, n-1).$$

Now grouping on the right the terms with the same summation indices μ and ν ($\mu+\nu \equiv r$) and again applying formula (19.17), we obtain

$$s_{jk} = a^{2(n-1)} \left(\frac{b}{a}\right)^{j+k} \left(\frac{\bar b}{b}\right)^{n-1} \sum_{\sigma=-(n-1)}^{n-1} c_\sigma \left(\frac{\bar b}{b}\right)^{\sigma} \times$$

$$\times \sum_{r=0}^{2(n-1)-(j+k)} \left(\sum_{\mu+\nu=r} C_{n-1-j}^{\mu} C_{n-1-k}^{\nu} \right) \left(\frac{\bar a b}{a \bar b}\right)^{r} C_{j+k}^{\sigma+(n-1)-r} =$$

$$= a^{2(n-1)} \left(\frac{b}{a}\right)^{j+k} \left(\frac{\bar b}{b}\right)^{n-1} \sum_{\sigma=-(n-1)}^{n-1} c_\sigma \left(\frac{\bar b}{b}\right)^{\sigma} \times$$

$$\times \sum_{r=0}^{2(n-1)-(j+k)} C_{2(n-1)-(j+k)}^{r} C_{r+k}^{\sigma+(n-1)-r} \left(\frac{\bar a b}{a \bar b}\right)^{r}, \qquad (19.18)$$

i.e., the coefficients s_{jk} of the transformed form (19.16) depend only on the sum $j+k$ of the indices j and k.

In particular, since the matrix $\|s_{jk}\|_{j,k=0}^{n-1}$ is Hermitian, it follows thence, that all coefficients s_{jk} are real (cf. Sec. 9.1; see also exercise 11 at the end of this section).

Thus is proved the proposition

5° *Every Hermitian Toeplitz form* (19.14) *is by a transformation* (19.4) *transformed into a Hermitian Hankel form.*

19.5 We now take an arbitrary Hankel form

$$H_{n-1}(y,y) = \sum_{j,k=0}^{n-1} s_{j+k} \eta_j \bar\eta_k \qquad (19.19)$$

with real coefficients

$$s_{j+k} = \bar s_{j+k} \qquad (j,k = 0,1,\cdots,n-1)$$

and we apply to it a transformation (19.6), induced by the identity (F.-F. bis) where $\Delta = a\bar{b} - \bar{a}b \neq 0$. We obtain

$$H_{n-1}(y,y) = \sum_{j,k=0}^{n-1} s_{j+k} \sum_{p=0}^{n-1} \beta_{jp}\xi_p \sum_{q=0}^{n-1} \bar{\beta}_{kq}\bar{\xi}_q = \sum_{p,q=0}^{n-1} c_{pq}\xi_p\bar{\xi}_q,$$

where

$$c_{pq} = \sum_{j,k=0}^{n-1} s_{j+k}\beta_{jp}\bar{\beta}_{kq} \quad (p,q = 0,1,\cdots,n-1).$$

Hence is clear at once, that the matrix $\|c_{pq}\|_{p,q=0}^{n}$ is Hermitian. Indeed,

$$c_{qp} = \sum_{j,k=0}^{n-1} s_{j+k}\beta_{jq}\bar{\beta}_{kp} = \overline{\sum_{j,k=0}^{n-1} s_{j+k}\bar{\beta}_{jq}\beta_{kp}} = \bar{c}_{pq} \quad (p,q = 0,1,\cdots,n-1).$$

Now we show that this matrix is a Toeplitz matrix. We apply at first proposition 4°, namely formula (19.13). We have

$$c_{pq} = \sum_{j,k=0}^{n-1} s_{j+k}\beta_{jp}\beta_{k,n-1-q} \quad (p,q = 0,1,\cdots,n-1).$$

Here we insert the values of the quantities β_{jp} (and correspondingly $\beta_{k,n-1-q}$), obtained in 3° (see (19.10)), and we transform the obtained expression through the same method as in the proof of proposition 5° (in particular, we use formula (19.17) twice). We obtain

$$c_{pq} = \frac{\bar{b}^{n-1-p}(-b)^p\bar{b}^q(-b)^{n-1-q}}{\Delta^{2(n-1)}} \sum_{j,k=0}^{n-1} s_{j+k}\left(\frac{a}{-b}\right)^{j+k} \times$$

$$\times \sum_{\mu=0}^{n-1-p} C^{\mu}_{n-1-p} C^{j-\mu}_{p} \left(\frac{\bar{a}b}{a\bar{b}}\right)^{\mu} \sum_{\nu=0}^{q} C^{\nu}_{q} C^{k-\nu}_{n-1-q} \left(\frac{\bar{a}b}{a\bar{b}}\right)^{\nu} = \frac{\bar{b}^{n-1-(p-q)}(-b)^{n-1+p-q}}{\Delta^{2(n-1)}} \times$$

$$\times \sum_{\mu=0}^{n-1-p} \sum_{\nu=0}^{q} C^{\mu}_{n-1-p} C^{\nu}_{q} \left(\frac{\bar{a}b}{a\bar{b}}\right)^{\mu+\nu} \sum_{j,k=0}^{n-1} s_{j+k}\left(\frac{a}{-b}\right)^{j+k} C^{j-\mu}_{p} C^{k-\nu}_{n-1-q} =$$

$$= \frac{\bar{b}^{n-1-(p-1)}(-b)^{n-1+p-q}}{\Delta^{2(n-1)}} \times$$

$$\times \sum_{\mu=0}^{n-1-p} \sum_{\nu=0}^{q} C^{\mu}_{n-1-p} C^{\nu}_{q} \left(\frac{\bar{a}b}{a\bar{b}}\right)^{\mu+\nu} \sum_{r=0}^{2n-2} \left(\sum_{j+k=r} C^{j-\mu}_{p} C^{k-\nu}_{n-1-q} \right) s_r \left(\frac{a}{-b}\right)^{r} =$$

$$= \frac{\bar{b}^{n-1-(p-q)}(-b)^{n-1+p-q}}{\Delta^{2(n-1)}} \times$$

$$\times \sum_{r=0}^{2(n-1)} s_r \left(\frac{a}{-b}\right)^{r} \sum_{\sigma=0}^{n-1-(p-1)} \left(\sum_{\mu+\nu=\sigma} C^{\mu}_{n-1-p} C^{\nu}_{q} \right) \left(\frac{\bar{a}b}{a\bar{b}}\right)^{\sigma} C^{r-\sigma}_{n-1+p-q} =$$

$$= \frac{\bar{b}^{n-1-(p-q)}(-b)^{n-1+p-q}}{\Delta^{2(n-1)}} \sum_{r=0}^{2(n-1)} s_r \left(\frac{a}{-b}\right)^{r} \times$$

$$\times \sum_{\sigma=0}^{n-1-(p-q)} C^{\sigma}_{n-1-(p-q)} C^{r-\sigma}_{n-1+(p-q)} \left(\frac{\bar{a}b}{a\bar{b}}\right)^{\sigma}$$

$$(p,q = 0,1,\cdots,n-1), \qquad (19.2o)$$

i.e., the coefficients c_{pq} depend only on the difference p-q of the indices p and q.

Thus is proved the proposition

6° *Every Hermitian Hankel form* (19.19) *is by a transformation* (19.6) *transformed into a Hermitian Toeplitz form.*

19.6. Summing up the proposition 5° and 6° we formulate the basic result of the present section, which is contained in them:

THEOREM 19.1. *Let* a *and* b *be complex numbers and* $\Delta = a\bar{b} - \bar{a}b \neq 0$. *Then the linear transformation* (19.4) *of Fischer-Frobenius, induced by the identity* (F.-F.), *is nonsingular. This transformation and its inverse transformation* (19.6) *(induced by the identity* (F.-F. bis)) *determine one - to - one relations between all Hermitian Toeplitz forms of order* n *and all Hermitian Hankel forms of the same order* n.

A result of such type was established for the first time by E. Fischer [18] (for a = 1/2, b = -i/2) for nonnegative Toeplitz forms of a special shape.

For arbitrary a and b G. Frobenius [2o] extended this result (precisely, the result, formulated above in proposition 5°) to arbitrary nonnegative Toeplitz forms. In its full generality Theorem 19.1 was established in [41] in two ways (both different from the straightforward calculation adduced above). We shall present one of these in Sec. 19.7., in view of its methodical interest. At this it will be necessary to rely on two classical theorems on the polynomial - and trigonometrical moment problem. For the sake of completeness the proof of these Theorems is presented below in Appendix II (see Theorem A.II.1 and A.II.2). There the reader also will find some helpful transformations of real quadratic Toeplitz forms into sums of Hankel forms (see Theorem A.II.3).

19.7. For the case of positive definite forms propositions 5° and 6° might be obtained much faster through the application of well-known results from the theory of moments (polynomial - and trigonometrical).

So, let be given a positive definite Toeplitz form

$$T_{n-1}(x,x) = \sum_{p,q=o}^{n-1} c_{p-q}\xi_p\bar{\xi}_q \ .$$

We perform on its independent parameters $\xi_o,\xi_1,\cdots,\xi_{n-1}$ the transformation (19.4), induced by the identity (F.-F.) (see Sec. 19.2). The coeficients c_p of the positive definite form $T_{n-1}(x,x)$ allow, by Theorem A.II.2, the following representation:

$$c_p = \sum_{k=1}^{n} r_k \varepsilon_k^p, \quad r_k > 0, \quad |\varepsilon_k| = 1 \qquad (k = 1,2,\cdots n; \; p = 0,\pm 1,\cdots,\pm(n-1)).$$

Introducing this expression into $T_{n-1}(x,x)$, and taking (F.-F.) into account, we obtain

$$\sum_{p,q=0}^{n-1} c_{p-q}\xi_p\bar{\xi}_q = \sum_{p,q=0}^{n-1}\sum_{k=1}^{n} r_k\varepsilon_k^{p-q}\xi_p\bar{\xi}_q = \sum_{k=1}^{n} r_k \sum_{p=0}^{n-1}\xi_p\varepsilon_k^p \sum_{q=0}^{n-1}\bar{\xi}_q\bar{\varepsilon}_k^q =$$

$$= \sum_{k=1}^{n} r_k \sum_{p,q=0}^{n-1} (a+\bar{a}\varepsilon_k)^{n-1-p}(b+\bar{b}\varepsilon_k)^p(\bar{a}+a\bar{\varepsilon}_k)^{n-1-q} \times$$

$$\times \quad (\bar{b}+b\bar{\varepsilon}_k)^q\eta_p\bar{\eta}_q = \sum_{p,q=0}^{n-1} s_{pq}\eta_p\bar{\eta}_q,$$

where

$$s_{pq} = \sum_{k=1}^{n} r_k\bar{\varepsilon}_k^{n-1}(a+\bar{a}\varepsilon_k)^{2n-2-(p+q)}(b+\bar{b}\varepsilon_k)^{p+q} \qquad (p,q = 0,1,\cdots,n-1),$$

i.e., each coefficient $s_{pq} \equiv s_{p+q}$ depends only on the sum p+q of the indices p and q and is, as one easily verifies, a real number. The latter, however, one can establish very simply, not using the positive definiteness of the form $T_{n-1}(x,x)$ and in general not resorting to any kind of formula for the coefficients c_p (see exercise 11 at the end of this section) [3].

Conversely, let

$$H_{n-1}(y,y) = \sum_{j,k=0}^{n-1} s_{j+k}\eta_j\eta_k, \quad s_k = \bar{s}_k \qquad (k = 0,1,\cdots,2n-2)$$

be a positive definite Hermitian Hankel form. We insert into it transformation (19.6), induced by identity (F.-F. bis), where $\Delta = a\bar{b} - \bar{a}b \neq 0$. By Theorem A.II.1 the positive definiteness of the form $H_{n-1}(y,y)$ is equivalent to the fact, that its coefficients allow the representation [4)]

$$s_k = \sum_{\nu=1}^{n} \rho_\nu\theta_\nu^k, \quad \rho_\nu > 0, \quad \theta_\nu = \bar{\theta}_\nu \quad (\nu = 1,2,\cdots,n; \; k = 0,1,\cdots,2n-2).$$

Introducing this expression into the form $H_{n-1}(y,y)$, and taking (F.-F. bis) into account, we have

$$\sum_{j,k=0}^{n-1} s_{j+k}\eta_j\bar{\eta}_k = \sum_{j,k=0}^{n-1}\sum_{\nu=1}^{n}\rho_\nu\theta_\nu^{j+k}\eta_j\bar{\eta}_k = \sum_{\nu=1}^{n}\rho_\nu\sum_{j=0}^{n-1}\eta_j\theta_\nu^j\sum_{k=0}^{n-1}\bar{\eta}_k\theta_\nu^k =$$

$$= \frac{1}{\Delta^{2(n-1)}}\sum_{\nu=1}^{n}\rho_\nu\sum_{p=0}^{n-1}(\bar{b}-\bar{a}\theta_\nu)^{n-1-p}(a\theta_\nu - b)^p\xi_p \quad \times$$

$$\times \sum_{q=0}^{n-1} (b-a\theta_\nu)^{n-1-q}(\bar{a}\bar{\theta}_\nu-\bar{b})^q\bar{\xi}_q = \sum_{p,q=0}^{n-1} c_{pq}\xi_p\bar{\xi}_q,$$

where

$$c_{pq} = \frac{(-1)^{p-q}}{\Delta^{2(n-1)}} \sum_{\nu=1}^{n} \rho_\nu (a\theta_\nu - b)^{n-1+p-q}(\bar{a}\bar{\theta}_\nu - \bar{b})^{n-1-(p-q)}$$

$$(p,q = 0,1,\cdots,n-1).$$

Hence is clear, that $c_{qp} = \bar{c}_{pq}$ $(\equiv c_{p-q})$ depends only on the difference p-q of the indices p and q.

The results established above allow to give a new proof of Theorem 19.1.

Indeed, the coefficients s_{jk} of an arbitrary form, obtained through a linear transformation of a Toeplitz form $T_{n-1}(x,x)$, are linear functions (forms) of the coefficients c_p of the original form. In particular, for the case of the linear transformation (19.4) these linear forms were written down explicitly above (see formula (19.16) and the next). Now we fix in the form $T_{n-1}(x,x)$ all coefficients c_p with $p \neq 0$, and the coefficient c_0 we choose positive and so large, that all successive principal minors

$$D_0 = c_0, D_1 = \begin{vmatrix} c_0 & c_{-1} \\ c_1 & c_0 \end{vmatrix}, \ldots, D_{n-1} = \begin{vmatrix} c_0 & c_{-1} & \cdots & c_{-n+1} \\ c_1 & c_0 & \cdots & c_{-n+2} \\ \cdots & \cdots & \cdots & \cdots \\ c_{n-1} & c_{n-2} & \cdots & c_0 \end{vmatrix}$$

of the form $T_{n-1}(x,x)$ become positive. This is possible, since $D_0, D_1, \cdots, D_{n-1}$ are polynomials in c_0 with leading coefficient equal to one. But then the form $T_{n-1}(x,x)$ becomes, because of Theorem 8.1 and Corollary 2 of Theorem 5.2, positive definite and, as was shown above, the transformation (19.4) turns it into a Hankel form with real coefficients. This means, that the real numbers s_{jk} (the coefficients of the obtained Hankel form) depend only on the sum j+k of the indices j and k, i.e.,

$$s_{j+1,k} = s_{j,k+1} \quad (j,k = 0,1,\cdots,n-1;\ j+k < 2n-2).$$

But the last relations remain, as equations between linear functions in c_0 (all other coefficients c_p are fixed), which hold for all sufficiently large c_0, evidently identities for arbitrary c_0, i.e., for arbitrary forms $T_{n-1}(x,x)$. Thus the first statement of Theorem 19.1 is proved.

Now we insert the transformation (19.6), which is the inverse with respect to (19.4), into the arbitrary Hankel form

$$H_{n-1}(y,y) = \sum_{j,k=0}^{n-1} s_{j+k}\eta_j\overline{\eta}_k.$$

We obtain

$$\sum_{j,k=0}^{n-1} s_{j+k}\eta_j\overline{\eta}_k = \sum_{j,k=0}^{n-1} s_{j+k} \sum_{p=0}^{n-1} \beta_{jp}\xi_p \sum_{q=0}^{n-1} \overline{\beta}_{kq}\overline{\xi}_q \equiv \sum_{p,q=0}^{n-1} c_{pq}\xi_p\overline{\xi}_q ,$$

where

$$c_{pq} = \sum_{j,k=0}^{n-1} s_{j+k}\beta_{jp}\overline{\beta}_{kq} \quad (p,q = 0,1,\cdots,n-1).$$

Hence is clear, that the matrix $\|c_{pq}\|_{p,q=0}^{n-1}$ is Hermitian:

$$\overline{c}_{qp} = c_{pq} \qquad (p,q = 0,1,\cdots,n-1),$$

and its entries are linear forms in the parameters $s_0,s_1,\cdots,s_{2n-2}$. Moreover, we know, that in the case of a positive definite form $H_{n-1}(y,y)$

$$c_{pq} = c_{p-1,q-1}, \quad (p,q = 1,2,\cdots,n-1). \tag{19.21}$$

We fix in the form $H_{n-1}(y,y)$ coefficients $s_1,s_3,\cdots,s_{2n-3}$, and the coefficients $s_0,s_2,\cdots,s_{2n-2}$ we replace so, that the form $H_{n-1}(y,y)$ becomes positive definite. Namely, we choose at first, and then we fix $(D_0=)\ s_0 > 0$. Since

$$D_1 = \begin{vmatrix} s_0 & s_1 \\ s_1 & s_2 \end{vmatrix} = s_0s_2 - s_1^2 ,$$

we have $D_1 > 0$ for sufficiently large (positive) s_2. We fix such a s_2, and a positive s_4 we choose so large, that

$$D_2 = \begin{vmatrix} s_0 & s_1 & s_2 \\ s_1 & s_2 & s_3 \\ s_2 & s_3 & s_4 \end{vmatrix} = s_4D_1 + \cdots > 0$$

and so on. As result we obtain for certain sufficiently large $s_0,s_2,\cdots,$ s_{2n-2}

$$D_0 > 0,\ D_1 > 0,\ \cdots,\ D_{n-1} > 0,$$

where it is clear from this reasoning, that such a set $s_0,s_2,\cdots,s_{2n-2}$ can be composed in infinitely many ways, such that all the values of all parameters $s_0,s_2,\cdots,s_{2n-2}$ are changed at once. But for all such sets hold, as we know, the relations (19.21), in which both sides (for fixed p and q) are linear function in $s_0,s_2,\cdots,s_{2n-2}$. So, these relations are realized identically relative to $s_0,s_2,\cdots,s_{2n-2}$, i.e., for arbitrary Hankel forms $H_{n-1}(y,y)$.

Theorem 19.1 is proved.

EXAMPLES AND EXERCISES

1. (F.I. Lander). Prove that transformation (19.4) can be rewritten in the equivalent "symbolic" shape

$$\xi_p = C_{n-1}^p (a+b\omega)^{n-p-1}(\bar{a}+\bar{b}\omega)^p \qquad (p = 0,1,\cdots,n-1),$$

where one must perform on the righthand side, after expanding the parentheses, the substitution

$$\omega^j = \frac{1}{C_{n-1}^j}\eta_j \qquad (j = 0,1,\cdots,n-1).$$

HINT. Put $\eta_j = C_{n-1}^j\omega^j$ $(j = 0,1,\cdots,n-1)$ in (F.-F.) and calculate the successive derivatives to ε in the point $\varepsilon = 0$.

2. Derive formulae (19.7) from the result of exercise 1.

3. Establish in analogy to exercise 1 for the transformation (19.6) the equivalent "symbolic" form

$$\eta_j = \frac{1}{\Delta^{n-1}}\frac{(n-1)!}{(n-j-1)!j!}(\bar{b}-b\varphi)^{n-1-j}(a\varphi-\bar{a})^j \qquad (j = 0,1,\cdots,n-1),$$

where $\Delta = a\bar{b}-\bar{a}b$, and after expanding the parentheses on the right the transformation

$$\varphi^p = \frac{1}{C_{n-1}^p}\xi_p \qquad (p = 0,1,\cdots,n-1)$$

is performed.

4. Derive the formulae (19.10) from the result of exercise 3.

5. Derive formulae (19.13) directly from (19.10).

HINT. Having substituted $n-1-p$ instead of p in (19.10), perform on the righthand side the transformation of the summation index μ to $\mu' = j-\mu$, and take into account (19.8).

6. Derive formulae (19.12) and (19.13), not using, in general, the direct expressions (19.7) and (19.10) for the coefficients α_{pj} and β_{jp}, respectively.

HINT. Use respective the identities (F.-F) and (F.-F. bis), assuming $|\varepsilon| = 1$, $\theta = \bar{\theta}$ in them.

7. (E. Fischer [19].) Prove, that the parameters $\eta_0,\eta_1,\cdots,\eta_{n-1}$ in the transformation (F.-F.) are real if and only if the parameters $\xi_0,\xi_1,\cdots,\xi_{n-1}$ satisfy the condition

$$\xi_{n-1-p} = \bar{\xi}_p \qquad (p = 0,1,\cdots,n-1).$$

HINT. Use formulae (19.12) and (19.13).

8. As one easily perceives directly from the identity (F.-F.), the formulae (19.4) appear as follows:

for n = 2

$$\xi_0 = a\eta_0 + b\eta_1,$$
$$\xi_1 = \bar{a}\eta_0 + \bar{b}\eta_1;$$

for n = 3

$$\xi_0 = a^2\eta_0 + ab\eta_1 + b^2\eta_2,$$
$$\xi_1 = 2a\bar{a}\eta_0 + (a\bar{b} + \bar{a}b)\eta_1 + 2b\bar{b}\eta_2,$$
$$\xi_2 = \bar{a}^2\eta_0 + \bar{a}\bar{b}\eta_1 + \bar{b}^2\eta_2.$$

Compare these equations with the direct formulae (19.7) for the coef-ficents α_{pj} of transformation (19.4).

9. In analogy to exercise 8 we find from the identity (F.-F. bis), that formulae (19.6) can be rewritten as ($\Delta = a\bar{b} - \bar{a}b$):

for n = 2

$$\eta_0 = \frac{1}{\Delta}(\bar{b}\xi_0 - b\xi_1),$$
$$\eta_1 = \frac{1}{\Delta}(-\bar{a}\xi_0 + a\xi_1);$$

for n = 3

$$\eta_0 = \frac{1}{\Delta^2}(\bar{b}^2\xi_0 - \bar{b}b\xi_1 + b^2\xi_2),$$
$$\eta_1 = \frac{1}{\Delta^2}(-2\bar{b}\bar{a}\xi_0 + [\bar{b}a + b\bar{a}]\xi_1 - 2ba\xi_2),$$
$$\eta_2 = \frac{1}{\Delta^2}(\bar{a}^2\xi_0 - \bar{a}a\xi_1 + a^2\xi_2).$$

Compare these equations with the direct formulae (19.1o) for the co-efficients β_{jp} of transformation (19.6) and also with note [2)] below.

1o. Verify that the matrix $\|s_{jk}\|_{j,k=0}^{n-1}$ is real and the matrix $\|c_{pq}\|_{p,q=0}^{n-1}$ Hermitian, where the matrices are defined by the formulae (19.18) and (19.2o), respectively, starting directly from these formulae (cf. with exercise 11).

11. In the begin of Sec. 19.4 was shown, that after the transformation (19.4) the form (19.14) turns into the Hermitian form (19.16): $\bar{s}_{jk} = s_{kj}$ $(j,k = 0,1,\cdots,n-1)$. Show, not using (19.18), that the coef-ficients are real: $\bar{s}_{jk} = s_{jk}$ $(j,k = 0,1,\cdots,n-1)$.

HINT. Use the relations (19.12).

NOTES

1) The relation (F.-F) was suggested by G. Frobenius [2o]. He generalized a transformation, introduced earlier by E. Fischer [18], which is obtained from (F.-F.) for the particular values $a = 1/2$, $b = -i/2$.

2) It suffices to compare the formulae (19.9) and (19.11) in order to observe, that for the transition from the first to the second (and so, also from (19.7) to (19.1o)) one must insert on the lefthand side the factor Δ^{n-1} and then replace the letters: ε for θ, α for β, j for p, p for j, a for $\bar{b}$, $\bar{a}$ for $(-\bar{a})$, b for (-b) and $\bar{b}$ for a.

3) From the remark to Theorem A.II.2 follows, that the whole followed argument is also applicable to nonnegative (degenerate) forms $T_{n-1}(x,x)$.

4) Here is everywhere, by definition, $0^0 = 1$. We note, that in [41] by mistake only nonnegativity of the form $H_{n-1}(y,y)$ was required (cf. below the remark to Theorem A.II.1).

REMARK. For new methods and numerical results concerning the inversion of Hankel- and Toeplitz matrices (both normal and block matrices), and also for the algebraic structure of spaces of such matrices and new transformations, defined over Hankel- and Toeplitz matrices, see [1], [4],[6],[1o],[11],[12],[13] and [16] of the additional list of references.

I. THE THEOREMS OF BORHARDT-JACOBI AND OF HERGLOTZ-M. KREĬN ON THE ROOTS OF REAL AND HERMITIAN-SYMMETRIC POLYNOMIALS.

1. We show here one of the spheres of direct application of the results concerning the signature of Hankel- and Toeplitz forms, presented in §§ 12 and 16. We mean the problem of the distribution of the roots of a real polynomial

$$P_n(\lambda) = a_0\lambda^n + a_1\lambda^{n-1} + \cdots + a_{n-1}\lambda + a_n \quad (\bar{a}_k = a_k;\ k = 0,1,\cdots,n) \qquad \text{(A.I.1)}$$

or a Hermitian-symmetric polynomial

$$Q_n(\lambda) = b_0\lambda^n + b_1\lambda^{n-1} + \cdots + b_{n-1}\lambda + b_n \quad (b_k = \bar{b}_{n-k};\ k = 0,1,\cdots,n) \qquad \text{(A.I.2)}$$

with respect to the real axis or the unit circle, respectively.

We recall, that the *roots of the polynomial* $P_n(\lambda)$ (see (A.I.1)) *are always symmetrically distributed with respect to the real axis*, where the nonreal complex-conjugated pairs (if they exist) consist of roots of the same multiplicity. *A quite analoguous picture is valid for the roots of the polynomial* $Q_n(\lambda)$ (see (A.I.2)), *but with the unit circle substituted for the real axis*, i.e., instead of real roots one must speak of roots with modulus one, and instead of complex-conjugated pairs - of pairs of the shape (β,β^*), situated mirror-like (symmetrically) with respect to the unit circle:

$$\beta^* = \frac{1}{\bar{\beta}}.$$

Both these statements, which are quite well-known in the algebraic theory of polynomials, are also easily verified directly (see the exercisis 1,2 below).

2. We consider the real polynomial $P_n(\lambda)$ (A.I.1), and let $\alpha_1,\alpha_2,\cdots,\alpha_n$ be all its roots (here each root is repeated according to its multiplicity). We put together the *Newton sums* [1)]

$$s_k = \alpha_1^k + \alpha_2^k + \cdots + \alpha_n^k \qquad (k = 0,1,\cdots) \qquad \text{(A.I.3)}$$

(here the symbol 0^0 is considered to be equal to 1), and use these as coefficients of the Hankel quadratic form

$$H_{n-1}(x,x) = \sum_{j,k=0}^{n-1} s_{j+k}\xi_j\xi_k \qquad \text{(A.I.4)}$$

(we recall, that because of the properties of the roots of the polynomial $P_n(\lambda)$ all sums s_k are real).

THEOREM A.I.1 (BORHARDT-JACOBI). *If* π *is the number of positive squares and* ν *the number of negative squares of the quadratic form* $H_{n-1}(x,x)$ *(see (A.I.4)), then the polynomial* $P_n(\lambda)$ *has* ν *different pairs of complex-conjugated roots and* $\sigma = \pi - \nu$ *different real roots.*

PROOF. The form $H_{n-1}(x,x)$ can easily be represented as a sum of squares

$$H_{n-1}(x,x) = \sum_{j,k=0}^{n-1} s_{j+k}\xi_j\xi_k = \sum_{j,k=0}^{n-1} (\alpha_1^{j+k} + \alpha_2^{j+k} + \cdots + \alpha_n^{j+k})\xi_j\xi_k =$$

$$= \sum_{j,k=0}^{n-1} (\alpha_1^j\xi_j\alpha_1^k\xi_k + \alpha_2^j\xi_j\alpha_2^k\xi_k + \cdots + \alpha_n^j\xi_j\alpha_n^k\xi_k),$$

i.e.,

$$H_{n-1}(x,x) = \sum_{k=1}^{n} (\xi_0 + \alpha_k\xi_1 + \alpha_k^2\xi_2 + \cdots + \alpha_k^{n-1}\xi_{n-1})^2. \qquad \text{(A.I.5)}$$

One cannot yet call the representation (A.I.5) canonical, (see Sec. 5.2), as among the roots $\alpha_1,\alpha_2,\cdots,\alpha_n$ there may be multiples, i.e., the linear forms corresponding to these are dependent (they simply coincide, and in representation (A.I.5) their squares are repeated as many times as the multiplicity of the corresponding root).

If among the roots $\alpha_1,\alpha_2,\cdots,\alpha_n$ there are p different real and q different pairs of complex-conjugated roots, then after the reduction of similar terms on the righthand side, the representation (A.I.5) turns out to be a sum of squares of $p+2q$ linear forms, which (over the field of complex numbers) are linearly independent; the matrix of their coefficients (see Sec. 5.1), now consisting of the nonnegative entire powers of the $p+2q$ different roots of the polynomial $P_n(\lambda)$, has a rank which is exactly equal to $\rho = p+2q$, since the corresponding Vandermonde determinant is nonzero.

But also the thus transformed representation (A.I.5) is, if $q \neq 0$, still not canonical, since the real quadratic form $H_{n-1}(x,x)$ is here represented as the sum of the squares of p real and 2q nonreal linear forms. To each of these 2q squares of nonreal forms of the shape

$$[M(x) + iN(x)]^2,$$

where $M(x)$ and $N(x)$ are real linear forms in the parameters $\xi_0, \xi_1, \cdots, \xi_{n-1}$ there corresponds in the same sum a square

$$[M(x) - iN(x)]^2.$$

Combining corresponding terms, we obtain

$$[M(x) + iN(x)]^2 + [M(x) - iN(x)]^2 = 2[M(x)]^2 - 2[N(x)]^2.$$

By performing this on all nonreal squares we "reorganize" representation (A.I.5) such, that it now will have $p+q$ positive and q negative squares. It is easy to understand, that all linear forms, which are used for these squares, are independent, since we obtained them from $p+2q$ independent forms through a simple transformation: we substituted for q pairs of independent forms of the shape

$$\{M(x) + iN(x), \; M(x) - iN(x)\}$$

their half-sums $M(x)$ and their - by $2i$ divided - half-differences $N(x)$.

So, if $\sigma = \pi - \nu$ is the signature of the form $H_{n-1}(x,x)$, then $\pi = p+q$ and $\nu = q$, whence

$$p = \pi - \nu = \sigma, \qquad q = \nu.$$

The theorem is proved.

3. In analogy to Theorem A.I.1 there arises

THEOREM A.I.2 (HERGLOTZ - M. KREĬN). *Let $\varepsilon_1, \varepsilon_2, \cdots, \varepsilon_p$ be all different roots of the Hermitian-symmetric polynomial $Q_n(\lambda)$ (see (A.I.2)), which lie on the unit circle, with the mulitplicities $\rho_1, \rho_2, \cdots, \rho_p$, respectively. Let further $\{\beta_1, \beta_1^*\}, \{\beta_2, \beta_2^*\}, \cdots, \{\beta_q, \beta_q^*\}$ be all different pairs of roots of the same polynomial $Q_n(\lambda)$, which are arranged symmetrically ($\beta_k^* = 1/\bar{\beta}_k$, $k = 1,2,\cdots,q$) with respect to the unit circle, with the corresponding multiplicities $\sigma_1, \sigma_2, \cdots, \sigma_q$. Through s_k ($k = 0,1,2,\cdots$) we denote, just as in Theorem A.I.1, the Newton sums of the roots (each repeated with its multiplicity).*

If the Hermitian Toeplitz form

$$T_{n-1}(x,x) = \sum_{j,k=0}^{n-1} s_{j-k}\xi_j\bar{\xi}_k \tag{A.I.6}$$

has π positive and ν negative squares, then the polynomial $Q_n(\lambda)$ has $\pi - \nu (=p)$ different roots ε_k with modulus 1, and $\nu(=q)$ different pairs of roots $\{\beta_k, \beta_k^*\}$ which lie symmetrically with respect to the circle $|\lambda| = 1$.

PROOF. Above all, we note that the Newton sums s_k now have the shape

$$s_k = \sum_{\mu=1}^{p} \rho_\mu \varepsilon_\mu^k + \sum_{\nu=1}^{q} \sigma_\nu (\beta_\nu^k + \beta_\nu^{*k}) \quad (k = 0,1,2,\cdots). \tag{A.I.7}$$

But since

$$\bar{\varepsilon}_\mu^k = \varepsilon_\mu^{-k},\ \overline{\beta^{*k}_\nu} = \beta_\nu^{-k} \qquad (\mu = 1,2,\cdots,p;\ \nu = 1,2,\cdots,q),$$

then

$$s_{-k} = \bar{s}_k \quad (k = 0,1,2,\cdots),$$

so that the form (A.I.6) is, in fact, Hermitian.

According to formula (A.I.7)

$$s_{j-k} = \sum_{\mu=1}^{p} \rho_\mu \varepsilon_\mu^{j-k} + \sum_{\nu=1}^{q} \sigma_\nu [\beta_\nu^{j-k} + \beta_\nu^{*j-k}] =$$

$$= \sum_{\mu=1}^{p} \rho_\mu \varepsilon_\mu^{j} \bar{\varepsilon}_\mu^{k} + \sum_{\nu=1}^{q} \sigma_\nu [\beta_\nu^{j} \bar{\beta}_\nu^{*k} + \bar{\beta}_\nu^{k} \beta_\nu^{*j}].$$

Hence

$$T_{n-1}(x,x) = \sum_{j,k=0}^{n-1} s_{j-k}\xi_j\bar{\xi}_k = \sum_{\mu=1}^{p} \rho_\mu |\xi_0 + \xi_1\varepsilon_\mu + \cdots + \xi_{n-1}\varepsilon_\mu^{n-1}|^2 +$$

$$+ \sum_{\nu=1}^{q} \sigma_\nu (\xi_0 + \xi_1\beta_\nu + \cdots + \xi_{n-1}\beta_\nu^{n-1})(\bar{\xi}_0 + \bar{\xi}_1\bar{\beta}_\nu^{*} + \cdots + \bar{\xi}_{n-1}\bar{\beta}_\nu^{*n-1}) \quad +$$

$$+ \sum_{\nu=1}^{q} \sigma_\nu (\xi_0 + \xi_1\beta_\nu^{*} + \cdots + \xi_{n-1}\beta_\nu^{*n-1})(\bar{\xi}_0 + \bar{\xi}_1\bar{\beta}_\nu + \cdots + \bar{\xi}_{n-1}\bar{\beta}_\nu^{n-1}).$$

Now we consider the $p+2q$ independent linear forms

$$X_\mu(x) = \xi_0 + \xi_1\varepsilon_\mu + \cdots + \xi_{n-1}\varepsilon_\mu^{n-1} \qquad (\mu = 1,2,\cdots,p),$$

$$\left.\begin{aligned} Y_\nu(x) &= \xi_0 + \xi_1\beta_\nu + \cdots + \xi_{n-1}\beta_\nu^{n-1} \\ Z_\nu(x) &= \xi_0 + \xi_1\beta_\nu^{*} + \cdots + \xi_{n-1}\beta_\nu^{*n-1} \end{aligned}\right\} \qquad (\nu = 1,2,\cdots,q)$$

We see, that

$$T_{n-1}(x,x) = \sum_{\mu=1}^{p} \rho_\mu |X_\mu(x)|^2 + \sum_{\nu=1}^{q} \sigma_\nu |Y_\nu(x)\bar{Z}_\nu(x) + Z_\nu(x)\bar{Y}_\nu(x)| =$$

$$= \sum_{\mu=1}^{p} \rho_\mu |X_\mu(x)|^2 + \sum_{\nu=1}^{q} 2\sigma_\nu |U_\nu(x)|^2 - \sum_{\nu=1}^{q} 2\sigma_\nu |V_\nu(x)|^2,$$

where

$$\left.\begin{aligned} U_\nu(x) &= \tfrac{1}{2}[Y_\nu(x) + Z_\nu(x)], \\ V_\nu(x) &= \tfrac{1}{2}[Y_\nu(x) - Z_\nu(x)] \end{aligned}\right. \qquad (\nu = 1,2,\cdots,q).$$

Thus, the form $T_{n-1}(x,x)$ is represented as sum of $p+q$ positive and q negative (independent!) squares, whence the full statement of the Theorem follows.

4. As historical information we note, that Theorem A.I.1 was initially established by Borhardt [15] in a different (more restricted) formulation: *the number of different pairs of complex-conjugated roots*

of the real polynomial (A.I.1) *is equal to the number*

$$V(1, D_1, D_2, \cdots, D_{n-1})$$

of sign variations in the sequence of successive principal minors of the form (A.I.4). Here it is assumed that all

$$D_k \neq 0 \quad (k = 1, 2, \cdots, n-1)$$

(cf. Theorem 8.1).

In the general formulation Theorem A.I.1 was established together with the proof adduced above, by Jacobi and only after his death published by Borhardt [16].

The history of Theorem A.I.2 is also not quite trivial. Being established at first (by a more complicated method) by Herglotz [27], it was a few years later proved independently by M.G. Kreĭn [44], moreover, with an elementary method, in complete analogy to the method of Jacobi and that presentd by us in the text.

We note, that we have restricted ourselves in the present Apppendix I to just the very first theorems on the application of Hankel- and Toeplitz forms in the theory of the distribution of the roots of algebraic equations. These applications are very multiform, and in the first third of our century there was an extensive literature devoted to them.

An extensive survey of this one can, probably, find only (at least in the Russian language) in the brochure of M. Kreĭn and M. Neĭmark [8], which became already long ago a bibliographical rarity.

EXAMPLES AND EXERCISES

1. The roots of the real polynomial

$$P_n(\lambda) = a_0\lambda^n + a_1\lambda^{n-1} + \cdots + a_{n-1}\lambda + a_n \quad (a_k = \bar{a}_k;\ k = 0, 1, \cdots, n)$$

are symmetrically distributed with respect to the real axis, since together with a root $\lambda = \mu$ the number $\bar{\mu}$ is also a root of the polynomial $P_n(\lambda)$:

$$P_n(\bar{\mu}) = \overline{P_n(\mu)} = 0.$$

The multiplicities of the nonreal roots μ and $\bar{\mu}$ coincide, since after division of $P_n(\lambda)$ by the real polynomial

$$(\lambda-\mu)(\lambda-\bar{\mu}) = \lambda^2 - 2(\mathrm{Re}\ \mu)\lambda + |\mu|^2$$

we obtain again a real polynomial.

Prove the analoguous statement in relation to the roots of a Hermitian-symmetric polynomial

$$Q_n(\lambda) = b_0\lambda^n + b_1\lambda^{n-1} + \cdots + b_{n-1}\lambda + b_n \quad (b_k = \bar{b}_{n-k},\ k = 0,1,\cdots,n)$$

substituting the pair $(\mu,\bar{\mu})$ of numbers, symmetrically with respect to the real axis, by the pair $(\mu,1/\bar{\mu})$ of numbers which lie mirror-like (symmetrically) with respect to the unit circle.

HINT. Use identity $\bar{Q}_n(\lambda) = \lambda^n Q_n(1/\lambda)$ (the symbol $\bar{R}(\lambda)$ denotes the substitution of all coefficients of a polynomial by their complex conjugates) and the fact, that (for $b_0 \neq 0$) $\lambda = 0$ is not a root of $Q_n(\lambda)$.

2. Verify that each of the two theorems which form together exercise 1, can be obtained from the other by means of linear-fractional transformation (cf. Sec. 19.2)

$$\theta = \frac{b+\bar{b}\varepsilon}{a+\bar{a}\varepsilon} \qquad (a\bar{b} - \bar{a}b \neq 0)$$

transforming the unit circle $|\varepsilon| = 1$ into the real line $\theta = \bar{\theta}$, and its inverse transformation

$$\varepsilon = \frac{a\theta-b}{\bar{b}-\bar{a}\theta}.$$

HINT. After performing the appropriate substitution in the polynomial $Q_n(\lambda)$, consider separately the case of even $(n = 2m)$ and odd $(n = 2m-1)$ degree of the polynomial $Q_n(\lambda)$.

3. We consider the real polynomial

$$P_3(\lambda) = \lambda^3 - 4\lambda^2 + 7\lambda + 1.$$

It is asked to determine how many among its roots $\alpha_1,\alpha_2,\alpha_3$ are real. Clearly, this number can be 3 or 1 (see exercise 1). Here (cf. exercise 4 below)

$$\alpha_1 + \alpha_2 + \alpha_3 = 4$$
$$\alpha_1\alpha_2 + \alpha_2\alpha_3 + \alpha_3\alpha_1 = 7$$
$$\alpha_1\alpha_2\alpha_3 = -1.$$

Hence the Newton sums of these roots are as follows:

$$s_0 = 3,\ s_1 = 4,\ s_2 = \alpha_1^2 + \alpha_2^2 + \alpha_3^2 =$$
$$= (\alpha_1+\alpha_2+\alpha_3)^2 - 2(\alpha_1\alpha_2+\alpha_2\alpha_3+\alpha_3\alpha_1) = 16 - 14 = 2.$$

Since

$$D_0 = s_0 = 3,\ D_1 = \begin{vmatrix} s_0 & s_1 \\ s_1 & s_2 \end{vmatrix} = \begin{vmatrix} 3 & 4 \\ 4 & 2 \end{vmatrix} = -10$$

it is here not necessary to calculate the further sums s_3,s_4 and the minor D_2.

Indeed, already

$$V(1,D_0,D_1) = V(1,3,-10) = 1,$$

i.e. (Theorems A.I.1 and 8.1), the polynomial $P_3(\lambda)$ has a pair of complex conjugates roots and, so, only one real root.

4. Let it be required to find out, whether the Hermitian-symmetric polynomial

$$Q_4(\lambda) = \lambda^4 - 5i\lambda^3 - \frac{33}{4}\lambda^2 + 5i\lambda + 1$$

has roots which lie on the unit circle. We note its roots by $\alpha_1,\alpha_2,\alpha_3,\alpha_4$. Their Newton sums are

$$\begin{aligned} s_0 &= 4,\ s_1 = 5i,\ s_2 = \alpha_1^2 + \alpha_2^2 + \alpha_3^2 + \alpha_4^2 = \\ &= (\alpha_1+\alpha_2+\alpha_3+\alpha_4)^2 - 2(\alpha_1\alpha_2+\alpha_1\alpha_3+\alpha_1\alpha_4+\alpha_2\alpha_3+\alpha_2\alpha_4+\alpha_3\alpha_4) = \\ &= s_1^2 - 2\left(-\frac{33}{4}\right) = -25 + \frac{33}{2} = -\frac{17}{2}\ . \end{aligned}$$

Hence

$$(D_{-1} \equiv 1),\ D_0 = 4,\ D_1 = \begin{vmatrix} 4 & -5i \\ 5i & 4 \end{vmatrix} = 16 - 25 = -9$$

$$D_2 = \begin{vmatrix} 4 & -5i & -17/2 \\ 5i & 4 & -5i \\ -17/2 & 5i & 4 \end{vmatrix} = \frac{1}{4}\,(81 - |-25+34|^2) = 0$$

(calculated through Sylvester's formula (2.6)).

The minor D_3 (and so the sum s_3) needs not to be calculated, as by rule (16.6) the signature of the form $\sum\limits_{p,q=0}^{3} s_{p-q}\xi_p\bar{\xi}_q$ is equal to

$$\sigma = \sum_{\nu=0}^{3} \text{sign}(D_{\nu-1}D_\nu) = \text{sign}(1.4) + \text{sign}[4\cdot(-9)] + \text{sign}[(-9)\cdot 0] +$$
$$+ \text{sign}[0\cdot D_3] = 0.$$

So (Theorem A.I.2), the polynomial $Q_4(\lambda)$ has no roots on the unit circle.

However, if one must find out whether this polynomial has one (multiple) pair or two (different) pairs of roots which lie mirror-like with respect to the unit circle, then one must also know the minor D_3, i.e., also calculate the sum s_3.

To this end we apply *Newton's formulae*, already mentioned in Note [1] below (cf., for example, [11], § 125, for their derivation).

If an arbitrary polynomial $f(\lambda)$ of degree n is written in the shape

$$f(\lambda) = \lambda^n - f_1\lambda^{n-1} + f_2\lambda^{n-2} - \cdots + (-1)^n f_n,$$

then there hold for the Newton sums s_k of its roots the recurrence relations

$$s_k - f_1 s_{k-1} + f_2 s_{n+k-2} - \cdots + (-1)^k k f_k = 0 \quad (k = 1,2,\cdots,n-1)$$

and

$$s_{n+k} - f_1 s_{n+k-1} + f_2 s_{n+k-2} - \cdots + (-1)^n f_n s_k = 0 \quad (k = 0,1,2,\cdots).$$

In particular, for n = 4 we obtain from these formulae

$$\begin{aligned} s_1 &= f_1, \\ s_2 &= f_1^2 - 2f_2, \\ s_3 &= f_1^3 - 3f_1 f_2 + 3f_3, \\ s_4 &= f_1^4 - 4f_1^2 f_2 + 4f_1 f_3 + 2f_2^2 - 4f_4 \end{aligned}$$

(we note, that we have already used the second of these formulae twice - in this and in the previous exercise).

Returning to the polynomial $Q_4(\lambda)$ we have

$$s_3 = (5i)^3 - 3(5i)(-\frac{33}{4}) + 3(-5i) = -125i + \frac{495}{4} i - 15i = -\frac{65}{4} i.$$

So (calculating it again by formula (2.6)),

$$D_3 = \begin{vmatrix} 4 & -5i & -\frac{17}{2} & \frac{65}{4} i \\ 5i & \boxed{\begin{matrix} 4 & -5i \\ 5i & 4 \end{matrix}} & & -\frac{17}{2} \\ -\frac{17}{2} & & & -5i \\ -\frac{65}{4} i & -\frac{17}{2} & 5i & 4 \end{vmatrix} =$$

$$= -\frac{1}{9} \left[0^2 - \left| \begin{vmatrix} -5i & -\frac{17}{2} & \frac{65}{4} i \\ 4 & -5i & -\frac{17}{2} \\ 5i & 4 & -5i \end{vmatrix} \right|^2 \right] =$$

$$= \frac{1}{9} \cdot \frac{i}{5} \left| (-25 + 34)^2 - \left(\frac{289}{4} - \frac{325}{4} \right) (16 - 25) \right|^2 = \frac{i}{45} (81 - 81) = 0.$$

Thus, in the present case the form (A.I.6) is degenerate and for its rank ρ we have: $2 \leqq \rho \leqq 3$. But since its signature $\sigma(=0)$ is even, one must have $\rho = 2$, i.e., $\pi = \nu = 1$, and according to Theorem A.I.2 the polynomial Q_4 has $(\nu = 1)$ one pair of roots (of multiplicity two), which lie mirror-like with respect to the circle $|\lambda| = 1$.

5. How many different real roots and different pairs of complex-conjugated roots has the polynomial

$$P_4(\lambda) = \lambda^4 + 4\lambda^3 + 2\lambda^2 + 12\lambda + 45?$$

Solution. One (double) real root and one pair of conjugated roots.

HINT. Use Newton's formulae.

6. How many different roots on the unit circle and different pairs of roots which lie mirror-like with respect to it has the polynomial

$$4i\lambda^4 + (10-10i)\lambda^3 - 25\lambda^2 + (10+10i)\lambda - 4i?$$

Solution. Two (different) pairs which lie mirror-like with respect to the unit circle.

7. Formulae (A.I.3) allow, since $k = 0,1,2,\cdots$, to construct besides the Hankel form (A.I.4) of order n (= the degree of the polynomial $P_n(\lambda)$), analoguous forms of arbitrary order. Prove that for all these forms

$$H_{m-1}(x,x) = \sum_{j,k=0}^{m-1} s_{j+k}\xi_j\xi_k \qquad (m = n,n+1,\cdots)$$

we have one and the same rank ρ and one and the same signature σ, i.e., $\rho = p+2q$ - the number of (all) different roots of the polynomial $P_n(\lambda)$, and $\sigma = p$ - the number of different real roots.

HINT. For arbitrary $m = n,n+1,\cdots$ one can repeat all arguments from the proof of Theorem A.I.1.

8. Formulate and prove the analogon of the result of exercise 7 for the forms (A.I.6) (replacing n in it by $m = n,n+1,\cdots$).

9. The *Cauchy index* $I_\alpha^\beta R(\lambda)$ of the real rational function

$$R(\lambda) = \frac{a_0\lambda^n + a_1\lambda^{n-1} + \cdots + a_{n-1}\lambda + a_n}{b_0\lambda^m + b_1\lambda^{m-1} + \cdots + b_{m-1}\lambda + b_m}$$

on the interval (α,β) is the name for the difference between the number of discontinuities of $R(\lambda)$ with a transition from $-\infty$ to $+\infty$ and the number of its discontinuities with a transition from $+\infty$ to $-\infty$, varying the argument λ from α to β $(\alpha < \lambda < \beta)$.

Prove, that for the rational function

$$R(\lambda) = \frac{P_n'(\lambda)}{P_n(\lambda)}$$

where $P_n(\lambda)$ is the real polynomial (A.I.1) and $P_n'(\lambda)$ is its derivative, $I_{-\infty}^{+\infty}R(\lambda) = \sigma$, where σ is the signature of the form (A.I.4).

HINT. Use Theorem A.I.1.

REMARK. The result, formulated in exercise 9, is a special case of a more general Theorem:

Let R(z) *be a rational function and*

$$R(z) = s_{-m-1}z^m + \cdots + s_{-2}z + s_{-1} + \frac{s_0}{z} + \frac{s_1}{z} + \cdots \quad (m \geq 0)$$

its series expansion (converging outside an arbitrary circle with centre in the point o and containing all poles of R(z) [1o]). *The sequence* $s_0, s_1, s_2, \cdots$ *defines an infinite Hankel matrix*

$$H_\infty = \| s_{j+k} \|_{j,k=0}^{\infty}$$

of finite rank ρ (cf. exercise 1o to § 11).

Then the signatures of all Hankel forms

$$H_{n-1}(x,x) = \sum_{j,k=0}^{n-1} s_{j+k}\xi_j\xi_k \quad (n \geq \rho)$$

coincide, and if we denote them by σ*, then*

$$I_{-\infty}^{+\infty} R(\lambda) = \sigma.$$

This theorem was proved by Hermit [28] for the case where all poles of R(z) are simple [1o], and in the general case by Hurwitz [29]. A different proof is presented in [3] (Ch. XVI, § 11, Theorem 9). There, see also (Ch. XVI, § 12) the application of this result to the well-known Raus-Hurwitz problem.

NOTE

1) In the theory of symmetrical functions (see, for example, [11], § 125) is proved that the sum (A.I.3) can be expressed by the well-know *Newton formulae* through basic symmetrical functions in $\alpha_1, \alpha_2, \cdots, \alpha_n$ and these, in turn, through the coefficients of the polynomial $P_n(\lambda)$ (see above, exercise 4).

II. THE FUNCTIONALS S AND C AND SOME OF THEIR APPLICATIONS.

1. Let be given a fixed set

$$s_0, s_1, s_2, \cdots, s_m \quad (s)$$

of real numbers. By means of this we define the linear functional S on the space of the (generally speaking, complex) polynomials

$$G_m(\lambda) = A_0 + A_1\lambda + \cdots + A_m\lambda^m$$

of degree not exceeding m through the formula

$$S\{G_m\} = A_0 s_0 + A_1 s_1 + \cdots + A_m s_m.$$

In a similar way one defines, by means of fixed complex numbers

$$c_o \; (=\bar{c}_o), c_1, c_2, \cdots, c_m \tag{c}$$

on the space of trigonometrical polynomials

$$T_m(z) = \sum_{k=-m}^{m} A_k z^k \qquad (z = e^{it})$$

of order not exceeding m the linear functional C:

$$C\{T_m\} = \sum_{k=-m}^{m} A_k c_k$$

where

$$c_{-k} = \bar{c}_k \qquad (k = 1,2,\cdots,m).$$

In particular, if $T_m(e^{it})$ is a *real trigonometrical polynomial*, i.e., $A_{-k} = \bar{A}_k$ $(k = 0,1,\cdots,m)$, then the value $C\{T_m\}$ of the functional C on this polynomial is a real number.

In our further considerations such functionals S and C will appear which are generated by sets (s) and (c) which consist of the coefficients of real Hankel- and Hermitian Toeplitz forms, respectively.

2. As a first application of the functionals S and C we present, albeit in an incomplete, but for our purpose sufficient form, two theorems from the classical moment problem (the polynomial- and the trigonometrical moment problem). This is the more useful, since the book of N.I. Ahiezer and M.G. Kreĭn [1], from which we have taken this material, did appear quite long ago (1938)[1).]

THEOREM A.II.1. *In order that the (real) Hankel quadratic form*

$$H_{n-1}(x,x) = \sum_{j,k=0}^{n-1} s_{j+k}\xi_j\xi_k \tag{A.II.1}$$

is positive, it is necessary and sufficient that its coefficiens s_k *allow a representation*

$$s_k = \sum_{\nu=1}^{n} \rho_\nu \theta_\nu^k \qquad (k = 0,1,\cdots,2n-2) \tag{A.II.2}$$

(here again $0^0 = 1$), *where*

$$\rho_\nu > 0, \; (\nu=1,2,\cdots,n), \; -\infty < \theta_1 < \theta_2 < \cdots < \theta_n < +\infty.$$

PROOF. SUFFICIENCY. From representation (A.II.2) it follows that

$$H_{n-1}(x,x) = \sum_{j,k=0}^{n-1} s_{j+k}\xi_j\xi_k = \sum_{j,k=0}^{n-1} \sum_{\nu=1}^{n} \rho_\nu \theta_\nu^{j+k}\xi_j\xi_k =$$

$$= \sum_{\nu=1}^{n} \rho_\nu \left(\sum_{k=0}^{n-1} \xi_k \theta_\nu^k \right)^2 > 0$$

for all $x = \{\xi_0, \xi_1, \cdots, \xi_{n-1}\} \neq 0$, since the simultaneous identities

$$\sum_{k=0}^{n-1} \xi_k \theta_\nu^k = 0 \quad (\nu = 1,2,\cdots,n)$$

are impossible. Indeed, in the opposite case the Vandermonde determinant

$$\begin{vmatrix} 1 & \theta_1 & \cdots & \theta_1^{n-1} \\ 1 & \theta_2 & \cdots & \theta_2^{n-1} \\ \cdot & \cdot & \cdots & \cdot \\ 1 & \theta_n & & \theta_n^{n-1} \end{vmatrix}$$

would (since $x \neq 0$) be equal to zero, which is impossible, as all $\theta_\nu (\nu = 1,2,\cdots,n)$ are different.

Thus, $H_{n-1}(x,x)$ is a positive definite form (Sec. 5.4).

NECESSITY. We start with a useful definition.

A polynomial $Q_k(\lambda) \not\equiv 0$ of degre $\leq k$ is called a *quasi-orthogonal polynomial of rank* k if it has the property

$$S\{Q_k(\lambda)\lambda^j\} = 0 \quad (j = 0,1,\cdots,k-2;\ k \leq n) \qquad \text{(A.II.3)}$$

where S is the functional defined in Sec. 1, and the sequence (s) which determines it is the sequence of coefficients $s_0, s_1, \cdots, s_{2n-2}$ of the form (A.II.1).

It is convenient to extract the following fact as an indenpendent proposition:

1° *For a real quasi-orthogonal polynomial* $Q_k(\lambda)$ *all roots are real and simple.*

Indeed, let $\theta_1, \theta_2, \cdots, \theta_p$ be all those real different roots of the quasiorthogonal polynomial $Q_k(\lambda)$ of rank k, which have odd multiplicity (if such exist). Then for a suitable choice of the factor $\varepsilon = \pm 1$ the product

$$G(\lambda) = \varepsilon(\lambda-\theta_1)(\lambda-\theta_2)\cdots(\lambda-\theta_p)Q_k(\lambda) \quad (\not\equiv 0)$$

is nonnegative for all real λ. As is well-know (see exercise 1 below), an arbitrary nonnegative real polynomial is the square of the modulus of some complex polynomial, in particular

$$G(\lambda) = \left|\sum_{\mu=0}^{n-1} (\xi_\mu + i\eta_\mu)\lambda^\mu\right|^2 = \sum_{\mu,\nu=0}^{n-1} \xi_\mu\xi_\nu\lambda^{\mu+\nu} + \sum_{\mu,\nu=0}^{n-1} \eta_\mu\eta_\nu\lambda^{\mu+\nu}$$

$$(\bar{\xi}_\mu = \xi_\mu, \bar{\eta}_\mu = \eta_\mu;\ \mu = 0,1,\cdots,n-1).$$

But then under the assumptions of the theorem

$$\mathcal{S}\{G(\lambda)\} = \sum_{\mu,\nu=0}^{n-1} s_{\mu+\nu}\xi_\mu\xi_\nu + \sum_{\mu,\nu=0}^{n-1} s_{\mu+\nu}\eta_\mu\eta_\nu > 0$$

(we recall, that $G(\lambda) \neq 0$, so that at least one of the systems of numbers $\{\xi_0,\xi_1,\cdots,\xi_{n-1}\}$ and $\{\xi_0,\xi_1,\cdots,\xi_{n-1}\}$ is nonzero)[2].

If we now assume, that $p \leqq k-2$, then, according to the definition of quasi-ortogonal polynomials (A.II.3), $\mathcal{S}(G(\lambda)) = 0$. So $p \geqq k-1$, and none or the roots $\theta_1,\theta_2,\cdots,\theta_p$ can be multiple, since in the opposite case, as by assumption these multiplicities are odd, the degree of the polynomial $Q_k(\lambda)$ would turn out to be larger than k. Since $Q_k(\lambda)$ is a real polynomial, the number p must be exactly equal to its degree.

Proposition 1° is proved.

Now we consider an arbitrary polynomial $G(\lambda)$ of degree $\leqq 2n-2$. Let $Q_n(\lambda)$ be a real quasi-orthogonal polynomial of degree n (such a polynomial exists, see Note [3] below) and $\theta_1 < \theta_2 < \cdots < \theta_n$ all its roots. We represent $G(\lambda)$ in the shape

$$G(\lambda) = Q_n(\lambda)q(\lambda) + r(\lambda) \tag{A.II.4}$$

where $q(\lambda)$ is a polynomial of degree $\leqq n-2$ and the remainder $r(\lambda)$ a polynomial of degree $\leqq n-1$.

Substituting $\lambda = \theta_\nu$, we obtain

$$r(\theta_\nu) = G(\theta_\nu) \qquad (\nu = 1,2,\cdots,n). \tag{A.II.5}$$

We represent $r(\lambda)$ through the interpolation formula of Lagrange ([11], § 62)[4]:

$$r(\lambda) = \sum_{\nu=1}^{n} \frac{Q_n(\lambda)}{(\lambda-\theta_\nu)Q_n'(\theta_\nu)}\, r(\theta_\nu). \tag{A.II.6}$$

In combination with (A.II.5) this yields

$$r(\lambda) = \sum_{\nu=1}^{n} \frac{Q_n(\lambda)}{(\lambda-\theta_\nu)Q_n'(\theta_\nu)}\, G(\theta_\nu). \tag{A,II.6'}$$

Now we apply the functional $\mathcal{S}$ to both sides of (A.II.4). Because of definition (A.II.3) we have, taking (A.II.6') into account,

$$\mathcal{S}\{G(\lambda)\} = \mathcal{S}\{r(\lambda)\} = \sum_{\nu=1}^{n} \rho_\nu G(\theta_\nu), \tag{A.II.7}$$

where the quantities

$$\rho_\nu = \mathcal{S}\left\{\frac{Q_n(\lambda)}{(\lambda-\theta_\nu)Q_n'(\theta_\nu)}\right\} \qquad (\nu = 1,2,\cdots,n)$$

do, evidently, not depend on the choice of the polynomial $G(\lambda)$. Using this, we insert into formula (A.II.7) in turn:

$$G(\lambda) = q_k^2(\lambda) \equiv \left|\frac{Q_n(\lambda)}{(\lambda-\theta_k)Q_n'(\lambda)}\right|^2 \qquad (k = 1,2,\cdots,n).$$

We have

$$S\{q_k^2(\lambda)\} = \sum_{\nu=1}^{n} \rho_\nu q_k^2(\theta_\nu) = \rho_k \qquad (k = 1,2,\cdots,n).$$

But the polynomials $G(\lambda) = q_k^2(\lambda)$ are nonnegative, and none of these is identically zero, and for such polynomials, one has, as we have seen (see the beginning of the proof of proposition 1°), because of the positive definiteness of the form (A.II.1)

$$\rho_k = S\{q_k^2(\lambda)\} > 0 \qquad (k = 1,2,\cdots,n).$$

It remains to note, that for $G(\lambda) = \lambda^k$ $(k = 0,1,\cdots,2n-2)$ formula (A.II.7) turns into (A.II.2).
Theorem A.II.1 is proved [5)].

REMARK. As follows from a theorem of E. Fischer [18] (see also [1], p. 13), Theorem A.II.1 can, for the case of nonnegative (degenerate) forms (A.II.1), i.e., such, for which $D_{n-1} = \det \|s_{j+k}\|_{j,k=0}^{n-1}$, be modified in the following way (see also below exercise 4).

Let for the nonnegative form (A.II.1) $D_{n-1} = 0$, *but* $D_{n-2} \neq 0$. *Then the coefficients* s_k $(k = 0,1,\cdots,2n-2)$ *allow a, moreover unique, representation*

$$s_k = \sum_{\nu=1}^{n-1} \rho_\nu \theta_\nu^k \quad (k = 0,1,\cdots,2n-2) \qquad \text{(A.II.8)}$$

where

$$\rho_\nu > 0 \ (\nu = 1,2,\cdots,n-1), -\infty < \theta_1 < \theta_2 < \cdots < \theta_{n-1} < +\infty.$$

For the converse of this proposition see below, exercise 5.

3. An analogon of Theorem A.II.1 holds also for Toeplitz forms. Its proof is in many aspects analoguous to the proof of Theorem A.II.1, and in connection with it we formulate, as a preliminary, some auxiliary propositions which are the analoga of the results, used above in Sec. 2.

Instead of the common polynomials, considered in Sec. 2, we now deal with real trigonometrical polynomials

$$T_n(e^{it}) = \sum_{k=-n}^{n} a_k e^{ikt} \quad (a_k = \bar{a}_k,\ k = 0,1,\cdots,n) \qquad \text{(A.II.9)}$$

of order n (if $a_n \neq 0$) and, in particular, with polynomials (A.II.9) which do not assume negative values.

2° (THEOREM OF FÉJER - F. RIESZ). *If the polynomial* (A.II.9) *is nonnegative* ($0 \leq t \leq 2\pi$) *then it allows the representation*

$$T_n(e^{it}) = \left| \sum_{k=0}^{n} \xi_k e^{ikt} \right|^2 = \sum_{p,q=0}^{n} e^{i(p-q)t} \xi_p \bar{\xi}_q . \tag{A.II.10}$$

For the proof we consider the "quasi-polynomial"

$$T_n(z) = \sum_{k=-n}^{n} a_k z^k$$

which, after multiplication by z^n turns, evidently, into the common polynomial

$$G(z) = z^n T_n(z) = a_{-n} + a_{-n+1} z + \cdots + a_{n-1} z^{2n-1} + a_n z^{2n} \tag{A.II.11}$$

which is, by assumption, Hermitian-symmetric ($a_{-k} = \bar{a}_k$, $k = 0,1,\cdots,n$). Hence (see exercise 1 to Appendix I) its roots are mirror-like distributed with respect to the unit circle with the only exception that in the present case some of its roots can be equal to zero: there are exactly r (≥ 0) such roots if $a_n = a_{n-1} = \cdots = a_{n-r+1} = 0$, $a_{n-r} \neq 0$.

Thus

$$G(z) = Cz^r \prod_{\mu=1}^{k} (z-\zeta_\mu) \prod_{\nu=1}^{\ell} (z-z_\nu) \left(z - \frac{1}{\bar{z}_\nu} \right), \tag{A.II.12}$$

where

$$|\zeta_\mu| = 1; \ \mu = 1,2,\cdots,k; \ 0 < |z_\nu| < 1,$$

$$\nu = 1,2,\cdots,\ell (r + k + 2\ell \leq 2n), \ C = \text{const},$$

and any two of the three polynomials following C in representation (A.II.12) may be lacking.

Representation (A.II.12) can be rewritten as

$$G(z) = C' z^{r+\ell} \prod_{\mu=1}^{k} (z-\zeta_\mu) \prod_{\nu=1}^{\ell} (z-z_\nu) \left(\frac{1}{z} - \bar{z}_\nu \right),$$

where $C' = (-1)^\ell C \prod_{\nu=1}^{\ell} \frac{1}{\bar{z}_\nu}$. Hence

$$T_n(z) = z^{-n} G(z) = C' z^s \prod_{\mu=1}^{k} (z-\zeta_\mu) \prod_{\nu=1}^{\ell} (z-z_\nu) \left(\frac{1}{z} - \bar{z}_\nu \right),$$

where $s = r+\ell-n$. As here the last polynomial $\left(\prod_{\nu=1}^{\ell} \right)$ is nonnegative on the circle $z = e^{it}$ ($0 \leq t \leq 2\pi$) and $T_n(e^{it})$ has the same property, it follows that also the function

$$f(t) = C' e^{ist} \prod_{\mu=1}^{k} (e^{it} - \zeta_\mu) \tag{A.II.13}$$

is nonnegative, whence follows, that the multiplicity of each of the roots ζ_μ is even. Indeed, if among them there is found a root, say $\zeta_1 = e^{it_1}$, with multiplicity $2m+1$, then the product $\left(\prod_{\mu=1}^{k}\right)$ on the right-hand side of (A.II.13) would contain the factor

$$\left(e^{it} - e^{it_1}\right)^{2m+1} = (2i)^{2m+1} e^{i(2m+1)\left(\frac{t+t_1}{2}\right)} \sin^{2m+1}\left(\frac{t-t_1}{2}\right) .$$

But then the real function

$$f(t) / \sin^{2m+1}\left(\frac{t-t_1}{2}\right)$$

is nonzero on a sufficiently small neighbourhood of the point t_1, and, because of continuity, it retains the same sign. But the numerator of this quotient is always nonnegative, and the denominator changes sign in a neighbourhood of the point t_1: a contradiction!

So, one can write

$$\prod_{\mu=1}^{k} (e^{it} - \zeta_\mu) = \prod_{\mu=1}^{k/2} (e^{it} - \zeta_\mu)^2. \tag{A.II.14}$$

But then it follows from (A.II.11), (A.II.12) and (A.II.14) that

$$0 \leq T_n(e^{it}) = |T_n(e^{it})| = |G(e^{it})| = |C| \prod_{\mu=1}^{k/2} |e^{it} - \zeta_\mu|^2 \prod_{\nu=1}^{\ell} \frac{|e^{it} - z_\nu|^2}{|z_\nu|} .$$

Now it suffices to form (recalling that $k/2 + \ell \leq n$)

$$\left(\frac{|C|}{\prod_{\nu=1}^{\ell} |z_\nu|}\right)^{1/2} \prod_{\mu=1}^{k/2} (e^{it} - \zeta_\mu) \prod_{\nu=1}^{\ell} (e^{it} - z_\nu) = \sum_{p=0}^{n} \xi_p e^{it},$$

in order to obtain representation (A.II.1o).

Proposition 2° is proved.

Let C be the functional, considered in Sec. 1, defined on the trigonometrical polynomials (of order not above n)

$$T_n(e^{it}) = \sum_{k=-n}^{n} a_k e^{ikt}$$

through the formula

$$C\{T_n\} = \sum_{k=-n}^{n} a_k c_k, \tag{A.II.15}$$

where $c_k = \bar{c}_{-k}$ $(k = 0,1,\cdots,n)$ are the coefficients of a given (fixed) Toeplitz form

$$\sum_{p,q=0}^{n} c_{p-q} \xi_p \bar{\xi}_q. \tag{A.II.16}$$

3° *In order that the value* $C\{T_n\}$ *is nonnegative (positive) for all nonnegative trigonometrical polynomials which are not identically zero, it is necessary and sufficient that the Toeplitz form* (A.II.16), *which determines* C, *is nonnegative (positive definite).*

Indeed, because of proposition 2° can each nonnegative trigonometrical polynomial $T_n(e^{it})$ be represented as

$$T_n(e^{it}) = \sum_{p,q=0}^{n} \xi_p \overline{\xi}_q e^{i(p-q)t}$$

so that

$$C\{T_n\} = \sum_{p,q=0}^{n} c_{p-q} \xi_p \overline{\xi}_q$$

whence statement 3° follows as well.

THEOREM A.II.2. *In order that the Hermitian Toeplitz form*

$$T_{n-1}(x,x) = \sum_{p,q=0}^{n} c_{p-q} \xi_p \overline{\xi}_q \qquad (A.II.17)$$

is positive definite, it is necessary and sufficient that its coefficients c_p $(=\overline{c}_{-p})$ *allow a representation*

$$c_p = \sum_{\nu=1}^{n} r_\nu \varepsilon_\nu^p \quad (p = 0,\pm 1,\cdots,\pm(n-1)), \qquad (A.II.18)$$

in which

$$r_\nu > 0, \quad |\varepsilon_\nu| = 1 \quad (\nu = 1,2,\cdots,n)$$

and all ε_ν *are different.*

PROOF. The SUFFICIENCY of the condition is clear from the fact, that a form (A.II.17) with coefficients (A.II.18) can be rewritten as

$$T_{n-1}(x,x) = \sum_{p,q=0}^{n-1} \left(\sum_{\nu=1}^{n} r_\nu \varepsilon_\nu^{p-q} \right) \xi_p \overline{\xi}_q = \sum_{\nu=1}^{n} r_\nu \left| \sum_{p=0}^{n-1} \varepsilon_\nu^p \xi_p \right|^2 ,$$

from which follows, that $T_{n-1}(x,x) > 0$ for $x = \{\xi_0,\xi_1,\cdots,\xi_{n-1}\} \neq 0$. Since all $r_\nu > 0$ and all ε_ν are different $(\nu = 1,2,\cdots,n)$ (cf. the proof of the sufficiency of Theorem A.II.1).

NECESSITY. For a positive definite form (A.II.17) is the determinant $D_{n-1} = \det \|c_{p-q}\|_{p,q=0}^{n-1} \neq 0$ (Corollary 1 of Theorem 5.2). We choose an arbitrary $\zeta = c_n$ on the circle (see Remark 1 to Theorem 13.1) so that

$$\begin{Vmatrix} c_0 & c_{-1} & \cdots & c_{-n+1} & \bar{\zeta} \\ c_1 & c_0 & \cdots & c_{-n+2} & c_{-n+1} \\ \cdot & \cdot & \cdots & \cdot & \cdot \\ c_{n-1} & c_{n-2} & \cdots & c_0 & c_{-1} \\ \zeta & c_{n-1} & \cdots & c_1 & c_o \end{Vmatrix} = 0.$$

Then the homogeneous linear system

$$\sum_{p=0}^{n} c_{p-q} u_q = 0 \qquad (q = 0,1,\cdots,n) \tag{A.II.19}$$

allows a nonzero solution $\{u_0, u_1, \cdots, u_{n-1}, u_n\}$. Replacing in this system p by n-p and q by n-q, we convince ourselves that the numbers $\{\bar{u}_n, \bar{u}_{n-1}, \cdots, \bar{u}_1, \bar{u}_0\}$ solve the same system

$$\sum_{p=0}^{n} c_{p-q} \bar{u}_{n-q} = 0 \qquad (q = 0,1,\cdots,n).$$

But since $D_{n-1} \neq 0$ and $c_n \neq 0$ it follows that $u_0 \neq 0$ and $u_n \neq 0$ and that the solution of the considered system is determined up to multiplicative constant, so that for some $\varepsilon \neq 0$

$$\bar{u}_n = \varepsilon u_0,\ \bar{u}_{n-1} = \varepsilon u_1,\ \cdots,\ \bar{u}_1 = \varepsilon u_{n-1},\ \bar{u}_0 = \varepsilon u_n.$$

Since $u_0, u_n \neq 0$ it follows that $|\varepsilon| = 1$. So, at the expense of multiplying all u_q $(q = 0,1,\cdots,n)$ with some factor one can achieve, that $\varepsilon = 1$. We shall assume, that this has been done, i.e., that

$$u_{n-q} = \bar{u}_q \qquad (q = 0,1,\cdots,n).$$

We consider the Hermitian-symmetric polynomial

$$U_n(z) = u_0 + u_1 z + \cdots + u_n z^n \qquad (u_0 = \bar{u}_n \neq 0).$$

If we multiply it with z^{-q} and apply for $z = e^{it}$ the functional C (see (A.II.15)) to it, then we obtain, because of (A.II.19)

$$C\{U_n(e^{it})e^{-qit}\} =$$

$$= c_{-q}u_0 + c_{1-q}u_1 + \cdots + c_{n-q}u_n = 0 \quad (q = 0,1,\cdots,n) \tag{A.II.20}$$

We shall prove, that all roots of the polynomial $U_n(z)$ are different and have modulus equal to 1. Let $e^{it_1}, e^{it_2}, \cdots, e^{it_m}$ be all different roots of odd multiplicity of the polynomial $U_n(z)$ which lie on the circle $|z| = 1$. We define a polynomial $G(z)$ of degree m through the formula

$$G(z) \begin{cases} = e^{\frac{\pi i m}{2}} e^{\frac{1}{2}(t_1+t_2+\cdots+t_m)}(z-e^{it_1})\cdots(z-e^{it_m}) & \text{for } m \geqq 1 \\ \equiv 1 \quad \text{for } m = 0 \end{cases}$$

If the polynomial $U_n(z)$ has roots which are not situated on the circle $|z| = 1$, then these lie mirror-like with respect to it in pairs $\{\beta_\ell, 1/\bar{\beta}_\ell\}(\ell = 1,\cdots,s)$, so that their total number, counted with their mulitplicity, is even. Even is, clearly, also the total number of roots of even multiplicity which lie on the circle $|z| = 1$ (each root counted with its mulitplicity). Hence one easily concludes that the number $n - m = 2k$ is even as well.

Now $U_n(z)$ can be represented in the shape

$$U_n(z) = CG(z)K(z)\bar{K}(\tfrac{1}{z})z^k \tag{A.II.21}$$

where $K(z)$ is a polynomial of degree k [6] and C a constant. But then the quasi-polynomial

$$H(z) = \frac{\bar{C}}{z^k}\bar{G}(\tfrac{1}{z})U_n(z) \tag{A.II.22}$$

turns for $z = e^{it}$ into a real trigonometrical polynomial. Indeed, because of (A.II.21) we have

$$H(z) = |C|^2 G(z)\bar{G}(\tfrac{1}{z})K(z)\bar{K}(\tfrac{1}{z})$$

i.e.

$$H(e^{it}) = |C|^2 G(e^{it})\overline{G(e^{it})}K(e^{it})\overline{K(e^{it})}.$$

Moreover, hence is clear, that the trigonometrical polynomial $H(e^{it})$ of order $m + k = n - k$ is nonnegative. At the same time, $H(e^{it}) \not\equiv 0$ if $k > 0$ (recall that $m \geqq 0$). But since the form (A.II.17) is positive definite, is because of proposition 3°

$$\mathcal{C}\{H\} > 0.$$

But this contradicts, if one considers the structure (A.II.22) of the polynomial $H(z)$, the identity (A.II.2o) (we recall that the degree of G is equal to m, and $m + k = n - k < n$).

So $k = 0$, i.e., $m = n$ and all roots $\varepsilon_\nu = e^{it_\nu}$ $(\nu = 1,2,\cdots,n)$ of the polynomial $U_n(z)$ are different and lie on the unit circle.

Now we consider an arbitrary polynomial $Q(z)$ of degree $\leqq n$ and represent it in the shape [7]

$$Q(z) = \sum_{\nu=1}^{n} Q(\varepsilon_\nu)h_\nu(z) + \text{const.}U(z),$$

where

$$h_\nu(z) = \frac{U_n(z)}{(z-\varepsilon_\nu)U_n'(\varepsilon_\nu)} \qquad (\nu = 1,2,\cdots,n) \tag{A.II.23}$$

are polynomials of degree $n-1$ (for $n=1$ we have $h_1 \equiv 1 \neq 0$). But since $\mathcal{C}\{U_n\} = 0$, because of (A.II.2o), one has

$$C\{Q\} = \sum_{\nu=1}^{n} r_\nu Q(\varepsilon_\nu), \tag{A.II.24}$$

where

$$r_\nu = C\{h_\nu\} \qquad (\nu = 1,2,\cdots,n).$$

If we insert $Q(z) = z^p$ into (A.II.24) $(p = 0,1,\cdots,n-1)$ we obtain

$$c_p = \sum_{\nu=1}^{n} r_\nu \varepsilon_\nu^p \qquad (p = 0,1,\cdots,n-1),$$

and it remains only to prove that

$$r_\nu > 0 \qquad (\nu = 1,2,\cdots,n).$$

To this end we note, that $h_\nu(\varepsilon_\nu) = 1$ and hence

$$\bar{h}_\nu\left(\frac{1}{\varepsilon_\nu}\right) = \bar{h}_\nu(\bar{\varepsilon}_\nu) = \overline{h(\varepsilon_\nu)} = 1 \quad (\nu = 1,2,\cdots,n).$$

So,

$$\bar{h}_\nu\left(\frac{1}{z}\right) - 1 = (z - \varepsilon_\nu) g_\nu\left(\frac{1}{z}\right),$$

where $g_\nu(z)$ is a polynomial of degree $n-1$ $(\nu = 1,2,\cdots,n)$. But then (see (A.II.23))

$$h_\nu(z)\bar{h}_\nu\left(\frac{1}{z}\right) - h_\nu(z) = \frac{1}{U_n'(\varepsilon_\nu)} U_n(z) g_\nu\left(\frac{1}{z}\right) \qquad (\nu = 1,2,\cdots,n).$$

Substituting here $z = e^{it}$ and applying to both sides the functional C, we obtain, with regard to (A.II.20),

$$r_\nu = C\{h_\nu(e^{it})\} = C\{h_\nu(e^{it})\bar{h}_\nu(e^{-it})\} \qquad (\nu = 1,2,\cdots,n).$$

But since the trigonometrical polynomials

$$h_\nu(e^{it})\bar{h}_\nu(e^{-it}) = |h_\nu(e^{it})|^2 \qquad (\nu = 1,2,\cdots,n)$$

are nonnegative and not identically zero, one has by means of 3°

$$r_\nu > 0 \qquad (\nu = 1,2,\cdots,n).$$

The theorem is proved [8)].

REMARK. One can prove, that in the case *where the form* (A.II.17) *is nonnegative and degenerate, and its rank is equal to* ρ $(<n)$, *the coefficients* c_p *allow a representation*

$$c_p = \sum_{\nu=1}^{\rho} r_\nu \varepsilon_\nu^p \qquad (p = 0,1,\cdots,n-1), \tag{A.II.25}$$

(which is unique at that), where $r_\nu > 0$, $|\varepsilon_\nu| = 1$ *and all* ε_ν $(\nu = 1, 2,\cdots,\rho)$ *are different* (see [1], part. I, Ch. I, Theorem 12).

As we see is here, in contrast to (A.II.8), the appropriate representation obtained without the additional requirement $D_{n-2} \neq 0$.

4. The functionals S and C, introduced in Sec. 1, which act in the spaces of polynomials and trigonometrical polynomials, respectively, allow in the case of real Toeplitz forms to construct another interesting transformation, which has a row of applications on the moment problem [9].

For the simplification of the notation we shall consider a Toeplitz form of order n+1, namely, the real quadratic form

$$\sum_{p,q=0}^{n} c_{p-q}\xi_p\xi_q \qquad (c_{-p} = c_p,\ p = 0,1,\cdots,n). \qquad \text{(A.II.26)}$$

We are interested in the transformation generated by the formulae

$$s_k = \frac{1}{2^k}\sum_{r=0}^{k} C_k^r c_{k-2r} \qquad (k = 0,1,\cdots,n). \qquad \text{(A.II.27)}$$

We recall, that the linear functional C, corresponding to the form (A.II.26), relates the trigonometrical polynomial

$$T_n(z) = \sum_{p=-n}^{n} A_p z^p \qquad (z = e^{it})$$

of order not above n to the number

$$C\{T_n(z)\} = \sum_{p=-n}^{n} A_p c_p.$$

Hence the formulae (A.II.27) can be rewritten as

$$s_k = C\{\frac{1}{2^k}(z+\frac{1}{z})^k\} \qquad (z = e^{it},\ k = 0,1,\cdots,n), \qquad \text{(A.II.28)}$$

since

$$(z+\frac{1}{z})^k = \sum_{r=0}^{k} C_k^r z^{k-2r} \qquad (k = 0,1,\cdots,n).$$

We recall, that the functional S, determined by some set of real numbers $s_0, s_1, \cdots, s_n$ (see Sec. 1), relates the polynomial

$$G(u) = \sum_{k=0}^{n} a_k u^k \qquad \text{(A.II.29)}$$

of degree not above n to the number

$$S\{G(u)\} = \sum_{k=0}^{n} a_k s_k.$$

Combining this identity with (A.II.28) and (A.II.29) we have

$$S\{G(u)\} = \sum_{k=0}^{n} a_k\, C\{(\frac{z+z^{-1}}{2})^k\} = C\{G(\frac{z+z^{-1}}{2})\} \qquad (z = e^{it}) \qquad \text{(A.II.30)}$$

Returning to the form (A.II.26), we note that

$$\sum_{p,q=0}^{n} c_{p-q}\xi_p\xi_q = \sum_{p,q=0}^{n} C\{z^{p-q}\}\xi_p\xi_q = C\{\sum_{p,q=0}^{n} \xi_p\xi_q z^{p-q}\} =$$

$$= C\{(\xi_0 + \xi_1 z^{-1} + \cdots + \xi_n z^{-n})(\xi_0 + \xi_1 z + \cdots + \xi_n z^n)\} \quad (z = e^{it}) \qquad \text{(A.II.31)}$$

We convert the trigonometrical polynomial which stands between the braces:

$$T_n(z) = \sum_{k=0}^{n} \xi_k e^{-ikt} \cdot \sum_{k=0}^{n} \xi_k e^{ikt} = \left|\sum_{k=0}^{n} \xi_k e^{-ikt}\right|^2 = \left|\sum_{k=0}^{n} \xi_k e^{i(n/2-k)t}\right|^2$$

(since $|e^{int/2}| = 1$).

First, we consider the case of even $n = 2m$. Then

$$T_n(z) = \left|\sum_{k=0}^{2m} \xi_k[\cos(m-k)t + i\sin(m-k)t]\right|^2 =$$

$$= |(\xi_0 + \xi_{2m})\cos mt + (\xi_1 + \xi_{2m-1})\cos(m-1)t + \cdots + (\xi_{m-1} + \xi_{m+1})\cos t + \xi_m +$$

$$+ i[(\xi_0 - \xi_{2m})\sin mt + (\xi_1 - \xi_{2m-1})\sin(m-1)t + \cdots + (\xi_{m-1} - \xi_{m+1})\sin t]|^2 =$$

$$= \left(\sum_{k=0}^{m} \eta_{m-k}\cos kt\right)^2 + \sin^2 t\left(\sum_{k=1}^{m} \eta_{m+k}\frac{\sin kt}{\sin t}\right)^2, \qquad \text{(A.II.32)}$$

where we have introduced the independent linear forms

$$\eta_m = \xi_m,\ \eta_{m-k} = \xi_{m-k} + \xi_{m+k},\ \eta_{m+k} = \xi_{m-k} - \xi_{m+k} \quad (k = 1,2,\cdots,m). \qquad \text{(A.II.33)}$$

Now we use the fact that for arbitrary k (=o,1,2,$\cdots$) cos kt is a real polynomial of degree k in cos t, and sin kt/sin t for k = 1,2,$\cdots$ also a real polynomial in cos t, but of degree k-1. Hence

$$\left.\begin{aligned} \sum_{k=0}^{m} \eta_{m-k}\cos kt &= \sum_{k=0}^{m} X_k\cos^k t, \\ \sum_{k=1}^{m} \eta_{m+k}\frac{\sin kt}{\sin t} &= \sum_{k=0}^{m-1} Y_k\cos^k t, \end{aligned}\right\} \qquad \text{(A.II.34)}$$

where

$$\left.\begin{aligned} X_k &= X_k(\eta_0,\eta_1,\cdots,\eta_m) \quad (k = 0,1,\cdots,m), \\ Y_k &= Y_k(\eta_{m+1},\eta_{m+2},\cdots,\eta_{2m}) \quad (k = 0,1,\cdots,m-1) \end{aligned}\right\} \qquad \text{(A.II.35)}$$

arc linear forms in their parameters.

Inserting the expressions (A.II.34) into (A.II.32), we obtain

$$T_{2m}(z) = \left(\sum_{k=0}^{m} X_k u^k\right)^2 + (1-u^2)\left(\sum_{k=0}^{m-1} Y_k u^k\right)^2,$$

where

$$z = e^{it},\ u = \frac{z+z^{-1}}{2} = \cos t.$$

Hence we have, taking (A.II.3o) into account

$$C\{T_{2m}(z)\} = S\left\{\left(\sum_{k=0}^{m} X_k u^k\right)^2\right\} + S\left\{(1-u^2)\left(\sum_{k=0}^{m-1} Y_k u^k\right)^2\right\}.$$

But the lefthand side of this identity coincides on the basis of (A.II.31) (for $2m = n$, $z = e^{it}$) with the original form (A.II.2o). The righthand side is equal to

$$S\left\{\sum_{j,k=0}^{m} u^{j+k} X_j X_k\right\} + S\left\{\sum_{j,k=0}^{m-1} u^{j+k} Y_j Y_k - \sum_{j,k=0}^{m-1} u^{j+k+2} Y_j Y_k\right\} =$$

$$= \sum_{j,k=0}^{m} s_{j+k} X_j X_k + \sum_{j,k=0}^{m-1} (s_{j+k} - s_{j+k+2}) Y_j Y_k,$$

wherein these two Hankel forms are *independent* in the sense, that the parameters X_k ($k = 0,1,\cdots,m$) are in no way connected with the parameters Y_k ($k = 0,1,\cdots,m-1$), which is clear from (A.II.35), and the forms η_j ($j = 0,1,\cdots,n$) are independent.

For odd $n = 2m-1$ the calculations are analoguous. Now (leaving out the details) we find for $z = e^{it}$

$$T_{2m-1}(z) = \cos^2\frac{t}{2}\left(\sum_{k=1}^{m} \zeta_{m-k}\frac{\cos\left(k-\frac{1}{2}\right)t}{\cos\frac{t}{2}}\right)^2 +$$

$$+ \sin^2\frac{t}{2}\left(\sum_{k=1}^{m} \zeta_{m+k}\frac{\sin\left(k-\frac{1}{2}\right)t}{\sin\frac{t}{2}}\right)^2,$$

where

$$\zeta_{m-k} = \xi_{m-k} + \xi_{m+k-1},\quad \zeta_{m+k} = \xi_{m-k} - \xi_{m+k+1}$$

$$(k = 0,1,\cdots,m). \tag{A.II.36}$$

The forms

$$X_k = X_k(\zeta_0,\zeta_1,\cdots,\zeta_{m-1}) \qquad (k = 0,1,\cdots,m-1)$$
$$Y_k = Y_k(\zeta_{m+1},\zeta_{m+2},\cdots,\zeta_{2m})$$

are here induced by the identities

$$\left.\begin{aligned} \sum_{k=1}^{m}\zeta_{m-k}\frac{\cos\left(k-\frac{1}{2}\right)t}{\cos\frac{t}{2}} &= \sum_{k=0}^{m-1} X_k\cos^k t \\ \sum_{k=1}^{m}\zeta_{m+k}\frac{\sin\left(k-\frac{1}{2}\right)t}{\sin\frac{t}{2}} &= \sum_{k=0}^{m-1} Y_k\cos^k t \end{aligned}\right\} \tag{A.II.37}$$

respectively. Then

$$T_{2m-1}(z) = \frac{1}{2}(1+u)\left(\sum_{k=0}^{m-1} X_k u^k\right)^2 + \frac{1}{2}(1-u)\left(\sum_{k=0}^{m-1} Y_k u^k\right)^2 \qquad \text{(A.II.38)}$$

where, as before,

$$u = \frac{z+z^{-1}}{2} = \cos t.$$

Thence and from (A.II.3o)

$$C\{T_{2m-1}(z)\} = S\left\{\frac{1+u}{2}\left(\sum_{k=0}^{m-1} X_k u^k\right)^2\right\} + S\left\{\frac{1-u}{2}\left(\sum_{k=0}^{m-1} Y_k u^k\right)^2\right\}.$$

Here the lefthand side coincides (for $n = 2m-1$, $z = e^{it}$) again, because of (A.II.31), with the original form (A.II.26), and the righthand side is equal to

$$S\left\{\frac{1}{2}\sum_{j,k=0}^{m-1} u^{j+k}X_jX_k + \frac{1}{2}\sum_{j,k=0}^{m-1} u^{j+k+1}X_jX_k\right\} +$$
$$+ S\left\{\frac{1}{2}\sum_{j,k=0}^{m-1} u^{j+k}Y_jY_k - \frac{1}{2}\sum_{j,k=0}^{m-1} u^{j+k+1}Y_jY_k\right\} =$$
$$= \frac{1}{2}\sum_{j,k=0}^{m-1}(s_{j+k}+s_{j+k+1})X_jX_k + \frac{1}{2}\sum_{j,k=0}^{m-1}(s_{j+k}-s_{j+k+1})Y_jY_k,$$

i.e., to the sum of two *independent* Hankel forms.

Thus we have proved.

THEOREM A.II.3. *The real Toeplitz quadratic form*

$$\sum_{p,q=0}^{n} c_{p-q}\xi_p\xi_q$$

can for $n = 2m$ *be transformed into a sum of two independent Hankel forms*

$$\sum_{j,k=0}^{m} s_{j+k}X_jX_k + \sum_{j,k=0}^{m-1}(s_{j+k}-s_{j+k+2})Y_jY_k,$$

where the parameters X_k $(k = 0,1,\cdots,m)$ *and* Y_k $(k = 0,1,\cdots,m-1)$ *depend on the original parameters* ξ_p $(p = 0,1,\cdots,2m)$ *through the transformations* (A.II.34) *and* (A.II.33); *for* $n = 2m-1$ *this forms is transformed into the sum of the two independent Hankel forms* [1o)

$$\frac{1}{2}\sum_{j,k=0}^{m-1}(s_{j+k}+s_{j+k+1})X_jX_k + \frac{1}{2}\sum_{j,k=0}^{m-1}(s_{j+k}-s_{j+k+1})Y_jY_k,$$

where the parameters X_k, Y_k $(k = 0,1,\cdots,m-1)$ *depend on the original parameters* ξ_p $(p = 0,1,\cdots,2m-1)$ *through the transformations* (A.II.37) *and* (A.II.36). *In both cases the coefficients* c_k $(=c_{-k})$ *and* s_k $(k = 0,1,\cdots,n)$ *are connected to each other through formulae* (A.II.27).

EXAMPLES AND EXERCISES.

1. Each real nonnegative polynomial $P_n(\lambda)$ is the square of the modulus of some (generally speaking, complex) polynomial. Verify that this fact, unlike its nontrivial analogon for trigonemetrical polynomials - the Theorem of Fêjer - F. Riesz (see above proposition 2°), directly follows from the symmetry of the roots of the polynomial $P_n(\lambda)$ with respect to the real axis (exercise 1 to Appendix I).

2. Let be given the real Hankel form

$$\sum_{j,k=o}^{n-1} s_{j+k}\xi_j\xi_k,$$

for which the successive principal minors $D_o, D_1, \cdots, D_{n-2}$ are different from zero. For this form we construct the corresponding functional S (see Sec. 1 above).

We consider the polynomials

$$P_k(\lambda) = \begin{vmatrix} s_o & s_1 & \cdots & s_k \\ s_1 & s_2 & \cdots & s_{k+1} \\ \cdot & \cdot & \cdots & \cdot \\ s_{k-1} & s_k & \cdots & s_{2k-1} \\ 1 & \lambda & \cdots & \lambda^k \end{vmatrix} \qquad (k = 1,2,\cdots,n-1), \; P_o(\lambda) \equiv 1.$$

Verify, that the $P_k(\lambda)$ are polynomials of degree exactly k, which satisfy the conditions

$$S\{P_k(\lambda)\lambda^j\} = 0 \qquad (j = 0,1,\cdots,k-1;\; k \leqq n-1).$$

Such polynomials are called *orthogonal polynomials of degree* k (on the real axis).

In particular, we see that each orthogonal polynomial is also a quasi-orthogonal polynomial of rank k in the sense of definition (A.II.3). The converse statement is not true - present an example.

3. Prove, that each orthogonal polynomial of degree k (see exercise 2) is equal to $\alpha\, P_k(\lambda)$, where α is a constant.

4. Obtain Theorem A.II.1 as corollary of the proposition, formulated in Sec. 2 above (in the Remark to Theorem A.II.1).

HINT. Use Theorem 9.1.

5. Show, that the proposition, stated in the Remark in Sec. 2 above, has the converse:

if

$$s_k = \sum_{\nu=1}^{n-1} \rho_\nu \theta_\nu^k \quad (k = 0,1,\cdots,2n-2)$$

where $\rho_\nu > 0$ $(\nu = 1,2,\cdots,n-1)$, $-\infty < \theta_1 < \theta_2 < \cdots < \theta_{n-1} < +\infty$, *then the Hankel form*

$$\sum_{j,k=0}^{n-1} s_{j+k}\xi_j\xi_k$$

is nonnegative, degenerate $(D_{n-1} = 0)$, *and* $D_{n-2} \neq 0$.

6. Convince yourself, that the real trigonometrical polynomial

$$T_4(e^{it}) = e^{4it} + 6 + e^{-4it}$$

is nonnegative, and represent it according to the Theorem of Féjer and F. Riesz (proposition 2°).

Solution (one of the possibilities)

$$T_4(e^{it}) = \left| \frac{\sqrt{2}}{2} - i\frac{\sqrt{2}}{2} \quad e^{2it} + \sqrt{2}\,(1+i) \quad + \right.$$

$$\left. + \left(\frac{\sqrt{2}}{2} - i\frac{\sqrt{2}}{2}\right) e^{-2it}\right|^2 .$$

HINT. Use the fact, that

$$T_4(e^{it}) = 8(\cos^4 t + \sin^4 t).$$

7. Let be given the Hermitian Toeplitz form

$$\sum_{p,q=0}^{n-1} c_{p-q}\xi_p\overline{\xi}_q,$$

where the minors $D_0, D_1, \cdots, D_{n-2}$ are nonzero. For this form we construct the corresponding functional C (see above Sec. 1). We consider the polynomials

$$W_k(z) = \begin{vmatrix} c_0 & c_1 & \cdots & c_k \\ c_{-1} & c_0 & \cdots & c_{k-1} \\ \cdot & \cdot & \cdots & \cdot \\ c_{-k+1} & c_{-k+2} & \cdots & c_1 \\ 1 & z & \cdots & z^k \end{vmatrix} \quad (k = 1,2,\cdots,n-1), W_0(z) \equiv 1.$$

Verify, that the $W_k(z)$ are polynomials of degree exactly k, which satisfy for $z = e^{it}$ the conditions

$$C\{W_k(z)z^j\} = 0 \qquad (j = 0,1,\cdots,k-1,\ k \leq n-1).$$

Such polynomials are called *orthogonal polynomials of degree* k *on the unit circle*. (In particular, it follows from (A.II.2o) that the polynomials $U_n(z)$, constructed for the proof of Theorem A.II.2, are polynomials of order n which are orthogonal on the unit circle with respect to the functional $\mathcal{C}$ constructed from the form

$$\sum_{p,q=o}^{n} c_{p,q}\xi_p\overline{\xi}_q \qquad (c_n = \zeta),$$

with the matrix $\|c_{p-q}\|_{p,q=o}^{n}$ which was defined above just before (A.II.19)).

8. Prove, that each polynomial of degree k which is orthogonal on the unit circle (see exercise 7) is equal to $\alpha\, W_k(z)$, where α is a constant.

9. Obtain Theorem A.II.2 as corollary of the proposition, formulated in Sec. 3 above (in the Remark to Theorem A.II.2).

HINT. Use Theorem 13.1.

1o. Show, that the proposition which was the subject of exercise 9, has the converse:

if

$$c_p = \sum_{\nu=1}^{\rho} r_\nu \varepsilon_n^p \quad (p = 0,1,\cdots,n-1), \qquad \rho < n,$$

where $r_\nu > 0$, $|\varepsilon_\nu| = 1$ *and all* ε_ν $(\nu = 1,2,\cdots,\rho)$ *are different, then the Toeplitz form*

$$\sum_{p,q=o}^{n-1} c_{p-q}\xi_p\overline{\xi}_q$$

is nonnegative, its rank is equal to ρ *and* $D_{\rho-1} \neq 0$.

11. We transform the Toeplitz form of order $n+1 = 5$ $(n = 2m,\ m = 2)$

$$T_4(x,x) = 2(\xi_0\xi_1 + \xi_0\xi_3 + \xi_1\xi_2 + \xi_1\xi_4 + \xi_2\xi_3 + \xi_3\xi_4)$$

in accordance with Theorem A.II.3. The matrix of this form has the shape

$$T_4 = \left\|\begin{matrix} o & 1 & 0 & 1 & 0 \\ 1 & 0 & 1 & 0 & 1 \\ 0 & 1 & 0 & 1 & 0 \\ 1 & 0 & 1 & 0 & 1 \\ 0 & 1 & 0 & 1 & o \end{matrix}\right\| ,$$

i.e., $c_o = c_{\pm 2} = c_{\pm 4} = 0$, $c_{\pm 1} = c_{\pm 3} = 1$. By formula (A.II.27) $s_o = 0$, $s_1 = \frac{1}{2}(c_1 + c_{-1}) = 1$, $s_2 = 0$, $s_3 = \frac{1}{8}(c_3 + 3c_1 + 3c_{-1} + c_{-3}) = 1$, $s_4 = 0$.

Hence (Theorem A.II.3)

$$T_4(x,x) = \sum_{j,k=0}^{2} s_{j+k}X_jX_k + \sum_{j,k=0}^{1}(s_{j+k}-s_{j+k+2})Y_jY_k = 2X_0X_1 + 2X_1X_2 = 0,$$

where the forms

$$X_0 = X_0(\eta_0,\eta_1,\eta_2),\ X_1 = X_1(\eta_0,\eta_1,\eta_2)\ X_2 = X_2(\eta_0,\eta_1,\eta_2)$$

are defined through the relations (see (A.II.34))

$$\eta_2+\eta_1\cos t+\eta_0\cos 2t = X_0 + X_1\cos t + X_2\cos^2 t.$$

As $\cos 2t = \cos^2 t - 1$, one has

$$X_0 = \eta_2-\eta_0,\ X_1 = \eta_1,\ X_2 = 2\eta_0.$$

Next (see (A.II.32))

$$\eta_2 = \xi_2,\ \eta_1 = \xi_1+\xi_3,\ \eta_0 = \xi_0+\xi_4,$$

so that

$$X_0 = \xi_2-\xi_0-\xi_4;\ X_1 = \xi_1+\xi_3,\ X_2 = 2\xi_0+2\xi_4.$$

Verification:

$$2(X_0X_1+X_1X_2) = 2[(\xi_2-\xi_0-\xi_4)(\xi_1+\xi_3) + (\xi_1+\xi_3) \times$$
$$\times (2\xi_0+2\xi_4)] = 2(\xi_1+\xi_3)(\xi_2+\xi_0+\xi_4) = T_4(x,x).$$

12. Transform in accordance to Theorem A.II.3 the real Toeplitz form of order $n+1 = 4$ $(n=2m-1, m=2)$

$$T_3(x,x) = 2(\xi_0\xi_2+\xi_0\xi_3+\xi_1\xi_3)$$

Solution. $T_3(x,x) = \frac{1}{2}\left(X_0X_1+\frac{3}{4}X_1^2\right)+\frac{1}{2}\left(\frac{1}{4}Y_1^2-Y_0Y_1\right),$

where $\begin{cases} X_0 = \xi_1+\xi_2-\xi_0-\xi_3;\ Y_0 = \xi_0-\xi_3+\xi_1-\xi_2; \\ X_1 = 2(\xi_0+\xi_3);\ Y_1 = 2(\xi_0-\xi_3). \end{cases}$

13. The transformation, considered in Theorem A.II.3, is based on formula (A.II.27), which makes sense also for an infinite sequence

$$c_0,c_1,c_2,\cdots\ (c_{-p} = c_p;\ p = 0,1,2,\cdots),$$

generating a corresponding infinite sequence of real numbers

$$s_0,s_1,s_2,\cdots .$$

Show ([1], p.34), that these two sequences are connected through the formal identity of the two series

$$\sqrt{x^2-1}\left\{\frac{s_0}{x}+\frac{s_1}{x^2}+\frac{s_2}{x^3}+\cdots\right\} = c_0 + 2c_1w + 2c_2w^2 + \cdots + \left(x = \frac{1+w^2}{2w}\right),$$

where $x > 1$ $(0 < w < 1)$, $\sqrt{x^2-1} > 0$.

HINT. Apply (for $k = 0,1,2,\cdots$) formulae (A.II.28), which are equivalent to (A.II.27), to the coefficients of the series standing on the lefthand side. Next decompose the quotient $\frac{(1-w^2)z}{(z-w)(1-wz)}$, which results in course of the calculation, into two partial fractions (with respect to the parameter z), and expand, in turn, each of these into a geometrical series.

NOTES.

1) Note, that an English translation of this book appeared in 1962.

2) Through this was, in fact, proved some general proposition (cf.[1], article I, Ch. 1, Theorem 1) which is the analogon of proposition 3°, presented above.

3) As $H_{n-1}(x,x)$ is positive, the matrix $\|s_{j+k}\|_{j,k=0}^{n-1}$ is invertible. So the system of equations $\sum_{\nu=0}^{n-1} s_{j+\nu}q_\nu = -s_{n+j}$, $j = 0,1,\cdots,n-1$ has a unique solution $\{q_0,q_1,\cdots,q_{n-1}\}$. Then the polynomial $Q_n(\lambda) = \lambda^n + q_{n-1}\lambda^{n-1} + \cdots + \lambda q_1 + q_0$ is quasi-orthogonal (cf. also exercise 2 above).

4) Besides, formula (A.II.6) is evident, since (see (A.II.5)) the polynomial $r(\lambda)$ of degree $\leq n-1$ coincides with the polynomial, standing on the righthand side of (A.II.6), in the n different points $\theta_1,\theta_2,\cdots,\theta_n$.

5) Theorem A.II.1 itself represents only a part of a much more complete theorem (see [1], article I, Ch. 1, Theorem 3).

6) Here $K(z) = (z-\alpha_1)^{\mu_1}\cdots(z-\alpha_r)^{\mu_r}(z-\beta_1)^{\nu_1}\cdots(z-\beta_s)^{\nu_s}$, where $|\alpha_j| = 1$ $(j = 1,2,\cdots,r)$, $0 < |\beta_\ell| < 1$ $(\ell = 1,2,\cdots,s)$, $\mu_1 + \mu_2 + \cdots + \mu_r + \nu_1 + \nu_2 + \cdots + \nu_s = k$. But then $z^k \bar{K}(1/z) = C_1(z-\alpha_1)^{\mu_1}\cdots(z-\alpha_r)^{\mu_r} \cdot (z-1/\bar{\beta}_1)^{\nu_1}\cdots(z-1/\bar{\beta}_s)^{\nu_s}$, where $C_1 = (-1)^k \bar{\alpha}_1\cdots\bar{\alpha}_r\bar{\beta}_1\cdots\bar{\beta}_s$.

7) Here the interpolation formula of Lagrange (cf. A.II.6) was used again; the obtained relation can, however, be verified directly.

8) Theorem A.II.2 itself represents only a part of a more complete statement (see [1], article I, Ch. 1, Theorem 9).

9) See [1], article I, Ch. 1, § 4, from where the transformation, presented in Sec. 4, was taken as well.

10) In the book [1] on p. 35 the coefficient 1/2 in front of both these formulae is omitted, which stems from a corresponding error in the formula which appears as (A.II.38) in our text.

REFERENCES

a) Textbooks and monographs

[1] N.I. AHIEZER and M.G. KREĬN: *Some questions in the theory of moments*, DNTVU, Har'kov, 1938 [Russian] ≡ Transl. Math. Monographs 2, Amer. Math. Soc., Providence, R.I., 1962.

[2] R. BELLMANN: *Introduction to matrix analysis*, Mc Graw-Hill, New York, 196o.

[3] F.R. GANTMAHER: *Theory of matrices*, 3rd ed., "Nauka", 1967 [Russian] (available translation of 2nd ed.: Chelsea, New York, 1964).

[4] F.R. GANTMAHER and M.G. KREĬN: *Oscillatory matrices and kernels and small vibrations of mechanical systems*, 2nd ed., Gostehizdat, 195o [Russian].

[5] K.F. GAUSS: *Werke, Bd. 1, Disquisitiones arithmetica*, § 271, Göttingen, 1863.

[6] I.C. GOHBERG and I.A. FEL'DMAN: *Convolution equations and projection methods for their solution*, "Nauka", 1971 [Russian] ≡ Amer. Math. Soc., Providence, R.I., 1974.

[7] U. GRENANDER and G.SZEGÖ: *Toeplitz forms and their applications*, University of California Press, 1958.

[8] M.G. KREĬN and M.A. NEĬMARK: *The method of symmetrical and Hermitian forms in the theory of the distribution of roots of algebraic equations*, DNTVU, Har'kov, 1936 [Russian].

[9] A.I. MAL'CEV: *Foundations of linear algebra*, 3rd ed., "Nauka", 197o [Russian] (available translation of 2nd ed: Freemann, San Francisco, 1963).

[1o] I.I. PRIVALOV: *Introduction to the theory of functions of a complex parameter*, 11th ed., "Nauka", 1967 [Russian].

[11] A.K. SUŠKEVIČ: *Foundations of higher algebra*, 4th ed., Gostehizdat, 1941 [Russian].

[12] N.Ya. VILENKIN: *Combinatorial Analysis*, "Nauka", 1969 [Russian].

b) papers in journals

[13] H. AKAIKE: *Block Toeplitz matrix inversion*, SIAM J. Appl. Math. 24 (1973), 234-241.

[14] G. BAXTER and I.I. HIRSCHMAN Jr.: *An explicit inversion formula for finite-section Wiener-Hopf operators*, Bull. Amer. Math. Soc. 7o (1964),82o-823.

[15] C. BORHARDT: *Développements sur l'équation à l'aide, de laquelle on détermine les inégalités séculaires du mouvement des planètes*, J.Math. pures appl. 12 (1847), 5o-67.

[16] C. BORHARDT: *Bemerkung über die beiden vorstehenden Aufsätze*, J. reine angew. Math. 53 (1857), 281-283.

[17] J. DURBIN: *The fitting of time-series models*, Rev.Inst. Internat. Statist.28 (196o), 233-244.

[18] E. FISCHER: *Über das Carathéodory'sche Problem, Potenzreihen mit positivem reellen Teil betreffend*, Rend. del Circ. Mat. die Palermo 31 (1911), 24o-256.

[19] G. FROBENIUS: *Über das Trägheitsgesetz der quadratischen Formen*, Sitzungsber. der Königl. Preuss. Akad. der Wiss. (1894), 241-256, 4o7-431.

[2o] G. FROBENIUS: *Ableitung eines Satzes von Carathéodory aus einer Formel von Kronecker*, Sitzungsber. der Königl. Preuss. Akad. der Wiss. (1912), 16-31.

[21] V.I. GORBAČUK: *On the integral representation of Hermitian-indefinite kernels (the case of several parameters)*, Ukrain. Mat. Ž. 16(2) (1964), 232-236 [Russian].

[22] V.I. GORBAČUK: *On the integral representation of Hermitian-indefinite kernels*, Ukrain. Mat. Ž. 17(3) (1965), 43-58 [Russian].

[23] V.I. GORBAČUK: *On extensions of real Hermitian-indefinite functions with one negative square*, Ukrain. Mat. Ž. 19(4) (1967), 119-125 [Russian].

[24] I.C. GOHBERG and N.Ya. KRUPNIK: *A formula for the inversion of finite Toeplitz matrices,* Mat. Issled. 7(2) (1972), 272-283 [Russian].

[25] I.C. GOHBERG and A.A. SEMENCUL: *On the inversion of finite Toeplitz matrices and their continual analoga,* Mat. Issled. 7(2) (1972), 2o1-223 [Russian].

[26] S. GUNDELFINGER: *Über die Transformation einer quadratischen Form in einer Summe von Quadraten,* J. reine angew. Math. 91 (1888), 221-237.

[27] G. HERGLOTZ: *Über die Wurzelanzahl algebraischer Gleichungen innerhalb und auf dem Einheitskreis,* Math. Zeitsch. 19 (1924), 26-34.

[28] C. HERMIT: *Sur le nombre des racines d'une équation algebrique comprise entre des limites données,* J. reine angew. Math. 52 (1856), 39-51.

[29] A. HURWITZ: *Über die Bedingungen, unter welchen eine Gleichung nur Wurzeln mit negativen reellen Teilen besitzt,* Math. Ann. 46 (1895), 273-284.

[3o] I.S. IOHVIDOV: *On the theory of indefinite Toeplitz forms,* Dokl. Akad. Nauk SSSR 1o1(2) (1955), 213-216 [Russian].

[31] I.S. IOHVIDOV: *Extension of Toeplitz forms with an invariant number of positive squares,* Ukrain. Mat. Ž. 18(3) (1966), 4o-5o [Russian] ≡ Amer. Math. Soc. Transl. (2) 76(1968), 1o3-115.

[32] I.S. IOHVIDOV: *On the signatures of Toeplitz forms,* Dokl. Akad. Nauk SSSR 169(6) (1966), 1258-1261 [Russian] ≡ Soviet Math. Dokl. 7(4) (1966), 1o74-1o77.

[33] I.S. IOHVIDOV: *Real sequences of class P_1,* Akad. Nauk Armjan. SSR Dokl. 42(5) (1966), 269-273 [Russian].

[34] I.S. IOHVIDOV: *On Hankel and Toeplitz matrices and the signatures of Toeplitz forms,* Ukrain. Mat. Ž. 19(1) (1967), 25-35 [Russian] ≡ Amer. Math. Soc. Transl. (2)76 (1968), 89-1o2.

[35] I.S. IOHVIDOV: *Unitary extensions of isometrical operators in a Π_1-space and extensions in the class P_1 of finite sequences of class $P_{1,n}$,* Dokl. Akad. Nauk SSSR 173(4) (1967), 758-761 [Russian] ≡ Soviet Math. Dokl. 8(2) (1967),471-474.

[36] I.S. IOHVIDOV: *On spectral trajectories, generated by unitary extensions of isometrical shift operators in a finite-dimensional* Π_1*-space*, Dokl. Akad. Nauk SSSR 173(5) (1967), 1oo2-1oo5 [Russian] ≡ Soviet Math. Dokl. 8(2) (1967), 512-516.

[37] I.S. IOHVIDOV: *On the rank of Toeplitz matrices*, Mat. Sb. 76(1) (1968), 26-38 [Russian] ≡ Math. USSR - Sb.5(1) (1968), 25-37.

[38] I.S. IOHVIDOV: *On the rules of Jacobi, Gundelfinger and Frobenius*, Mat. Issled. 3(4) (1968), 162-165 [Russian].

[39] I.S. IOHVIDOV: *On the (r,k)-characteristic of a Hankel matrix*, Uspehi Mat. Nauk 24(4) (1969), 199-2oo [Russian].

[4o] I.S. IOHVIDOV: *On Hankel matrices and forms*, Mat. Sb. 8o(2) (1969),241-252 [Russian] ≡ Math. USSR - Sb. 9(2) (1969), 229-24o.

[41] I.S. IOHVIDOV: *On the transformation of Fischer-Frobenius*, Teor. Funkciĭ Funkcional. Anal. i Priložen. Resp. Sb., Nr. 15 (1972), 2o3-212 [Russian].

[42] I.S. IOHVIDOV and M.G. KREĬN: *Spectral theory of operators on spaces with an indefinite metric*, II, Trudy Moskov. Mat. Obšč. 8 (1959), 413-496 [Russian] ≡ Amer. Math. Soc. Transl. (2)34 (1963), 283-373.

[43] I.S. IOHVIDOV and M.G. KREĬN: *A letter to the editors*, Trudy Moskov. Mat. Obšč. 15 (1966), 452-454 [Russian] ≡ Trans. Moscow Math. Soc. 15 (1966), 5o5-5o8.

[44] M.G. KREĬN: *On the theory of symmetrical polynomials*, Mat. Sb. 4o(3) (1933), 271-283 [Russian].

[45] M.G. KREĬN: *On the integral representation of a continuous Hermitian-indefinite function with a finite number of negative squares*, Dokl. Akad. Nauk SSSR 125(1) (1959), 31-34 [Russian].

[46] M.G. KREĬN: *On the distribution of the roots of polynomials, which are orthogonal on the unit circle with respect to an alternating weight*, Teor. Funkciĭ Funkcional. Anal. i Priložen. Resp. Sb. Nr. 2 (1966), 131-137 [Russian].

[47] L. KRONECKER: *Bemerkungen zur Determinanten-Theorie*, J. reine angew. Math. 72 (187o), 152-175.

[48] L. KRONECKER: *Zur Theorie der Elimination einer Variabeln aus zwei algebraischen Gleichungen*, Monatsber. der Königl.Preuss. Akad.

der Wiss. (1881), 535-6oo.

[49] L.M. KUTIKOV: *On the structure of matrices, which are the inverses of correlation matrices of vector random processes,* Ž.Vyčisl. Mat. i Mat. Fiz. 7(4) (1967), 764-773 [Russian] ≡ U.S.S.R. Computational Math. and Math. Phys. 7(4) (1967), 58-71.

[5o] F.I. LANDER: *Extensions of shift operators in the Pontryagin space* $\Pi_\kappa^{(n)}$, Dokl. Akad. Nauk SSSR 199(3) (1971), 529-532 [Russian] ≡ Soviet Math. Dokl. 12(4) (1971), 11o5-111o.

[51] F.I. LANDER: *Extensions of isometrical shift operators in the Pontryagin space* $\Pi_\kappa^{(n)}$, Akad. Nauk Armjan. SSR Dokl. 53(4) (1971), 193-198 [Russian].

[52] F.I. LANDER: *The Bezoutiant and the inversion of Hankel and Toeplitz matrices,* Mat. Issled. 9(2) (1974), 69-87 [Russian].

[53] N. LEVINSON: *The Wiener RMS (root mean square) error in filter design,* J. Math. and Phys. 25 (1947), 261-278.

[54] V.I. PLYUŠČEVA (GORBAČUK): *On the integral representation of Hermitian indefinite functions with* κ *negative squares,* Ukrain. Mat. Ž. 14(1) (1962), 3o-39 [Russian].

[55] A.L. SAHNOVIČ: *On a method of inversion of finite Toeplitz matrices,* Mat. Issled. 8(4) (1973), 18o-186 [Russian].

[56] L.A. SAHNOVIČ: *On the similarity of operators,* Sibirsk. Mat. Ž. 13(4) (1972), 868-883 [Russian] ≡ Siberian Math. J. 13(4) (1972), 6o4-615.

[57] L.A. SAHNOVIČ: *On an integral equation with a kernel, depending on the difference of the arguments,* Mat. Issled. 8(2) (1973), 138-146 [Russian].

[58] A.A. SEMENCUL: *Inversion of finite Toeplitz matrices and their continual analoga,* Appendix II in: Projection methods in the solution of Wiener-Hopf equations, by I.C. Gohberg and I.A. Fel'dman, R.I.O. Akad. Nauk MSSR, Kišinev, 1967, 14o-156 [Russian].

[59] V.A. ŠTRAUS: *On continuous Hermitian-indefinite functions,* Mat. zametki 13(2) (1973), 3o3-31o [Russian] ≡ Math. Notes 13(2) (1973) 183-188.

[6o] W.F. TRENCH: *An algorithm for the inversion of finite Toeplitz matrices,* J. Soc. Indust. Appl. Math. 12(3) (1964), 515-522.

[61] P. WITTLE: *The analysis of multiple stationary time series,* J. Roy. Statist. Soc., ser. B (1953), 125-139.

[62] P. WITTLE: *On the fitting of multivariate autoregressions and the approximate factorization of a special density matrix,* Biometrika 5o (1963), 129-134.

[63] S. ZOHAR: *Toeplitz matrix inversion: the algorithm of W.F. Trench,* J. Assoc. Comput. Math. 16 (1967), 592-6o1.

ADDITIONAL REFERENCES

published in the years 1974-198o

[1] A.I. BALINSKIĬ and LI-GYUN-Y: *On the inversion of Hankel and Toeplitz matrices,* Mat. Metody i Fiz.-Meh. Polja 9 (1979), 31-37 [Russian].

[2] R.R. BITMEAD and B.D.O. ANDERSON: *The matrix Cauchy index: properties and applications,* SIAM J. Appl. Math. 33(4) (1977), 655-672.

[3] B.N. DATTA: *A relationship between a Hankel matrix of Markov parameters and the associated matrix polynomial with some applications,* Czecheslowak. Math. J. 3o (198o), 71-79.

[4] I. GOHBERG and G. HEINIG: *Inversion of finite Toeplitz matrices, consisting of elements of a noncommutative algebra,* Rev. Roum. math. pures et appl. 19(5) (1974), 623-663 [Russian].

[5] G. HEINIG: *Endliche Toeplitzmatrizen und zweidimensionale Wiener-Hopf-Operatoren mit homogenem Symbol,* Math. Nachr. 82 (1978), 29-52, 53-68.

[6] G. HEINIG: *Transformationen von Toeplitz- und Hankelmatrizen,* Wiss. Zeitschr. der Techn. Hochschule Karl-Marx-Stadt 21(7) (1979), 859-864.

[7] I.S. IOHVIDOV: *On the (r,k,)-characteristics of rectangular Hankel matrices,* Dokl. Akad. Nauk SSSR 233(2) (1977), 276-279 [Russian] ≡ Soviet Math. Dokl. 18(2) (1977), 316-32o.

[8] I.S. IOHVIDOV: *On extensions of rectangular Hankel matrices,* Dokl. Akad. Nauk SSSR 233(3) (1977), 285-289 [Russian] ≡ Soviet

Math. Dokl. 18(2) (1977), 378-382.

[9] I.S. IOHVIDOV and O.D. TOLSTYH: *On the (r,k,l)-characteristics of rectangular Toeplitz matrices*, Ukrain. Mat. Ž. 32(4) (1980), 477-482 [Russian] ≡ Ukrainian Math. J. 32 (1980), 327-331.

[10] T. KAILATH, A. VIERA and M.MORF: *Inverses of Toeplitz operators, innovations and orthogonal polynomials*, SIAM Rev. 20(1) (1978), 106-119.

[11] L.D. PUSTYL'NIKOV: *On the algebraic structure of spaces of Toeplitz and Hankel matrices*, Dokl. Akad. Nauk SSSR 250(3) (1980), 556-559 [Russian] ≡ Soviet Math. Dokl. 21(1)(1980), 141-144.

[12] L.D. PUSTYL'NIKOV: *On fast calculations in some problems of linear algebra, connected with Toeplitz and Hankel matrices*, Uspehi Mat. Nauk 35(5) (1980), 241-242 [Russian] ≡ Russian Math. Surveys 35(5) (1980), 271-272.

[13] E.M. RUSSAKOVSKIĬ: *On Hankel and Toeplitz matrices and the Bezoutiant*, Teor. Funkciĭ Funkcional. Anal.i Priložen. 32 (1979), 77-72 and 33(1980), 119-124 [Russian].

[14] A.I. SAUDARGAS: *An approach to the problem of partial realization*, Avtomat. i Telemeh. 2 (1977), 73-80 [Russian] ≡ Automat. Remote Control 2 (1977).

[15] O.D. TOLSTYH: *On extensions of rectangular Toeplitz matrices*, in: Funkcional'nyĭ analiz, No. 13, Ul'yanovsk, 1979, 144-155 [Russian].

[16] E.E. TYRTYŠNIKOV: *On some problems, connected with Toeplitz matrices*, in: Cislennyĭ analiz na FORTRANe, Moskva, 1979, No. 25, 105-113 [Russian].

INDEX